Adipose Tissue and Gene Expression

Adipose Tissue and Gene Expression

Guest Editors

Margherita Maffei
Gaia Scabia

MDPI

Basel • Beijing • Wuhan • Barcelona • Belgrade • Novi Sad • Cluj • Manchester

Guest Editors
Margherita Maffei
Institute of Clinical Physiology
National Research Council (CNR)
Pisa
Italy

Gaia Scabia
Institute of Clinical Physiology
National Research Council (CNR)
Pisa
Italy

Editorial Office
MDPI AG
Grosspeteranlage 5
4052 Basel, Switzerland

This is a reprint of the Special Issue, published open access by the journal *International Journal of Molecular Sciences* (ISSN 1422-0067), freely accessible at: https://www.mdpi.com/journal/ijms/special_issues/LX1OZ5S7WU.

For citation purposes, cite each article independently as indicated on the article page online and as indicated below:

Lastname, A.A.; Lastname, B.B. Article Title. *Journal Name* **Year**, *Volume Number*, Page Range.

ISBN 978-3-7258-7330-2 (Hbk)
ISBN 978-3-7258-7331-9 (PDF)
https://doi.org/10.3390/books978-3-7258-7331-9

Cover image courtesy of Margherita Maffei

Contents

Preface

Understanding fat depot-specific variations in gene expression profiles is essential for clarifying pathological mechanisms that remain only partially understood. Moreover, dissecting these molecular differences may reveal novel biomarkers and identify promising therapeutic targets to be explored in future research.

Adipose tissue is no longer viewed merely as a passive reservoir of stored energy but rather as a highly dynamic and plastic organ with profound endocrine and metabolic functions. Both white adipose tissue (WAT) and brown adipose tissue (BAT) actively participate in systemic homeostasis, responding to nutritional inputs, hormonal signals, aging, and environmental challenges through finely tuned regulatory programs. Their capacity to adapt structurally and functionally relies on coordinated transcriptional networks and post-transcriptional mechanisms that shape cellular identity, metabolic flexibility, and intercellular communication.

This Special Issue was conceived with the aim of gathering novel insights into the regulation of gene expression in WAT and BAT, particularly in the context of dysfunctions or alterations induced by metabolic challenges such as obesity and diabetes, and by nutritional stimuli, aging, and environmental cues. The contributions collected here explore the molecular pathways that underpin adipose tissue remodeling and maladaptation, offering perspectives that span different biological models, experimental approaches, and pathological conditions.

Together, these studies highlight the central role of gene regulatory mechanisms in determining adipose tissue functionality and adaptability. By integrating diverse models and cross-disciplinary expertise, this volume reflects the strong and continuing interest in adipose biology and exemplifies the collaborative spirit required to advance the field. We hope that the insights presented here will stimulate further investigation, foster innovative therapeutic strategies, and contribute to a deeper understanding of adipose tissue physiology and its impact on health and disease.

Margherita Maffei and Gaia Scabia
Guest Editors

International Journal of
Molecular Sciences

MDPI

Editorial

Special Issue "Adipose Tissue and Gene Expression"

Margherita Maffei * and Gaia Scabia *

CNR, Institute of Clinical Physiology, 56124 Pisa, Italy
* Correspondence: margherita.maffei@cnr.it (M.M.); gaia.scabia@cnr.it (G.S.)

Since the recognition of white adipose tissue (WAT) as an endocrine organ rather than merely an energy reservoir, interest in this complex ensemble of cell types, characterized by extraordinary and, in many respects, unique plasticity, has steadily increased [1]. Beyond its initially recognized role in the regulation of long-term energy homeostasis, it has become increasingly clear that WAT profoundly influences fertility, immune responses, glucose metabolism, bone remodeling, and blood pressure regulation [2,3]. It is now reasonable to conclude that virtually no physiological function is exempt from WAT control, a concept closely linked to the central role of energy availability in individual survival. Concomitantly, WAT itself is highly responsive to a wide range of stimuli, acting as a dynamic sensor of environmental and systemic changes [4]. The strong response to this Special Issue underscores the scientific relevance and broad interest surrounding this field. The original aim has been fully achieved, bringing together studies that highlight both the multifaceted nature of WAT and its remarkable plasticity—an adaptive feature fundamentally driven by changes in gene-expression regulation.

Indeed, although at first glance the nine studies featured in this issue may appear to have little in common, they share a clear mechanistic core centered on lipid metabolic pathways and gene-expression programs that shape cellular growth and function. Together, they span species ranging from closely related mammals to a phylogenetically distant invertebrate (*Bombyx mori*); address diverse conditions, including COVID-19, mastitis, obesity, prostate cancer, and diabetes; and dissect regulatory layers encompassing nutritional cues, microRNA-mediated control, transcriptional networks, and RNA-splicing mechanisms.

Regarding gene expression, a fil rouge running through most of the studies, Pérez-Gómez J.M. and colleagues (Contribution 1) focus on the importance of optimizing molecular tools to enhance the reliability and precision of WAT gene-expression analyses. The authors applied two widely used gene-expression methodologies in combination with multiple analytical algorithms to validate a panel of previously reported internal reference genes (IRGs) in two independent cohorts of periprostatic WAT, a specialized adipose depot positioned at the pathophysiological interface between obesity and prostate cancer. They identify *LRP10*, *PGK1*, and *RPLP0* as the genes with the highest stability to be employed in gene expression studies involving human periprostatic WAT.

A considerable number of the studies rely on omic approaches to unravel how metabolic pathways and gene regulation shape tissue physiology across species. Poklukar K. and collaborators (Contribution 2) show that a protein-restricted diet in pigs affects WAT composition and transcriptomic responses in ways that depend on rearing conditions. Their results highlight how nutrition and environment converge on lipid synthesis, stress responses, and immune pathways, further emphasizing the plasticity of adipose biology.

A similar metabolic sensitivity emerges in dairy cows. Zhou B. and co-authors (Contribution 3) investigated the relationship between fatty acid metabolism and clinical mastitis (CM) through the identification and characterization of differentially expressed proteins in healthy and CM-affected animals. They found that the enzyme ACOT7 is positively correlated with CM onset and progression, linking long-chain fatty acid elongation and unsaturated lipid biosynthesis to inflammatory pathology.

Alazaabi M. and co-authors (Contribution 4) provide a developmental view of adipocyte formation. Using mouse embryonic stem cells, they mapped the transcriptional choreography underlying adipogenesis and, through a Robust Rank Aggregation approach, identified Atf5, Ccnd1, and Nr4a1 as potential fine regulators of this process.

Two contributions illustrate the importance of microRNAs as fine-tuners of adipose cell biology. In obese rabbits, Shao J. and co-authors (Contribution 5) show that miR-let-7a-5p dampens PI3K–AKT–mTOR signaling, acting as a modulator of lipid accumulation. Using primary rabbit white adipocytes, Sun W. and collaborators (Contribution 6) demonstrate that miR-889-3p facilitates the differentiation toward a brown-adipocyte phenotype and promotes the expression of thermogenic markers, likely through the inhibition of the DNA-binding protein SON.

Using genetic manipulation and molecular assays, Qin X. and co-authors (Contribution 7) shift the attention to invertebrates and demonstrated that the domestication of *Bombyx mori*, and the improvement of its economic traits for the silk market, are associated with the loss of a transposable element upstream of *Mlx*. Silencing *Mlx* decreases the expression of glucose-metabolism and lipogenic genes, whereas its overexpression in the silkworm fat body increases larval body size, cocoon-shell weight, and reproductive output.

Gene expression dynamics in adipose tissue could have a broad clinical implication in diverse pathologies. Salazar M. and co-authors (Contribution 8) demonstrate how obesity, aging, and metabolic history influence viral susceptibility at the tissue level. By profiling the differential expression of SARS-CoV-2 cell infection mediators (ACE2, TMPRSS2, ADAM17, and NRP1) in different adipose depots from middle-aged and older women with distinct metabolic backgrounds, they show that visceral adipose tissue may act both as a sentinel and a source of vulnerability in the context of COVID-19 pathogenesis, in a manner dependent on other risk factors such as obesity and aging. This is relevant as obesity and aging lead to a significantly elevated risk of developing severe and fatal forms of the disease [5,6]. Lui A. and co-authors (Contribution 9) uncover how alternative splicing affects adipocyte dysfunction in type 2 diabetes. Working with human-visceral-adipose-derived stem cells, the authors reveal that in the context of diabetes, WAT expresses a truncated, alternatively spliced form of sortilin that interferes with GLUT4 trafficking and impairs glucose uptake. They further show that GLP-1 can suppress this aberrant isoform by modulating TRA2B-dependent splicing, a regulatory mechanism that opens a therapeutic window for rescuing metabolic function in insulin-resistant adipose tissue.

Taken together, these studies form a coherent narrative: despite differences in species, techniques, and biological questions, they converge on the same underlying principle, i.e., the metabolic and regulatory versatility of adipose tissue. Whether shaping viral entry, responding to diet and environment, driving inflammation, enabling developmental transitions, directing domestication, or fine-tuning lipid storage and thermogenesis, adipose tissue consistently emerges as a central node of biological regulation.

We sincerely thank all the authors, reviewers, and editors whose collective efforts have shaped this Special Issue. By collating diverse models and cross-species perspectives, their contributions demonstrate the strong and continuing interest in this topic and exemplify the collaborative and integrative spirit needed to advance our understanding of adipose biology in health and disease.

https://doi.org/10.3390/ijms27031479

Conflicts of Interest: The author declares no conflicts of interest.

List of Contributions:

1. Pérez-Gómez, J.M.; Porcel-Pastrana, F.; De La Luz-Borrero, M.; Montero-Hidalgo, A.J.; Gómez-Gómez, E.; Herrera-Martínez, A.D.; Guzmán-Ruiz, R.; Malagón, M.M.; Gahete, M.D.; Luque, R.M. LRP10, PGK1 and RPLP0: Best Reference Genes in Periprostatic Adipose Tissue under Obesity and Prostate Cancer Conditions. *Int. J. Mol. Sci.* **2023**, *24*, 15140.
2. Poklukar, K.; C″andek-Potokar, M.; Vrecl, M.; Brankovič, J.; Uršič, M.; Škrlep, M. The Effect of Reduced Dietary Protein on Adipose Tissue in Local Krškopolje Pigs. *Int. J. Mol. Sci.* **2025**, *26*, 4440.
3. Zhou, B.; Zhang, B.; Han, J.; Zhang, J.; Li, J.; Dong, W.; Zhao, X.; Zhang, Y.; Zhang, Q. Role of Acyl-CoA Thioesterase 7 in Regulating Fatty Acid Metabolism and Its Contribution to the Onset and Progression of Bovine Clinical Mastitis. *Int. J. Mol. Sci.* **2024**, *25*, 13046.
4. Alzaabi, M.; Khalili, M.; Sultana, M.; Al-Sayegh, M. Transcriptional Dynamics and Key Regulators of Adipogenesis in Mouse Embryonic Stem Cells: Insights from Robust Rank Aggregation Analysis. *Int. J. Mol. Sci.* **2024**, *25*, 9154.
5. Shao, J.; Jiang, G.; Li, Y.; Wang, M.; Tang, T.; Wang, J.; Jia, X.; Lai, S. Let-7a-5p Regulates Animal Lipid Accumulation by Targeting Srebf2 and Thbs1 Signaling. *Int. J. Mol. Sci.* **2024**, *25*, 894.
6. Sun, W.; Zhang, X.; Bai, X.; Du, K.; Chen, L.; Wang, H.; Jia, X.; Lai, S. miR-889-3p Facilitates the Browning Process of White Adipocyte Precursors by Targeting the SON Gene. *Int. J. Mol. Sci.* **2023**, *24*, 17580.
7. Qin, X.; Jiang, L.; Zhao, P.; Lin, Y.; Zhang, Y.; Xia, Q. Domestication Gene Mlx and Its Partner Mondo Are Involved in Controlling the Larval Body Size and Cocoon Shell Weight of Bombyx mori. *Int. J. Mol. Sci.* **2024**, *25*, 3427.
8. Salazar, M.; Ferreira, M.; Oliveira, S.M.; Saraiva, F.; Pinho, C.; Jarnalo, M.; Correia-Sá, I.; Falcão-Pires, I.; Leite-Moreira, A.; Neves, D.; et al. Impact of Obesity and Ageing on the Expression of Key Mediators of SARS-CoV-2 Infection in Human Adipose Tissue. *Int. J. Mol. Sci.* **2025**, *26*, 7313.
9. Lui, A.; Patel, R.S.; Krause-Hauch, M.; Sparks, R.P.; Patel, N.A. Regulation of Human Sortilin Alternative Splicing by Glucagon-like Peptide-1 (GLP1) in Adipocytes. *Int. J. Mol. Sci.* **2023**, *24*, 14324.

References

1. Maffei, M.; Barone, I.; Scabia, G.; Santini, F. The Multifaceted Haptoglobin in the Context of Adipose Tissue and Metabolism. *Endocr. Rev.* **2016**, *37*, 403–416. [CrossRef] [PubMed]
2. Huang, L.; Li, C. Leptin: A multifunctional hormone. *Cell Res.* **2000**, *10*, 81–92. [CrossRef] [PubMed]
3. Iacobellis, G. Epicardial adipose tissue in contemporary cardiology. *Nat. Rev. Cardiol.* **2022**, *19*, 593–606. [CrossRef] [PubMed]
4. Corvera, S.; Rajan, A.; Townsend, K.L.; Shamsi, F.; Wu, J.; Svensson, K.J.; Zeltser, L.M.; Collins, S.; Reis, T.; Tseng, Y.H.; et al. Advances in Adipose Tissue Biology. *Endocr. Rev.* **2026**, *47*, 75–92. [CrossRef] [PubMed]
5. GBD 2023 Causes of Death Collaborators. Global burden of 292 causes of death in 204 countries and territories and 660 subnational locations, 1990-2023: A systematic analysis for the Global Burden of Disease Study 2023. *Lancet* **2025**, *406*, 1811–1872. [CrossRef] [PubMed]
6. Quaranta, P.; Scabia, G.; Storti, B.; Dattilo, A.; Quintino, L.; Perrera, P.; Di Primio, C.; Costa, M.; Pistello, M.; Bizzarri, R.; et al. SARSCoV-2 Infection Alters the Phenotype and Gene Expression of Adipocytes. *Int. J. Mol. Sci.* **2024**, *25*, 2086. [CrossRef] [PubMed]

https://doi.org/10.3390/ijms27031479

International Journal of
Molecular Sciences

MDPI

Article

LRP10, PGK1 and RPLP0: Best Reference Genes in Periprostatic Adipose Tissue under Obesity and Prostate Cancer Conditions

Jesús M. Pérez-Gómez [1,2,3,4], Francisco Porcel-Pastrana [1,2,3,4], Marina De La Luz-Borrero [1,2,3,4], Antonio J. Montero-Hidalgo [1,2,3,4], Enrique Gómez-Gómez [1,3,5], Aura D. Herrera-Martínez [1,3,4,6], Rocío Guzmán-Ruiz [1,2,3,4], María M. Malagón [1,2,3,4], Manuel D. Gahete [1,2,3,4] and Raúl M. Luque [1,2,3,4,*]

1 Maimonides Biomedical Research Institute of Cordoba (IMIBIC), 14004 Cordoba, Spain; b42pegoj@uco.es (J.M.P.-G.); z12popaf@uco.es (F.P.-P.); marina8noviembre@gmail.com (M.D.L.L.-B.); b42mohia@uco.es (A.J.M.-H.); enriquegomezgomez@yahoo.es (E.G.-G.); auritadhm@gmail.com (A.D.H.-M.); bc2gurur@uco.es (R.G.-R.); bc1mapom@uco.es (M.M.M.); bc2gaorm@uco.es (M.D.G.)
2 Department of Cell Biology, Physiology, and Immunology, University of Cordoba, 14004 Cordoba, Spain
3 Reina Sofia University Hospital (HURS), 14004 Cordoba, Spain
4 CIBER Physiopathology of Obesity and Nutrition (CIBERobn), 14004 Cordoba, Spain
5 Urology Service, Reina Sofia University Hospital, 14004 Cordoba, Spain
6 Endocrinology and Nutrition Service, Reina Sofia University Hospital, 14004 Cordoba, Spain
* Correspondence: bc2luhur@uco.es; Tel.: +34-957213740

Abstract: Obesity (OB) is a metabolic disorder characterized by adipose tissue dysfunction that has emerged as a health problem of epidemic proportions in recent decades. OB is associated with multiple comorbidities, including some types of cancers. Specifically, prostate cancer (PCa) has been postulated as one of the tumors that could have a causal relationship with OB. Particularly, a specialized adipose tissue (AT) depot known as periprostatic adipose tissue (PPAT) has gained increasing attention over the last few years as it could be a key player in the pathophysiological interaction between PCa and OB. However, to date, no studies have defined the most appropriate internal reference genes (IRGs) to be used in gene expression studies in this AT depot. In this work, two independent cohorts of PPAT samples ($n = 20/n = 48$) were used to assess the validity of a battery of 15 literature-selected IRGs using two widely used techniques (reverse transcription quantitative PCR [RT-qPCR] and microfluidic-based qPCR array). For this purpose, ΔCt method, GeNorm (v3.5), BestKeeper (v1.0), NormFinder (v.20.0), and RefFinder software were employed to assess the overall trends of our analyses. *LRP10*, *PGK1*, and *RPLP0* were identified as the best IRGs to be used for gene expression studies in human PPATs, specifically when considering PCa and OB conditions.

Keywords: periprostatic adipose tissue (PPAT); gene expression analysis; reference genes; RT-qPCR; prostate cancer (PCa); weight-related disorders

1. Introduction

Adipose tissue (AT) is a complex and dynamic organ mainly composed of adipocytes and the stromal vascular fraction, including fibroblast, endothelial cells, preadipocytes, adipocyte-derived stem cells, and a variable infiltrated immune cell population. AT has been classified into white adipose tissue (WAT) and brown adipose tissue (BAT) [1]. While BAT is related to non-shivering thermogenesis through the regulation of lipidic β-oxidation, WAT has been typically conceived as an organ specialized in energy storage. However, rather than acting as a simple lipidic depot, WAT has been recognized as an important contributor to endocrine homeostasis [2]. Through the secretion of the termed adipokines (i.e., any signaling molecule secreted by AT), fat depots are capable of controlling metabolism, inflammation, immune function, and energy homeostasis [3]. For those reasons, studies focused on AT have quickly gained interest, pinpointing it as an active contributor to the onset and progression of several types of endocrine/metabolic disorders such as obesity (OB).

OB is defined as a pathological metabolic condition characterized by an abnormal and excessive lipid accumulation and whose prevalence has drastically increased during the last decades, reaching pandemic proportions. OB is characterized by AT dysfunction and a systemic chronic low-grade inflammatory state which increases the risk for several associated disorders, including insulin resistance, type 2 diabetes, atherosclerosis, cardiovascular diseases, hypertension, and importantly, several endocrine-related cancer types [4]. In fact, according to the International Agency for Research on Cancer (IARC), up to 13 cancer types have been related to overweight/OB conditions, including breast, uterus, colon, kidney, gallbladder, pancreas, rectum, and prostate cancer (PCa), the most studied [5,6]. Regarding PCa, in addition to the mentioned increased risk, OB (along with clinical alterations typically associated with metabolic syndrome such as insulin resistance, elevated blood pressure, and high triglyceride plasma levels) has been linked to a more aggressive disease [7–9]. Despite that, the molecular bases linking OB and PCa have not been fully elucidated to date but the number of studies focusing on this pathological association has increased considerably over the last few years.

In this scenario, a specific depot of WAT that surrounds two-thirds of the prostate, typically known as periprostatic adipose tissue (PPAT), has been postulated as a putative active player in the interaction between OB and PCa, not only being a source of bioactive molecules, such as lipids, adipokines, and growth factors (that contribute to microenvironment modulation) but also acting as the first tissue susceptible to be invaded by tumor cells in the early stages of metastasis [10,11]. However, information about the molecular bases underlying that interaction is scarce. In this regard, to explore such a relationship, it is crucial to perform gene expression analyses in PPAT samples derived from patients with different metabolic conditions (e.g., normo-weight [NW] and OB).

Specifically, reverse transcription quantitative PCR (RT-qPCR) is the gold standard technique for quantifying mRNA expression due to its high accuracy and sensitivity even using small amounts of samples. However, the identification and validation of reliable internal reference genes (IRGs) for normalization of RT-qPCR data are critically needed to correct bias from multiple variables (such as RNA extraction yielding, retrotranscription performance, or primers efficiency, among others) before conclusions can be drawn. In fact, the choice of appropriate IRGs must be confirmed in a tissue-dependent manner and under the specific experimental conditions used in the study. Noteworthily, several studies using multiple rodent models have proven that the choice of IRGs in different AT subtypes depends on the metabolic status [12–14]. Regarding humans, although various studies evaluating the use of IRGs in different AT depots have been published [15,16], to the best of our knowledge, there are no reports regarding the identification, evaluation, selection, and validation of suitable IRGs for studying PPAT in expression analyses. Therefore, the objective of this work was to identify and validate, for the first time, reliable IRGs for the normalization of gene expression analyses in PPAT samples under health and disease (PCa and OB) conditions.

2. Results

2.1. *Evaluation of Gene Expression Stability in Periprostatic Adipose Tissue (PPAT) of Patients with Benign Prostate Hyperplasia (BPH) and Prostate Cancer (PCa)*

The evaluation of the potential stability in the expression levels of the 15 selected internal reference genes (IRGs) in two independent cohorts of periprostatic adipose tissue (PPAT) samples [i.e., analysis of all samples together from cohort 1 and all samples from cohort 2; Table 1] revealed that the top five IRGs with higher stability (based on changes in the Standard Deviation [SD]) using the RT-qPCR method were *LRP10* (27.99 ± 1.47), *B2M* (23.11 ± 1.50), *RPLP0* (25.97 ± 1.71), *PGK1* (26.64 ± 1.76), and *KDM2B* (31.66 ± 1.78) (Figure 1a; expression levels expressed as SD of Cycle threshold [Ct] values). Statistics details for all IRGs are summarized in Table S1. Similarly, the top five IRGs with the higher stability using the microfluidic-based qPCR array method were *RPS13* (19.21 ± 1.91), *PGK1* (20.19 ± 1.99), *RPLP0* (18.42 ± 2.01), *IPO8* (25.01 ± 2.02), and *LRP10* (19.84 ± 2.17) [Figure 1b (expression levels expressed as SD of Ct values) and Table S1)]. It should be mentioned that the use of the SD of the Ct value in these types of analyses/representations is widely used as a first approach since this offers a more visual and direct comparison of the genes by placing them over a narrow dynamic range of values [13,14,17]. In this regard, a representation of the linearized values of the Ct of each of the 15 IRGs (expressed in absolute mRNA copy number) for both techniques used (RT-qPCR and microfluidic-based qPCR array) are also represented in Figure S1, showing similar results. For this latter purpose, standard curves for each selected gene were run in parallel to quantify the absolute mRNA copy number across all the samples analyzed.

Table 1. Periprostatic adipose tissue (PPAT) human cohorts and methodologies employed. BMI—Body Mass Index; BPH—Benign Prostate Hyperplasia; PCa—Prostate Cancer; PSA—Prostate-Specific Antigen; SD—Standard Deviation. Data represent the mean ± SD of each group.

Methodology	BMI	Parameter	BPH Group	PCa Group
RT-qPCR Cohort 1 (n = 20)	Normo-weight	n	5	5
		BMI (kg/cm^2)	23.76 ± 1.71	23.47 ± 1.39
		PSA (ng/mL)	4.28 ± 5.67	6.73 ± 4.28
		Gleason score	-	7 ± 0
		Age	69.9 ± 5.59	60.8 ± 4.96
		PI-RADS	2.67 ± 1.15	4.25 ± 0.50
	Obesity	n	5	5
		BMI (kg/cm^2)	31.05 ± 2.09	33.63 ± 1.25
		PSA (ng/mL)	9.60 ± 17.04	14.29 ± 8.14
		Gleason score	-	7 ± 0
		Age	69.4 ± 7.66	62.2 ± 3.42
		PI-RADS	2 ± 0	4.20 ± 1.30
Microfluidic-based qPCR array Cohort 2 (n = 48)	Normo-weight	n	6	14
		BMI (kg/cm^2)	24.161 ± 3.20	23.80 ± 1.12
		PSA (ng/mL)	3.99 ± 5.12	8.43 ± 7.35
		Gleason score	-	7 ± 0
		Age	69.33 ± 5.04	59.45 ± 5.15
		PI-RADS	2.33 ± 0.81	4.51 ± 0.53
	Obesity	n	14	14
		BMI (kg/cm^2)	31.26 ± 2.07	31.13 ± 1.01
		PSA (ng/mL)	6.99 ± 10.31	7 ± 4.95
		Gleason score	-	6.85 ± 0.37
		Age	69.22 ± 6.62	58.85 ± 5.33
		PI-RADS	2 ± 0	4.27 ± 0.90

On the other hand, the less stable IRGs in terms of variability were *HMBS* (29.02 ± 3.17), *PPIG* (30.17 ± 2.61), *G6PD* (23.89 ± 2.32), *YWHAZ* (26.73 ± 2.21), and *GUSB* (29.54 ± 2.16) using the RT-qPCR method and *HMBS* (23.73 ± 3.46), *PPIG* (24.78 ± 3.03), *GUSB* (22.92 ± 3.01), *HSP90AB1* (20.02 ± 2.91), and *ACTB* (19.71 ± 2.80) using the microfluidic-based qPCR array method (Table S1).

These results suggest that more stable (less variable) IRG genes using both techniques are *LRP10*, *PGK1*, and *RPLP0* while the worst IRGs (with higher variability) were *HMBS*, *PPIG*, and *GUSB*.

2.2. Influence of BMI (Body Mass Index) and PCa Presence on the Expression of IRGs in PPATs

To evaluate the effect of BMI and PCa presence on the stability of the IRG expression levels previously analyzed in PPATs, we stratified the data derived from the two available cohorts of samples under these conditions (Figure 2). Firstly, when BMI was considered (patients with NW vs. OB) we noted that the IRGs previously observed to be more stable (less variable) by both techniques did not exhibit statistically significant differences in their expression levels (i.e., *LRP10*, *PGK1*, and *RPLP0*; Figure 2a), further suggesting that these IRGs are also appropriated candidates to be used in these analyses. In contrast, statistical differences were consistently found in the expression levels of *ACTB*, *B2M*, *GAPDH*, *GUSB*, *HMBS*, and *PPIG* when the samples were classified according to their BMI using both methodologies (RT-qPCR and microfluidic-based qPCR array; Figure 2a), suggesting that these IRGs are not good candidates to be used in these expression analyses.

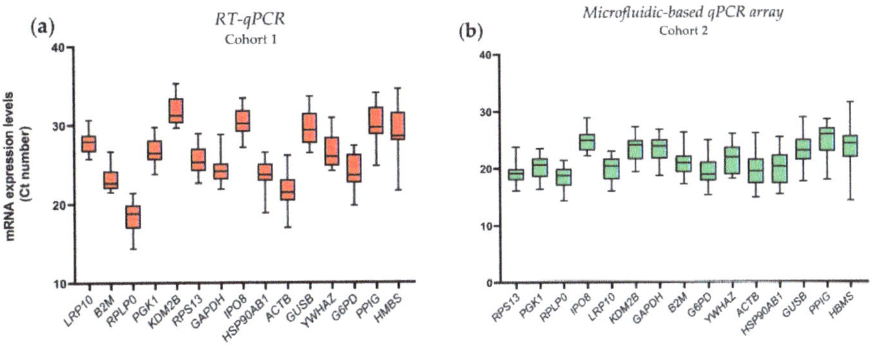

Figure 1. mRNA levels of 15 evaluated internal reference genes (IRGs) in human PPATs. Data (mean ± SD of Cycle threshold (Ct) levels) obtained by the reverse transcription quantitative PCR (RT-qPCR technique) (**a**) or by the microfluidic-based qPCR array technique (**b**). Genes are ordered ascendingly according to the standard deviation.

Secondly, (Figure 2b), when the presence of PCa was considered (patients with BPH vs. PCa), we again observed that the IRGs previously reported to be more stable (less variable) in the different analyses showed until this point (i.e., *LRP10*, *PGK1*, and *RPLP0*; data showed in Figures 1 and 2a) did not exhibit differences in their expression levels (Figure 2a). In contrast, statistical differences were consistently found in the expression levels of *GUSB*, *HMBS*, *IPO8*, *KDM2B*, and *PPIG* (using both RT-qPCR and microfluidic-based qPCR array, methodologies) as well as *ACTB*, *HSP90AB1*, *RPS13*, and *YWHAZ* (using one of the methodologies) when classifying patients in BPH vs. PCa (Figure 2b), suggesting that all these IRGs are not good candidates to be used in these expression analyses.

In order to further explore the influence of the different conditions previously studied [i.e., according to BMI status (NW vs. OB) or presence of PCA (BPH vs. PCa)], different methods specialized in the evaluation of the suitability of IRGs were used [ΔCt, NormFinder, GeNorm, and BestKeeper algorithms]. The results are summarized in Figure S2 which shows how the evaluated IRGs are ranked across the two techniques employed by the different software. It should be noted that differences were observed in these rankings depending on the tool used, which was due to the specific algorithms and parameters evaluated in each case. Moreover, it is important to clarify that, in all cases, the three algorithms displayed slightly worse parameters for the data obtained by the microfluidic-based qPCR array compared to the RT-qPCR. This may be due to the fact that, regarding protocol differences, this methodology uses a much smaller amount of starting genetic material; so, a previous pre-amplification step is necessary, which may introduce some kind of bias. Likewise, the pre-amplification step may also explain why the Ct values obtained on the microfluidic-based qPCR array are substantially lower compared to those of RT-qPCR. Nonetheless, in order to solve the discrepancies mentioned, we next used the web tool RefFinder to calculate the geometric mean of ranking values from the weighted values of each algorithm score for each of the IRGs evaluated. Specifically, a comprehensive ranking was generated (Figure 3a) which confirmed previous findings indicating that *LRP10*, *PGK1*, and *RPLP0* are in the top ranking of the most stable IRGs, which reinforces the idea that these three IRGs are the most appropriate to be used in expression analysis of PPATs, at least when considering the presence PCa and OB. Additionally, a bootstrap step strategy without replacement and size 100,000 was performed using the weighted scores previously calculated, confirming that *LRP10* is the best reference gene, followed by *PGK1* and *RPLP0* (Figure 3b).

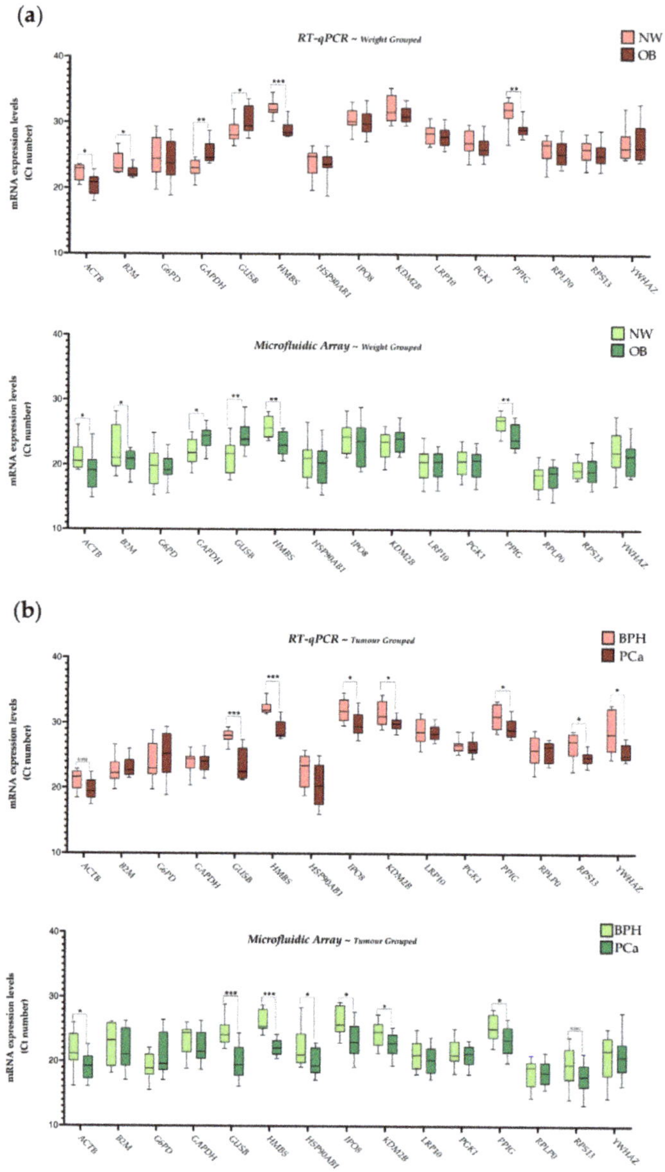

Figure 2. mRNA levels of the 15 evaluated IRGs in human PPAT samples from the two available cohorts of samples stratified by BMI or by the presence of PCa. Data obtained by the RT-qPCR technique (**a**) or by the microfluidic-based qPCR array technique (**b**). Asterisks (* $p < 0.05$; ** $p < 0.01$, and *** $p < 0.001$) indicate statistically significant differences between groups. Normo-weight—NW; Obesity—OB.

(a)

Nº	RT-qPCR	Microarray	OVERALL
1	*LRP10*	*PGK1*	*LRP10*
2	*RPS13*	*LRP10*	*PGK1*
3	*PGK1*	*RPLPO*	*RPLPO*
4	*KDM2B*	*B2M*	*RPS13*
5	*RPLPO*	*GAPDH*	*KDM2B*
6	*B2M*	*RPS13*	*B2M*
7	*G6PD*	*KDM2B*	*G6PD*
8	*IPO8*	*G6PD*	*GAPDH*
9	*YWHAZ*	*HSP90AB1*	*YWHAZ*
10	*HSP90AB1*	*YWHAZ*	*HSP90AB1*
11	*GAPDH*	*IPO8*	*IPO8*
12	*PPIG*	*ACTB*	*ACTB*
13	*ACTB*	*PPIG*	*PPIG*
14	*GUSB*	*GUSB*	*GUSB*
15	*HMBS*	*HMBS*	*HMBS*

(b)

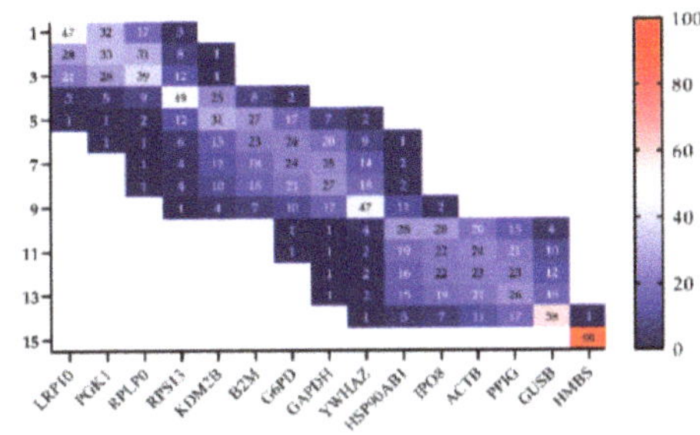

Figure 3. Ranking of the most stable IRGs in human periprostatic adipose tissues (PPAT) derived from the two cohorts of patients BPH and PCa with NW or OB. (**a**) Weighted gene ranking suitability of IRGs evaluated across the ΔCt method, GeNorm (v3.5), BestKeeper (v1.0), and NormFinder (v.20.0) software. (**b**) Empirical distribution of ranks shown in percentage (%) for the 15 evaluated IRGs in bootstrap.

Importantly, it should be also indicated that a more detailed view of the value distribution of the top three best-ranked IRGs (*LRP10*, *PGK1*, and *RPLPO*) throughout all experimental subgroups of PPATs indicated no statistically significant changes across the different groups (Figure 4). Specific software parameters for all evaluated genes are also listed in Table S1.

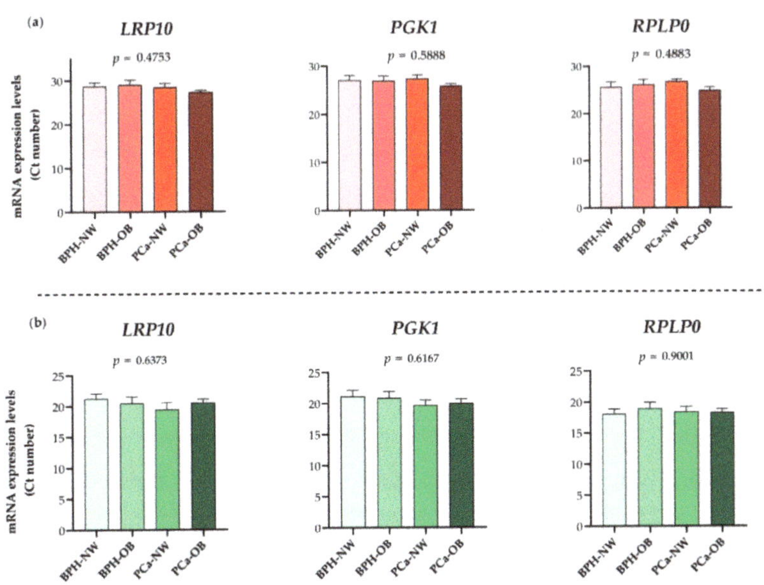

Figure 4. Ct value distribution of the top 3 best-ranked internal reference genes (IRGs) according to Body Mass Index (BMI) or by the presence of prostate cancer (PCa). Expression levels (Cts) were obtained by the RT-qPCR technique (**a**) or by the microfluidic-based qPCR array technique (**b**). Normo-weight—NW; Obesity—OB; Benign Prostate Hyperplasia (BPH).

2.3. Validation of RPLP0, PGK1, and LRP10 as Reference Genes

Finally, in order to empirically confirm that the selected IRGs were suitable to be used in expression analyses of PPATs obtained under different experimental conditions (i.e., patients with different BMIs or with and without PCa), we measured the expression levels of six metabolic- and tumor-related genes well-known to be dysregulated in the AT of humans under cancer-related conditions (i.e., *SPP1*, *IL6R*, *CD68*, *AGT*, *FNDC5*, and *RETN* [18–23]; primers sequences are listed in Table S2). Specifically, gene expression levels of *SPP1*, *IL6R*, *CD68*, *AGT*, *FNDC5*, and *RETN* were determined using a microfluidic qPCR-based approach and are represented before and after the adjustment by a normalization factor generated by the GeNorm software based on the expression levels of the three selected IRGs (*LRP10*, *PGK1*, and *RPLP0*) (Figure 5). As expected, the expression levels of *SPP1*, *IL6R*, *CD68*, and *AGT* were found to be up-regulated while *FNDC5* and *RETN* levels were down-regulated in PPAT samples from patients with PCa vs. BPH after data normalization. Importantly, these comparisons (non-normalized vs. normalized data) clearly indicated the necessity of using suitable IRGs for the normalization of gene expression levels.

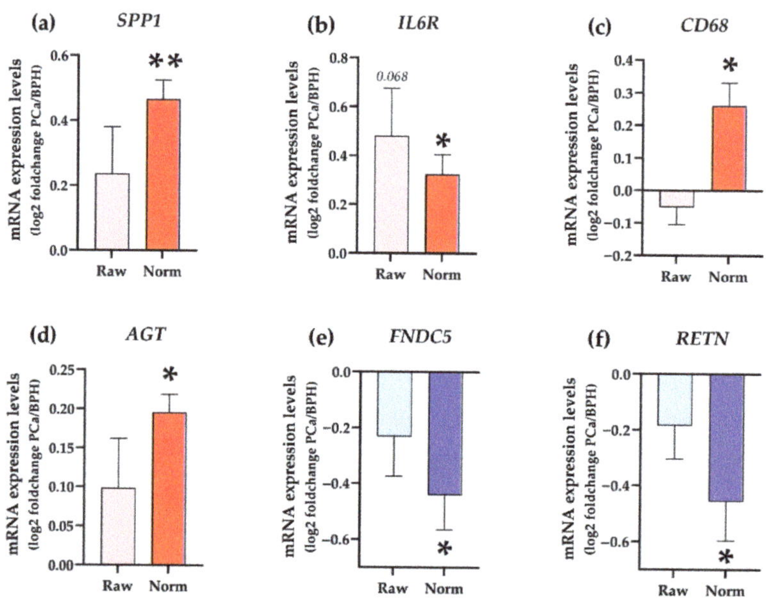

Figure 5. Changes in the expression levels of (**a**) *SPP1*, (**b**) *IL6R*, (**c**) *CD68*, (**d**) *AGT*, (**e**) *FNDC5*, and (**f**) *RETN* in human PPAT from patients with PCa compared with patients with BPH before normalization (raw data) and after normalization (norm) with selected IRGs (*LRP10*, *PGK1*, and *RPLP0*) are shown. Normalization of genes was performed using a normalization factor obtained with the GeNorm software (v3.5) based on the expression levels of *LRP10*, *PGK1*, and *RPLP0*. Asterisks (* $p < 0.05$, and ** $p < 0.01$) indicate statistically significant differences between groups.

Additionally, to further validate the selection of *LRP10*, *PGK1*, and *RPLP0* as suitable IRGs, we empirically determined the expression levels of two genes (*ANGPT1* and *LEP*) that were previously described to be dysregulated in PPATs of patients with PCa and OB [24]. Specifically, Ribeiro and colleagues indicated that *LEP* and *ANGPT1* were overexpressed in PPATs of patients with BPH under overweight/obese vs. normo-weight conditions, which was confirmed in our analyses using RT-qPCR methodology and normalizing the data with a normalization factor based on the expression levels of *LRP10*, *PGK1*, and *RPLP0*. Likewise, Ribeiro and colleagues demonstrated that the expression of two metalloproteases (*MMP2* and *MMP9*) was increased in conditioned media derived from PPAT explants of overweight PCa patients as compared with normo-weight patients with BPH [25] which was also confirmed in our study after normalization of the data by IRGs (Figure 6c,d). Importantly, these changes observed in the expression levels of *ANGPT1*, *LEP*, *MMP2*, and *MMP9* after normalization were not observed when the data were not normalized (raw data) (Figure 6a–d), which again clearly indicates the necessity of using suitable IRGs for normalization of gene expression levels in PPATs under different experimental conditions (i.e., presence or absence of Ob and PCa).

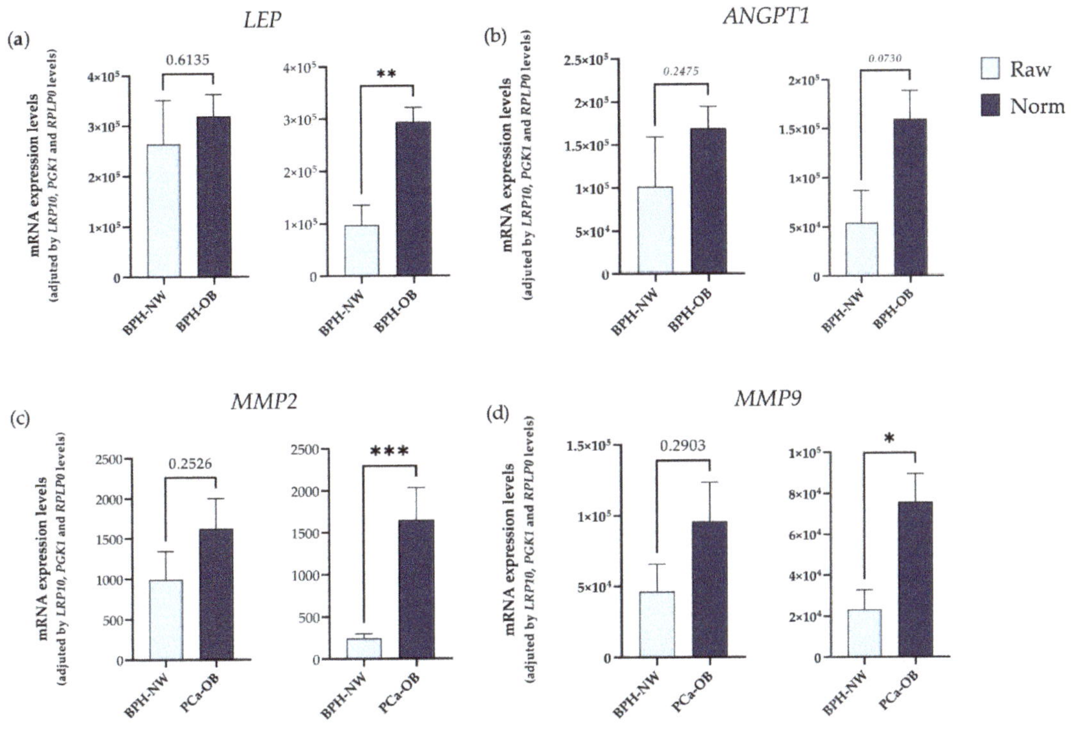

Figure 6. Empirical confirmation demonstrates that the selected IRGs (*LRP10*, *PGK1*, and *RPLP0*) were suitable to be used in expression analyses in human PPAT obtained under different experimental conditions. Expression levels of (**a**) *LEP*, (**b**) *ANGPT1*, (**c**) *MMP2*, and (**d**) *MMP9* before normalization (raw data) and after normalization with selected IRGs (*LRP10, PGK1*, and *RPLP0*) are shown. Normalization of genes was performed using a normalization factor obtained with the GeNorm software (v3.5) based on the expression levels of *LRP10*, *PGK1*, and *RPLP0*. Asterisks (* $p < 0.05$; ** $p < 0.01$, and *** $p < 0.001$) indicate statistically significant differences between groups.

3. Discussion

Gene expression analysis is one of the most used strategies in molecular biology research. However, selecting inappropriate IRGs for the normalization of the data can result in misleading or biased findings. Classically, *GAPDH* or *ACTB* were two of the most commonly used IRGs, but it is well known that these genes can exhibit significant variation in specific cells, tissues, and experimental conditions (such as the metabolic status and/or tumor presence) [15,26], highlighting the need for rigorous selection of IRGs for accurate gene expression analyses and reliable results.

In this study, we evaluated the suitability of 15 previously documented IRGs in order to determine the most appropriate IRGs for standardizing the mRNA data obtained from PPATs derived from two independent cohorts of samples, using two widely used methodologies (RT-qPCR and microfluidic-based qPCR array) and different algorithms (ΔCt method, GeNorm, BestKeeper, and NormFinder) and also considering the presence or absence of OB and PCa. We believe that this study is relevant because although the relationship between OB and cancer is well known in some specific cancer types, it is still ambiguous in the case of PCa [27] and also because the PPAT has been gaining important attention in recent years as a metabolic- and tumor-driver tissue associated with the prostate gland [28]. However, to the best of our knowledge, no works have evaluated the suitability of different IRGs in PPATs. Moreover, this is also the first study addressing this validation in PPATs from patients with and without PCa and/or OB.

Firstly, we found that certain IRGs (including *HMBS*, *PPIG*, *GUSB*, and *G6PD*) presented a high variability in their expression levels of PPATs. Indeed, the genes with the greatest variability corresponded to those that presented higher differences in expression levels when the cohorts were distributed according to their BMI and the presence/absence of PCa. Among them, some genes were highly dependent on BMI, such as *ACTB*, *B2M*, and *GAPDH*; some were highly dependent on the PCa existence, such as *GUSB*, *IPO8*, and *KDM2B*; and others varied in both conditions (e.g., *HMBS* and *PPIG*) and were, therefore, considered as the worst IRGs in PPATs under these different experimental conditions.

To evaluate the best IRGs among all the possible candidates, we used four methods based on different statistical parameters: the ΔCt method, GeNorm, BestKeeper, and NormFinder. These analyses revealed some inconsistencies in the results obtained by each software which might be due to different considerations when calculating the scores. For example, the ΔCt method is exclusively based on the average of the SD through the relative comparison of the expression of candidate genes by pairs, GeNorm is based on the arithmetic mean of the pair-wise variation, NormFinder relies on an intra- and inter-group variation of inputted genes, and BestKeeper considers the standard deviation and coefficient variance to calculate the stability factor. Thus, we used a fourth software, called RefFinder, that takes into consideration the weighted contribution score from the previously mentioned software to calculate a geometric mean ranking of values. As a result, we found that the highest stability for RT-qPCR determinations corresponded to the genes *LRP10*, *RPS13*, *PGK1*, *KDM2B*, and *RPLP0* while *PGK1*, *LRP10*, *RPLP0*, *B2M*, and *GAPDH* were the most appropriate for microfluidic-based qPCR arrays; with *LRP10*, *PGK1*, and *RPLP0* being common to both analyses. In fact, we performed a bootstrapping analysis based on the weighted contribution of each software for each gene to their global score which revealed and confirmed that the best genes to be used as IRGs for gene expression studies are *LRP10*, *PGK1*, and *RPLP0*.

Specifically, *LRP10* encodes a low-density lipoprotein receptor family protein but its role in humans is poorly explored. Most available studies associate *LRP10* with neurodegenerative disorders, especially Parkinson's disease [29]. In agreement with Gabrielsson and colleagues who analyzed a battery of potential IRGs in human AT biopsies in the context of a wide range of commonly studied physiological parameters (including BMI), *LRP10* was found to be the IRG with the highest stability [30]. In the case of *PGK1*, it encodes for a glycolytic enzyme that catalyzes the conversion of 1,3-diphosphoglycerate to 3-phosphoglycerate and exerts an important role in the control of glucose metabolism. Moreover, this enzyme has been widely explored in various fields, including cancer [31–33], although it has also been considered a potential IRG in different human tissues under several experimental conditions and even in distinct animal models [34,35]. In this sense, and regarding the adipose tissue depots, Neville and co-workers demonstrated that *PGK1* was the most stable gene among a set of several cohorts of human AT samples and cells, although PPAT was not considered in that work [36]. Lastly, *RPLP0* encodes a ribosomal protein that is a component of the 60S subunit of the ribosomes and has been already proposed to be a suitable IRG across a wide variety of tissues and experimental conditions. Consistent with our results, Zhang et al. evaluated the potential of some genes in three different ATs (epididymal white, inguinal beige, and brown adipose tissue) under different metabolic conditions by using the same software as that used in the present study. Specifically, they showed that the *RPLP0* gene had the lowest variability across all situations evaluated suggesting that could actually be used as a universal IRG in AT [14].

In summary, this study confirms the importance of selecting suitable IRGs for normalization to obtain reliable results using different gene expression techniques. It also provides fundamental information regarding a list of potential IRGs that can be used for the analysis of gene expression patterns in PPATs under different experimental conditions. Specifically, this study demonstrates, for the first time, that the expression stability of many IRGs widely used by research laboratories is not valid for the normalization of expression data since their expression significantly varies in PPATs among different experimental settings. Importantly, we solidly demonstrated that *LRP10* has the lowest variability, the highest stability, and more consistency across two widely used techniques (RT-qPCR and microfluidic-based qPCR array); throughout all the analytical software employed, we showed highly stable expression levels in PPAT samples that are independent of the BMI and the presence PCa. In addition, in accordance with the minimum information for publication of quantitative (MIQE) guidelines which recommend the use of multiple reference genes for normalization [37], we provide compelling evidence demonstrating the use of *LRP10*, *PGK1*, and *RPLP0* as the best IRGs for gene expression analyses in human PPATs under different experimental conditions, especially when considering the metabolic status (OB) and the presence of PCa.

4. Materials and Methods

4.1. Subjects and Samples

This study was approved by the Reina Sofia University Hospital Ethics Committee and conducted following the principles of the Declaration of Helsinki. Cordoba Biobank Node (through Andalusian Biobank) coordinated the collection, processing, and management of the PPAT samples used. Microdisected samples were immediately frozen in liquid nitrogen and then stored at −80 °C following the standard protocols established for this purpose. Written informed consent was obtained from all participants. PPAT samples were obtained from patients who were diagnosed with PCa and immediately treated by radical prostatectomy. All patients were diagnosed with clinically localized and hormone-naive PCa, representing a homogenous cohort of patients (Table 1).

Two different cohorts were used in this study. The first one (Cohort 1, n = 20) was used as a discovery cohort for qPCR determinations. A second larger and independent cohort (Cohort 2, n = 48) was used to implement the microfluidic-based qPCR array. Available clinical history details from all patients included in these cohorts are summarized in Table 1, including BMI, Prostate-Specific Antigen (PSA), Gleason score, age, and PI-RADS (prostate imaging reporting and data system). Briefly, four different experimental conditions were analyzed: patients with BPH or PCa and both NW and OB conditions. PI-RADS defines the standards of high-quality for multi-parametric magnetic resonance imaging (mpMRI), including image creation and reporting [38]. Briefly, this system is stratified into five categories which are PIRADS 1, very low (clinically significant cancer is highly unlikely to be present); PIRADS 2, low (clinically significant cancer is unlikely to be present); PIRADS 3, intermediate (the presence of clinically significant cancer is equivocal); PIRADS 4, high (clinically significant cancer is likely to be present); and PIRADS 5, very high (clinically significant cancer is highly likely to be present).

4.2. Total RNA Isolation and cDNA Synthesis

Total RNA was isolated with TRIzol Reagent (ThermoFisher Scientific, Bedford, MA, USA) from 100 mg of PPAT following supplier instructions. After tissue disruption, an additional step consisting of two consecutively centrifugations (5 min at 12,200× *g*-force) and supernatant discarding was performed to remove cellular debris and fat excess. A DNAse treatment (RNase-Free DNase Kit—Qiagen, Madrid, Spain) was applied previously to RNA concentration and purity determination on a Nanodrop One Microvolume UV–Vis Spectrophotometer (ThermoFisher Scientific, Bedford, MA, USA). A total amount of 1 µg per sample was then retrotranscribed using random hexamer primers and the RevertAid RT reverse transcription kit (ThermoFisher Scientific, Bedford, MA, USA). Retrotranscribed cDNA was then stored at −20 °C until use.

4.3. Potential Reference Gene Selection and Primer Design

For the final selection of the genes included in this study, a combinatorial integration of different bibliographic approaches was carried out. First, an intensive literature search was performed in PubMed, SciELO, and Medline databases using the terms "reference gene" and "housekeeping gene". Second, we also screened by the terms "adipose tissue", "cancer", and "diet" and evaluated two types of articles: (i) those articles whose main objective was to specifically assess the validity of reference genes in human adipose tissues under different cancer and/or diet conditions and (ii) general articles wherein different reference genes were commonly and routinely used in adipose tissues. Finally, and due to the clear limitation in finding information/references regarding human PPATs, we also checked the available information/references of related adipose tissues from rodent models (mouse and rat), including inguinal or perigonadal adipose tissues. Based on all these bibliographic searches, as well as preliminary results from our group using PPATs (unpublished), a list of 15 candidate genes was selected to be studied as potential IRGs in PPATs. The IRGs consulted in this bibliographic approach are included in Table S3 while the 15 candidate IRGs selected, primer sequences, amplicon size, and gene functions are listed in Table 2. Primers that specifically target these genes were designed with the Primer3 software v.0.4.0 while the validation for the suitability for RT-qPCR and microfluidic-based qPCR array technology was assessed by checking the specificity (agarose gel band size, sanger sequencing, and melting curves) as previously reported by our laboratory [39,40]. Specific custom oligonucleotide sequences were purchased from Metabion (Munich, Germany).

Table 2. Gene symbol, primers sequences, Amplicon Size (A.S.), official name, and function of evaluated genes.

Gene	Sequence	A.S (bp)	Official Name	Function
ACTB	ACTCTTCCAGCCTTCCTTCCT CAGTGATCTCCTTCTGCATCCT	176	Actin Beta	Highly conserved proteins involved in cell motility, structure, and integrity
B2M	GCTCGCGCTACTCTCTCTTT TCCATTCTCTGCTGGATGAC	88	Beta-2-Microglobulin	MHC class I associated protein that is found on nearly all cellular surfaces
G6PD	GCAAACAGAGTGAGCCCTTC GCCAGCCACATAGGAGTTG	89	Glucose-6-Phosphate Dehydrogenase	Cytosolic enzyme whose main function is to produce NADPH using glucose
GAPDH	GCCTCAAGATCATCAGCAATG CTTCCACGATACCAAAGTTGT	90	Glyceraldehyde-3-Phosphate Dehydrogenase	Enzyme related to carbohydrate metabolism and DNA glycosylation
GUSB	TGACCGCTATGGGATTGT CTACGCACCACTTCTTCCA	120	Glucuronidase Beta	Enzyme able to hydrolyze glycosaminoglycans
HMBS	TTGCTATGTCCACCACAGG CCAGGTCCACTTCATTCTTCT	117	Hydroxymethylbilane Synthase	Enzyme involved in the condensation of porphobilinogen
HSP90AB1	ATGGAAGAGAGCAAGGCAAAG GCAGCAAGGTGAAGACACAA	114	Heat Shock Protein 90α Family Class B Member 1	Constitutive form of the cytosolic 90 kDa heat-shock protein
IPO8	CCAAGGGGTGGTTCATTCT TCTTGCCACAGCTCTTCATC	120	Importin 8	GTPase involved in the nuclear import of proteins
KDM2B	TGGAGGGCAAAGATTTCAAC TCCCAGTCCATCCTTTTCTC	86	Lysine Demethylase 2B	Component of SCFs complex involved in ubiquitination

Table 2. *Cont.*

Gene	Sequence	A.S (bp)	Official Name	Function
LRP10	GGCAACGTCACCATCACTT AATCTTGGCTGTAGGAGAGCA	82	LDL Receptor Related Protein 10	Encodes for a low-density lipoprotein receptor family protein
PGK1	GGTGGAATGGCTTTTACCTT ATCTTGGCTCCCTCTTCATC	83	Phosphoglycerate Kinase 1	Glycolytic enzyme that catalyzes the conversion of 1,3-diphosphoglycerate
PPIG	TTGCGGAGGTACGGATACT TGAGGAGGAGGAAGATGATG	104	Peptidylprolyl Isomerase G	Protein involved in protein peptidyl-prolyl isomerization
RPLP0	AACTCTGCATTCTCGCTTCC GGACTCGTTTGTACCCGTTG	119	Ribosomal Protein Lateral Stalk Subunit P0	Ribosomal protein that is a component of the 60S subunit
RPS13	CCCACTTGGTTGAAGTTGAC CCGATCTGTGAAGGAGTAAGG	86	Ribosomal Protein S13	Ribosomal protein that is a component of the 40S subunit
YWHAZ	TACCGTTACTTGGCTGAGGTT GATGTGTTGGTTGCATTTCC	118	Y3-Monooxygenase/ W5-M.oxygenase Act. Prot. Z	Signal transduction mediation by binding to phosphoserine-containing proteins

4.4. RT-qPCR and Microfluidic-Based qPCR Array

Quantitative RT-qPCR was used to determine the mRNA levels of the selected IRGs in the discovery cohort 1 of PPATs. Specifically, primers for the selected IRGs were used in combination with Brilliant III Ultra-Fast SYBR Green (ThermoFisher Scientific, Bedford, MA, USA) using the Stratagene Mx3000P system (Agilent Technologies, Madrid, Spain) following the manufacturer's instructions. mRNA levels were determined to analyze the putative steady levels or the variability in the expression across the different groups of patients [i.e., with BPH or PCa and both under NW and OB conditions] using the MxPro 3.0 qPCR software (Agilent Technologies, Madrid, Spain).

Likewise, a microfluidic-based qPCR array (Standard BioTools, San Francisco, CA, USA) was used to simultaneously determine the gene expression levels of the 15 candidate IRGs in the second cohort of PPAT samples using a 48.48 Dynamic Array Plate (GE 48.48 Dynamic Array Reagent Kit with Control Line Fluid, Standard BioTools, San Francisco, CA, USA). Preamplification, exonuclease I treatment, and microfluidic-based qPCR array were implemented as previously described [41,42] following the manufacturer's instructions using the Biomark HD System and the Fluidigm Real-Time PCR Analysis v4.8.2 Software.

4.5. Evaluation of Putative IRGs

To study the suitability of the selected IRGs to be used for the normalization of gene expression analyses as well as to establish a prioritized ranking of these IRGs, a combination of different software analysis (including ΔCt method, NormFinder v.20.0, GeNorm v3.5, and BestKeeper v1.0) were used. Specifically, the ΔCt method is based on the relative comparison expression of 'pairs of genes' within each sample. On this basis, stability of the candidate housekeeping genes was ranked according to repeatability of the gene expression differences [43]. The NormFinder algorithm [44] relies on intra- and inter-group variation to estimate gene expression stability enabling the ranking of the candidates so that the less stable genes represent the worse as reference genes. GeNorm [45] calculates the called 'M' value using the arithmetic mean of the pair-wise variations of each candidate gene. A low M value (default value threshold set on 1.5) represents a stable gene expression. In addition, GeNorm also calculates the normalization factor according to the number of reference genes inputted. BestKeeper [46] considers the SD as a measure of estimated variability along with the coefficient of variance (CV) from the direct Ct values of each gene. The Ct is established as the number of cycles needed to reach the fluorescent signal threshold so that the lower the Ct level, the greater amount of nucleic sample in an inversely proportional way. Furthermore, RefFinder (http://blooge.cn/RefFinder/), (accessed on 2 September 2023) a web-based tool, was also used to assign a weighted and individual gene score by calculating the geometric mean of their weights for the overall final ranking, allowing the integration of the three computational programs cited below [47]. In addition, a bootstrap step with replacement [48] was performed to evaluate the confidence of the ranking (size of 100,000) across those weighted scores by using the R (v. 3.5.2) package Boot (v. 1.3-28.1).

4.6. Statistical Analyses

Statistical differences were assessed by a *t*-test or by two-way ANOVA followed by Fisher's correction exact test according to the comparison. All statistical analyses and plots were generated using Prism software 9.0 (GraphPad Software, La Jolla, CA, USA). The p-value of < 0.05 was considered statistically significant. Data represent the median (interquartile range) or means ± SD, wherein asterisks (* p <0.05; ** p < 0.01; and *** p < 0.001) indicate statistically significant differences between different conditions. The Metaboanalyst 5.0 online platform was used to perform normalization and density plot visualization.

4.7. Study Limitations

Limitations of this study include the selection of a defined set of genes mainly based on a literature search. Additionally, the number of samples (n = 20 in the discovery cohort-1 and n = 48 in the validation cohort-2) might be considered limited. However, it should be mentioned that the work herein presented represents a collection of PPAT tissues over a 5-year period where the availability of PPAT is dependent on the written informed consent from the participants which, in our experience, is not easy to obtain in many cases. Therefore, it is very difficult to obtain human PPAT samples for research purposes and this is probably the reason why the publications that include PPAT samples are very limited and include similar or lower numbers of samples than the one used in our study [49–51]. Furthermore, across the selected patients, other confounding factors such as androgen dependence, tumor behavior, disease duration, or previous anti-tumor treatments, could not be considered in this study. Finally, regarding the validation of the three genes selected as the most suitable IRGs among all the candidates analyzed, part of this validation is based on gene alterations that were described in other types of adipose tissues. Given the peculiar nature of periprostatic adipose tissue, the changes observed may not necessarily occur in the same way as in other depots. Therefore, results herein obtained should be corroborated in additional cohorts of samples.

Supplementary Materials: The supporting information can be downloaded at https://www.mdpi.com/article/10.3390/ijms242015140/s1. References [52–67] are cited in the Supplementary Material.

Author Contributions: J.M.P.-G.: Software, Validation, Formal Analysis, Investigation, Writing, Visualization. F.P.-P.: Investigation, Writing—Review and Editing. M.D.L.L.-B.: Writing—Review and Editing. A.J.M.-H.: Writing—Review and Editing. E.G.-G.: Writing—Review and Editing. A.D.H.-M.: Writing—Review and Editing. R.G.-R.: Writing—Review and Editing. M.M.M.: Conceptualization, Writing—Review and Editing. M.D.G.: Writing—Review and Editing. R.M.L.: Conceptualization, Writing—Review and Editing, Supervision, Project administration, Funding acquisition. All authors have read and agreed to the published version of the manuscript.

Funding: This research was funded by the Spanish Ministry of Science, Innovation, and Universities (Research-Grant: PID2022-1381850B-I00; PID2019-105564RB-I00; Predoctoral contracts: FPU18-06009, PRE2020-094225, FPU18-02485); Junta de Andalucia (BIO-0139); CIBERobn (CIBER is an initiative of Instituto de Salud Carlos III, Ministerio de Sanidad, Servicios Sociales e Igualdad, Spain).

Institutional Review Board Statement: The study was conducted in accordance with the Declaration of Helsinki, and approved by the Reina Sofia University Hospital Ethics Committee (5570, approved on 29 May 2023).

Informed Consent Statement: Informed consent was obtained from all subjects involved in the study.

Data Availability Statement: The data presented in this study are available in the article and Supplementary Material.

Conflicts of Interest: The authors declare no conflict of interest.

References

1. Ibrahim, M.M. Subcutaneous and visceral adipose tissue: Structural and functional differences. *Obes. Rev.* **2010**, *11*, 11–18. [CrossRef] [PubMed]
2. Scheja, L.; Heeren, J. The endocrine function of adipose tissues in health and cardiometabolic disease. *Nat. Rev. Endocrinol.* **2019**, *15*, 507–524. [CrossRef] [PubMed]
3. Unamuno, X.; Gómez-Ambrosi, J.; Rodríguez, A.; Becerril, S.; Frühbeck, G.; Catalán, V. Adipokine dysregulation and adipose tissue inflammation in human obesity. *Eur. J. Clin. Investig.* **2018**, *48*, e12997. [CrossRef]
4. Blüher, M. Obesity: Global epidemiology and pathogenesis. *Nat. Reviews. Endocrinol.* **2019**, *15*, 288–298. [CrossRef] [PubMed]
5. Arnold, M.; Lam, F.; Ervik, M.; Soerjomataram, I. *Cancer and Obesity: Global Burden of Cancer Attributable to Excess Weight*; International Agency for Research on Cancer: Lyon, France, 2015. Available online: http://gco.iarc.fr/obesity (accessed on 1 October 2022).
6. Arnold, M.; Pandeya, N.; Byrnes, G.; Renehan, P.A.G.; Stevens, G.A.; Ezzati, P.M.; Ferlay, J.; Miranda, J.J.; Romieu, I.; Dikshit, R.; et al. Global burden of cancer attributable to high body-mass index in 2012: A population-based study. *Lancet Oncol.* **2015**, *16*, 36–46. [CrossRef]
7. Di Sebastiano, K.M.; Pinthus, J.H.; Duivenvoorden, W.C.M.; Mourtzakis, M. Glucose impairments and insulin resistance in prostate cancer: The role of obesity, nutrition and exercise. *Obes. Rev.* **2018**, *19*, 1008–1016. [CrossRef] [PubMed]
8. Allott, E.H.; Masko, E.M.; Freedland, S.J. Obesity and prostate cancer: Weighing the evidence. *Eur. Urol.* **2013**, *63*, 800–809. [CrossRef] [PubMed]
9. Gómez-Gómez, E.; Carrasco-Valiente, J.; Campos-Hernández, J.P.; Blanca-Pedregosa, A.M.; Jiménez-Vacas, J.M.; Ruiz-García, J.; Valero-Rosa, J.; Luque, R.M.; Requena-Tapia, M.J. Clinical association of metabolic syndrome, C-reactive protein and testosterone levels with clinically significant prostate cancer. *J. Cell. Mol. Med.* **2019**, *23*, 934–942. [CrossRef]
10. Laurent, V.; Toulet, A.; Attané, C.; Milhas, D.; Dauvillier, S.; Zaidi, F.; Clement, E.; Cinato, M.; Le Gonidec, S.; Guérard, A.; et al. Periprostatic Adipose Tissue Favors Prostate Cancer Cell Invasion in an Obesity-Dependent Manner: Role of Oxidative Stress. *Mol. Cancer Res.* **2019**, *17*, 821–835. [CrossRef] [PubMed]
11. Zhang, Q.; Sun, L.J.; Qi, J.; Yang, Z.G.; Huang, T.; Huo, R.C. Periprostatic adiposity measured on magnetic resonance imaging correlates with prostate cancer aggressiveness. *Urol. J.* **2014**, *11*, 1793–1799. [CrossRef]
12. Almeida-Oliveira, F.; Leandro, J.G.B.; Ausina, P.; Sola-Penna, M.; Majerowicz, D. Reference genes for quantitative PCR in the adipose tissue of mice with metabolic disease. *Biomed. Pharmacother.* **2017**, *88*, 948–955. [CrossRef] [PubMed]
13. Gong, H.; Sun, L.; Chen, B.; Han, Y.; Pang, J.; Wu, W.; Qi, R.; Zhang, T.M. Evaluation of candidate reference genes for RT-qPCR studies in three metabolism related tissues of mice after caloric restriction. *Sci. Rep.* **2016**, *6*, 38513. [CrossRef]
14. Zhang, W.X.; Fan, J.; Ma, J.; Rao, Y.S.; Zhang, L.; Yan, Y.E. Selection of Suitable Reference Genes for Quantitative Real-Time PCR Normalization in Three Types of Rat Adipose Tissue. *Int. J. Mol. Sci.* **2016**, *17*, 968. [CrossRef] [PubMed]
15. Perez, L.J.; Rios, L.; Trivedi, P.; D'Souza, K.; Cowie, A.; Nzirorera, C.; Webster, D.; Brunt, K.; Legare, J.F.; Hassan, A.; et al. Validation of optimal reference genes for quantitative real time PCR in muscle and adipose tissue for obesity and diabetes research. *Sci. Rep.* **2017**, *7*, 3612. [CrossRef] [PubMed]
16. Taube, M.; Andersson-Assarsson, J.C.; Lindberg, K.; Pereira, M.J.; Gäbel, M.; Svensson, M.K.; Eriksson, J.W.; Svensson, P.A. Evaluation of reference genes for gene expression studies in human brown adipose tissue. *Adipocyte* **2015**, *4*, 280–285. [CrossRef]
17. Krautgasser, C.; Mandl, M.; Hatzmann, F.M.; Waldegger, P.; Mattesich, M.; Zwerschke, W. Reliable reference genes for expression analysis of proliferating and adipogenically differentiating human adipose stromal cells. *Cell. Mol. Biol. Lett.* **2019**, *24*, 14. [CrossRef]
18. Chang, M.C.; Eslami, Z.; Ennis, M.; Goodwin, P.J. Crown-like structures in breast adipose tissue of breast cancer patients: Associations with CD68 expression, obesity, metabolic factors and prognosis. *NPJ Breast Cancer* **2021**, *7*, 97. [CrossRef] [PubMed]
19. Di Zazzo, E.; Polito, R.; Bartollino, S.; Nigro, E.; Porcile, C.; Bianco, A.; Daniele, A.; Moncharmont, B. Adiponectin as Link Factor between Adipose Tissue and Cancer. *Int. J. Mol. Sci.* **2019**, *20*, 839. [CrossRef]
20. Lee, J.O.; Kim, N.; Lee, H.J.; Lee, Y.W.; Kim, S.J.; Park, S.H.; Kim, H.S. Resistin, a fat-derived secretory factor, promotes metastasis of MDA-MB-231 human breast cancer cells through ERM activation. *Sci. Rep.* **2016**, *6*, 18923. [CrossRef]
21. Maalouf, G.E.; El Khoury, D. Exercise-Induced Irisin, the Fat Browning Myokine, as a Potential Anticancer Agent. *J. Obes.* **2019**, *2019*, 6561726. [CrossRef]
22. Rasha, F.; Ramalingam, L.; Gollahon, L.; Rahman, R.L.; Rahman, S.M.; Menikdiwela, K.; Moustaid-Moussa, N. Mechanisms linking the renin-angiotensin system, obesity, and breast cancer. *Endocr. Relat. Cancer* **2019**, *26*, R653–R672. [CrossRef] [PubMed]
23. Sturtz, L.A.; Deyarmin, B.; van Laar, R.; Yarina, W.; Shriver, C.D.; Ellsworth, R.E. Gene expression differences in adipose tissue associated with breast tumorigenesis. *Adipocyte* **2014**, *3*, 107–114. [CrossRef]
24. Ribeiro, R.; Monteiro, C.; Catalán, V.; Hu, P.; Cunha, V.; Rodríguez, A.; Gómez-Ambrosi, J.; Fraga, A.; Príncipe, P.; Lobato, C.; et al. Obesity and prostate cancer: Gene expression signature of human periprostatic adipose tissue. *BMC Med.* **2012**, *10*, 108. [CrossRef] [PubMed]

25. Ribeiro, R.; Monteiro, C.; Cunha, V.; Oliveira, M.J.; Freitas, M.; Fraga, A.; Príncipe, P.; Lobato, C.; Lobo, F.; Morais, A.; et al. Human periprostatic adipose tissue promotes prostate cancer aggressiveness in vitro. *J. Exp. Clin. Cancer Res.* **2012**, *31*, 32. [CrossRef] [PubMed]
26. Jo, J.; Choi, S.; Oh, J.; Lee, S.-G.; Choi, S.Y.; Kim, K.K.; Park, C. Conventionally used reference genes are not outstanding for normalization of gene expression in human cancer research. *BMC Bioinform.* **2019**, *20*, 245. [CrossRef] [PubMed]
27. Bandini, M.; Gandaglia, G.; Briganti, A. Obesity and prostate cancer. *Curr. Opin. Urol.* **2017**, *27*, 415–421. [CrossRef] [PubMed]
28. Finley, D.S.; Calvert, V.S.; Inokuchi, J.; Lau, A.; Narula, N.; Petricoin, E.F.; Zaldivar, F.; Santos, R.; Tyson, D.R.; Ornstein, D.K. Periprostatic adipose tissue as a modulator of prostate cancer aggressiveness. *J. Urol.* **2009**, *182*, 1621–1627. [CrossRef] [PubMed]
29. Chen, Y.; Cen, Z.; Zheng, X.; Pan, Q.; Chen, X.; Zhu, L.; Chen, S.; Wu, H.; Xie, F.; Wang, H.; et al. LRP10 in autosomal-dominant Parkinson's disease. *Mov. Disord.* **2019**, *34*, 912–916. [CrossRef]
30. Gabrielsson, B.G.; Olofsson, L.E.; Sjögren, A.; Jernås, M.; Elander, A.; Lönn, M.; Rudemo, M.; Carlsson, L.M. Evaluation of reference genes for studies of gene expression in human adipose tissue. *Obes. Res.* **2005**, *13*, 649–652. [CrossRef]
31. Hu, H.; Zhu, W.; Qin, J.; Chen, M.; Gong, L.; Li, L.; Liu, X.; Tao, Y.; Yin, H.; Zhou, H.; et al. Acetylation of PGK1 promotes liver cancer cell proliferation and tumorigenesis. *Hepatology* **2017**, *65*, 515–528. [CrossRef]
32. Lu, W.; Gao, J.; Yang, J.; Cao, Y.; Jiang, L.; Li, M.; Zhang, Y.; Zhou, J.; Liu, Y. Down-Regulated Phosphoglycerate Kinase 1 Expression Is Associated With Poor Prognosis in Patients With Gallbladder Cancer. *Medicine* **2015**, *94*, e2244. [CrossRef] [PubMed]
33. Zhang, Y.; Yu, G.; Chu, H.; Wang, X.; Xiong, L.; Cai, G.; Liu, R.; Gao, H.; Tao, B.; Li, W.; et al. Macrophage-Associated PGK1 Phosphorylation Promotes Aerobic Glycolysis and Tumorigenesis. *Mol. Cell* **2018**, *71*, 201–215.e7. [CrossRef] [PubMed]
34. Falkenberg, V.R.; Whistler, T.; Murray, J.R.; Unger, E.R.; Rajeevan, M.S. Identification of Phosphoglycerate Kinase 1 (PGK1) as a reference gene for quantitative gene expression measurements in human blood RNA. *BMC Res. Notes* **2011**, *4*, 324. [CrossRef]
35. Panina, Y.; Germond, A.; Masui, S.; Watanabe, T.M. Validation of Common Housekeeping Genes as Reference for qPCR Gene Expression Analysis During iPS Reprogramming Process. *Sci. Rep.* **2018**, *8*, 8716. [CrossRef]
36. Neville, M.J.; Collins, J.M.; Gloyn, A.L.; McCarthy, M.I.; Karpe, F. Comprehensive human adipose tissue mRNA and microRNA endogenous control selection for quantitative real-time-PCR normalization. *Obesity* **2011**, *19*, 888–892. [CrossRef] [PubMed]
37. Bustin, S.A.; Benes, V.; Garson, J.A.; Hellemans, J.; Huggett, J.; Kubista, M.; Mueller, R.; Nolan, T.; Pfaffl, M.W.; Shipley, G.L.; et al. The MIQE guidelines: Minimum information for publication of quantitative real-time PCR experiments. *Clin. Chem.* **2009**, *55*, 611–622. [CrossRef] [PubMed]
38. Turkbey, B.; Rosenkrantz, A.B.; Haider, M.A.; Padhani, A.R.; Villeirs, G.; Macura, K.J.; Tempany, C.M.; Choyke, P.L.; Cornud, F.; Margolis, D.J.; et al. Prostate Imaging Reporting and Data System Version 2.1: 2019 Update of Prostate Imaging Reporting and Data System Version 2. *Eur. Urol.* **2019**, *76*, 340–351. [CrossRef]
39. Fuentes-Fayos, A.C.; Vázquez-Borrego, M.C.; Jiménez-Vacas, J.M.; Bejarano, L.; Pedraza-Arévalo, S.; L.-López, F.; Blanco-Acevedo, C.; Sánchez-Sánchez, R.; Reyes, O.; Ventura, S.; et al. Splicing machinery dysregulation drives glioblastoma development/aggressiveness: Oncogenic role of SRSF3. *Brain* **2020**, *143*, 3273–3293. [CrossRef]
40. Jiménez-Vacas, J.M.; Herrero-Aguayo, V.; Montero-Hidalgo, A.J.; Gómez-Gómez, E.; Fuentes-Fayos, A.C.; León-González, A.J.; Sáez-Martínez, P.; Alors-Pérez, E.; Pedraza-Arévalo, S.; González-Serrano, T.; et al. Dysregulation of the splicing machinery is directly associated to aggressiveness of prostate cancer. *EBioMedicine* **2020**, *51*, 102547. [CrossRef]
41. Del Río-Moreno, M.; Alors-Pérez, E.; González-Rubio, S.; Ferrín, G.; Reyes, O.; Rodríguez-Perálvarez, M.; Sánchez-Frías, M.E.; Sánchez-Sánchez, R.; Ventura, S.; López-Miranda, J.; et al. Dysregulation of the Splicing Machinery Is Associated to the Development of Nonalcoholic Fatty Liver Disease. *J. Clin. Endocrinol. Metab.* **2019**, *104*, 3389–3402. [CrossRef]
42. Gahete, M.D.; Del Rio-Moreno, M.; Camargo, A.; Alcala-Diaz, J.F.; Alors-Perez, E.; Delgado-Lista, J.; Reyes, O.; Ventura, S.; Perez-Martínez, P.; Castaño, J.P.; et al. Changes in Splicing Machinery Components Influence, Precede, and Early Predict the Development of Type 2 Diabetes: From the CORDIOPREV Study. *EBioMedicine* **2018**, *37*, 356–365. [CrossRef]
43. Silver, N.; Best, S.; Jiang, J.; Thein, S.L. Selection of housekeeping genes for gene expression studies in human reticulocytes using real-time PCR. *BMC Mol. Biol.* **2006**, *7*, 33. [CrossRef] [PubMed]
44. Andersen, C.L.; Jensen, J.L.; Ørntoft, T.F. Normalization of real-time quantitative reverse transcription-PCR data: A model-based variance estimation approach to identify genes suited for normalization, applied to bladder and colon cancer data sets. *Cancer Res.* **2004**, *64*, 5245–5250. [CrossRef]
45. Vandesompele, J.; De Preter, K.; Pattyn, F.; Poppe, B.; Van Roy, N.; De Paepe, A.; Speleman, F. Accurate normalization of real-time quantitative RT-PCR data by geometric averaging of multiple internal control genes. *Genome Biol.* **2002**, *3*, Research0034. [CrossRef]
46. Pfaffl, M.W.; Tichopad, A.; Prgomet, C.; Neuvians, T.P. Determination of stable housekeeping genes, differentially regulated target genes and sample integrity: BestKeeper--Excel-based tool using pair-wise correlations. *Biotechnol. Lett.* **2004**, *26*, 509–515. [CrossRef] [PubMed]
47. Xie, F.; Wang, J.; Zhang, B. RefFinder: A web-based tool for comprehensively analyzing and identifying reference genes. *Funct. Integr. Genom.* **2023**, *23*, 125. [CrossRef] [PubMed]
48. Boos, D.D. Introduction to the Bootstrap World. *Stat. Sci.* **2003**, *18*, 168–174. [CrossRef]
49. Altuna-Coy, A.; Ruiz-Plazas, X.; Sánchez-Martin, S.; Ascaso-Til, H.; Prados-Saavedra, M.; Alves-Santiago, M.; Bernal-Escoté, X.; Segarra-Tomás, J.; Chacón, M.R. The lipidomic profile of the tumoral periprostatic adipose tissue reveals alterations in tumor cell's metabolic crosstalk. *BMC Med.* **2022**, *20*, 255. [CrossRef]

50. Roumiguié, M.; Estève, D.; Manceau, C.; Toulet, A.; Gilleron, J.; Belles, C.; Jia, Y.; Houël, C.; Pericart, S.; LeGonidec, S.; et al. Periprostatic Adipose Tissue Displays a Chronic Hypoxic State that Limits Its Expandability. *Am. J. Pathol.* **2022**, *192*, 926–942. [CrossRef]
51. Zhang, Q.; Sun, L.J.; Yang, Z.G.; Zhang, G.M.; Huo, R.C. Influence of adipocytokines in periprostatic adipose tissue on prostate cancer aggressiveness. *Cytokine* **2016**, *85*, 148–156. [CrossRef]
52. Cahyadi, D.D.; Warita, T.; Irie, N.; Mizoguchi, K.; Tashiro, J.; Hosaka, Y.Z.; Warita, K. Housekeeping gene expression variability in differentiating and non-differentiating 3T3-L1 cells. *Adipocyte* **2023**, *12*, 2235081. [CrossRef] [PubMed]
53. Anderson, J.; Hollifield, G. Glucose-6-phosphate dehydrogenase activity in adipose tissue from two types of obese mice during starvation. *Metabolism* **1966**, *15*, 1092–1097. [CrossRef] [PubMed]
54. Pedersen, M.H.; Lauritzen, L.; Hellgren, L.I. Fish oil combined with SCFA synergistically prevent tissue accumulation of NEFA during weight loss in obese mice. *Br. J. Nutr.* **2011**, *106*, 1449–1456. [CrossRef]
55. Hoggard, N.; Cruickshank, M.; Moar, K.M.; Barrett, P.; Bashir, S.; Miller, J.D. Inhibin betaB expression in murine adipose tissue and its regulation by leptin, insulin and dexamethasone. *J. Mol. Endocrinol.* **2009**, *43*, 171–177. [CrossRef]
56. Fromm-Dornieden, C.; Lytovchenko, O.; von der Heyde, S.; Behnke, N.; Hogl, S.; Berghoff, J.; Köpper, F.; Opitz, L.; Renne, U.; Hoeflich, A.; et al. Extrinsic and intrinsic regulation of DOR/TP53INP2 expression in mice: Effects of dietary fat content, tissue type and sex in adipose and muscle tissues. *Nutr. Metab.* **2012**, *9*, 86. [CrossRef]
57. Wu, Y.; Kim, J.Y.; Zhou, S.; Smas, C.M. Differential screening identifies transcripts with depot-dependent expression in white adipose tissues. *BMC Genom.* **2008**, *9*, 397. [CrossRef]
58. Wargent, E.T.; O'Dowd, J.F.; Zaibi, M.S.; Gao, D.; Bing, C.; Trayhurn, P.; Cawthorne, M.A.; Arch, J.R.; Stocker, C.J. Contrasts between the effects of zinc-α2-glycoprotein, a putative β3/2-adrenoceptor agonist and the β3/2-adrenoceptor agonist BRL35135 in C57Bl/6 (ob/ob) mice. *J. Endocrinol.* **2013**, *216*, 157–168. [CrossRef] [PubMed]
59. Chechi, K.; Gelinas, Y.; Mathieu, P.; Deshaies, Y.; Richard, D. Validation of reference genes for the relative quantification of gene expression in human epicardial adipose tissue. *PLoS ONE* **2012**, *7*, e32265. [CrossRef] [PubMed]
60. Fan, X.; Yao, H.; Liu, X.; Shi, Q.; Lv, L.; Li, P.; Wang, R.; Tang, T.; Qi, K. High-Fat Diet Alters the Expression of Reference Genes in Male Mice. *Front. Nutr.* **2020**, *7*, 589771. [CrossRef]
61. Hurtado del Pozo, C.; Calvo, R.M.; Vesperinas-García, G.; Gómez-Ambrosi, J.; Frühbeck, G.; Corripio-Sánchez, R.; Rubio, M.A.; Obregon, M.J. IPO8 and FBXL10: New reference genes for gene expression studies in human adipose tissue. *Obesity* **2010**, *18*, 897–903. [CrossRef]
62. Catalán, V.; Gómez-Ambrosi, J.; Rotellar, F.; Silva, C.; Rodríguez, A.; Salvador, J.; Gil, M.J.; Cienfuegos, J.A.; Frühbeck, G. Validation of endogenous control genes in human adipose tissue: Relevance to obesity and obesity-associated type 2 diabetes mellitus. *Horm. Metab. Res.* **2007**, *39*, 495–500. [CrossRef] [PubMed]
63. Yuan, J.C.; Yogarajah, T.; Lim, S.K.; Tee, G.B.Y.; Khoo, B.Y. Pilot study and bioinformatics analysis of differentially expressed genes in adipose tissues of rats with excess dietary intake. *Mol. Med. Rep.* **2020**, *21*, 2063–2072. [CrossRef] [PubMed]
64. Rashid, M.; Shah, S.G.; Natu, A.; Verma, T.; Rauniyar, S.; Gera, P.B.; Gupta, S. RPS13, a potential universal reference gene for normalisation of gene expression in multiple human normal and cancer tissue samples. *Mol. Biol. Rep.* **2021**, *48*, 7967–7974. [CrossRef] [PubMed]
65. de Jonge, H.J.; Fehrmann, R.S.; de Bont, E.S.; Hofstra, R.M.; Gerbens, F.; Kamps, W.A.; de Vries, E.G.; van der Zee, A.G.; Meerman, G.J.T.; Elst, A.T. Evidence based selection of housekeeping genes. *PLoS ONE* **2007**, *2*, e898. [CrossRef] [PubMed]
66. Sharan, R.N.; Vaiphei, S.T.; Nongrum, S.; Keppen, J.; Ksoo, M. Consensus reference gene(s) for gene expression studies in human cancers: End of the tunnel visible? *Cell Oncol.* **2015**, *38*, 419–431. [CrossRef]
67. Caracausi, M.; Piovesan, A.; Antonaros, F.; Strippoli, P.; Vitale, L.; Pelleri, M.C. Systematic identification of human housekeeping genes possibly useful as references in gene expression studies. *Mol. Med. Rep.* **2017**, *16*, 2397–2410. [CrossRef]

International Journal of
Molecular Sciences

MDPI

Article

The Effect of Reduced Dietary Protein on Adipose Tissue in Local Krškopolje Pigs

Klavdija Poklukar [1], Marjeta Čandek-Potokar [1], Milka Vrecl [2], Jana Branković [2], Matjaž Uršič [2] and Martin Škrlep [1,*]

[1] Agricultural Institute of Slovenia, Hacquetova 17, 1000 Ljubljana, Slovenia; klavdija.poklukar@kis.si (K.P.); meta.candek-potokar@kis.si (M.Č.-P.)
[2] Institute of Preclinical Sciences, Veterinary Faculty, University of Ljubljana, Gerbičeva ulica 60, 1000 Ljubljana, Slovenia; milka.vrecl@vf.uni-lj.si (M.V.); jana.brankovic@vf.uni-lj.si (J.B.)
* Correspondence: martin.skrlep@kis.si

Abstract: The Slovenian autochthonous breed, Krškopolje pig, is known for high fatness and better adaptability to different environmental conditions and feed resources. However, the metabolic processes underlying these adaptations, especially in response to different diets, have not yet been studied. A deeper understanding of these mechanisms could provide valuable insights into the breed's adaptability to different environmental conditions. Therefore, the main objective of this study was to evaluate the effect of a low-protein (LP) diet on adipose tissue in Krškopolje pigs reared in either organic outdoor ($n = 2 \times 12$) or conventional indoor ($n = 2 \times 14$) systems. In the outdoor system, the LP diet had no effect on adipocyte size compared to the control (high-protein) diet, while it increased lipogenic enzyme activities and monounsaturated fatty acid content, and decreased polyunsaturated fatty acid content ($p < 0.05$). RNA sequencing revealed the upregulation of 28 genes and the downregulation of 37 genes. The upregulated genes were mainly involved in lipid metabolism (*ACLY*, *FASN*, *ACACA*, *MOGAT2*), oxidative stress, and mitochondrial function. In the indoor system, pigs on the LP diet had smaller adipocytes ($p < 0.05$), whereas no differences were detected in the lipogenic enzyme activities or fatty acid composition ($p > 0.10$). RNA sequencing revealed 30 upregulated and 28 downregulated genes. In the indoor system, heat shock proteins (*HSP70.2*, *HSPA6*) were upregulated in pigs on the LP diet, while genes involved in the innate immune system (*MSR1*, *TREM2*, *CSF3R*) were downregulated. To conclude, the present study showed that LP diet affected adipose tissue metabolism and gene expression in Krškopolje pigs, with different transcriptomic responses observed in outdoor and indoor rearing conditions.

Keywords: pig; local breed; low-protein diet; indoor system; outdoor system; subcutaneous adipose tissue; RNA-sequencing; nCounter gene expression assay

1. Introduction

Local pig breeds can play an important role in sustainable pig production due to their lower nutritional demands, despite their lower growth performance and higher fat content. The Krškopolje pig, the only preserved autochthonous Slovenian pig breed, has not undergone genetic selection and consequently displays lower growth performance. Currently, Krškopolje pigs are raised using a wide variety of husbandry practices, ranging from intensive indoor farming to fully extensive outdoor systems. The majority are reared on small-scale farms, which usually use organic or extensive production methods [1]. Compared to modern, lean pig breeds, Krškopolje pigs exhibit greater fat deposition [2] and moderate protein deposition potential [3]. A recent study by Škrlep et al. [4] has

https://doi.org/10.3390/ijms26094440

shown that dietary protein deficiency has a greater impact on the performance of modern, lean breeds than fatty Krškopolje pigs. This indicates the adaptability of the breed, and lower dietary protein requirements. A pilot study by Brossard et al. [3] also showed that the nutrition of local pig breeds may not be adequately adapted to their needs. Several studies on other local pig breeds have investigated the effects of various diets on growth performance, carcass characteristics, and meat quality. In these studies, it was reported that a controlled reduction of the protein levels in the diet is a suitable strategy to influence intramuscular fat content and adipose tissue characteristics without significantly affecting growth [5–9]. Our recent research on Krškopolje pigs [10] showed that a protein-reduced diet affected carcass fatness only in the organic outdoor system, but not in the conventional indoor system, highlighting the interaction with the production system. Nevertheless, the mechanisms driving this underlying response, particularly in adipose tissue, are still poorly understood. Previous studies reported variable response of lysine-deficient or low-protein (LP) diets. For instance, research on Chinese local breeds [11,12] suggested an increase in lipogenic enzyme activity, while studies on Mediterranean breeds [5,6,13] found no notable effect on lipogenic enzymes or the gene expression of lipogenic genes. Given that LP diets may affect a broader range of biological processes beyond lipogenesis, including transcriptional regulation, it becomes important to explore their impact on the global gene expression landscape using RNA sequencing (RNA-seq). While the RNA-seq method is widely used in metabolic and nutrigenomic studies of modern pig breeds [9,14,15], it has never been used in local Krškopolje pig, highlighting the novelty and significance of the current study. Therefore, the main objective of this study was to elucidate how the adipose tissue transcriptomic profile responds to a reduced-protein diet in Krškopolje pigs raised in two different production systems: conventional indoor and organic outdoor systems.

2. Results

2.1. Histomorphological Traits of Adipose Tissue

In the indoor system, the LP diet was associated with smaller adipocyte size in the outer layer ($p < 0.05$) and showed a tendency toward smaller adipocytes in the inner layer ($p < 0.10$). Additionally, a tendency toward a higher adipocyte number per region of interest (ROI) was observed in the inner layer ($p < 0.10$). In the outdoor system, the LP diet group showed a tendency toward increased backfat thickness; however, no significant effect on adipocyte size or number was observed (Table 1).

Table 1. Histomorphological characteristics of subcutaneous adipose tissue in Krškopolje pigs according to feeding in outdoor and indoor rearing systems.

	HP Diet	LP Diet	RMSE	p-Value
Outdoor rearing system				
Number of pigs	12	12		
Backfat thickness, mm	49.4	53.7	5.21	0.058
Adipocyte surface area, μm^2				
Inner layer	7828	8055	873.1	0.540
Outer layer	8596	8645	723.1	0.872
Number of adipocytes per ROI				
Inner layer	136	132	24.0	0.686
Outer layer	149	144	24.3	0.622

Table 1. *Cont.*

	HP Diet	LP Diet	RMSE	*p*-Value
Indoor rearing system				
Number of pigs	14	14		
Backfat thickness, mm	45.1	42.7	5.24	0.236
Adipocyte surface area, μm^2				
Inner layer	7330	6767	855.5	0.094
Outer layer	8048	7321	792.1	0.023
Number of adipocytes per ROI				
Inner layer	160	180	28.8	0.079
Outer layer	185	190	23.7	0.548

HP = control high-protein diet; LP = low-protein diet; RMSE = root mean square error; ROI = region of interest.

2.2. Lipogenic Enzyme Activities of Adipose Tissue

The results of lipogenic enzyme activities primarily showed an effect of the diet on the adipose tissue in outdoor-reared pigs (Table 2). The LP diet significantly increased the activity of all four lipogenic enzymes tested in the subcutaneous adipose tissue ($p < 0.05$), with values ranging from 1.4- to over 2-fold differences. In contrast, no diet-related differences ($p > 0.10$) were observed in the indoor system.

Table 2. Lipogenic enzyme activities in subcutaneous adipose tissue of Krškopolje pigs according to the dietary treatment in outdoor and indoor rearing systems.

	HP Diet	LP Diet	RMSE	*p*-Value
Outdoor rearing system				
Number of pigs	12	12		
ME1	298.5	416.4	108.24	0.014
G6PD	500.7	628.4	127.69	0.023
CCL	167.4	256.0	71.3	0.006
FASN	19.5	39.5	19.8	0.034
Indoor rearing system				
Number of pigs	14	14		
ME1	211.6	240.8	60.27	0.212
G6PD	491.3	495.1	102.7	0.922
CCL	123.9	124.1	53.59	0.992
FASN	18.1	15.0	8.77	0.456

HP = control high-protein diet; LP = low-protein diet; RMSE = root mean square error; ME1 = malic enzyme; G6PD = glucose-6-phosphate dehydrogenase; CCL = citrate cleavage enzyme; FASN = fatty acid synthase. The activities of ME1 and G6PD are expressed as nmol of reduced nicotinamide adenine dinucleotide phosphate per minute per gram of wet tissue, and the activities of CCL and FASN are expressed as nmol of oxidized nicotinamide adenine dinucleotide phosphate per minute per gram of wet tissue.

2.3. Fatty Acid Composition of Adipose Tissue

The content of 20 different fatty acids was measured in the adipose tissue of Krškopolje pigs. The subcutaneous adipose tissue of animals fed the LP diet (Table 3) contained more monounsaturated fatty acids (MUFAs) ($p < 0.05$) and less polyunsaturated fatty acids (PUFAs) ($p < 0.01$) compared to those fed the control high-protein (HP) diet. This was particularly reflected in a higher proportion of oleic acid (C18:1) and a lower proportion of linoleic acid (C18:2n−6), together with some less representative PUFAs (e.g., C18:3n−3, C20:2n−6, C20:2n−3). In the indoor system, no notable differences were observed, except for alpha-linolenic (C18:3n−3, $p = 0.001$) and palmitoleic acid (16:1n−7, $p < 0.10$), both of which were lower in pigs fed the LP diet.

Table 3. The fatty acid composition of subcutaneous adipose tissue of Krškopolje pigs according to the dietary treatment in outdoor and indoor rearing systems. Results are in g/100 g of fatty acids.

	HP Diet	LP Diet	RMSE	*p*-Value
Outdoor rearing system				
Number of pigs	12	12		
SFA	43.47	43.52	1.570	0.937
MUFA	45.55	46.73	1.273	0.034
PUFA	10.98	9.75	0.859	0.002
C10:00	0.05	0.05	0.016	0.787
C12:00	0.07	0.07	0.004	0.606
C14:00	1.36	1.31	0.060	0.066
C15:00	0.01	0.00	0.010	0.204
C16:00	25.79	25.93	0.724	0.637
C16:1n−7	1.8	1.86	0.267	0.616
C17:00	0.13	0.12	0.014	0.015
C17:1n−7	0.21	0.19	0.031	0.080
C18:00	15.83	15.8	1.248	0.950
C18:1n−9	42.18	43.26	1.144	0.031
C18:2n−6	9.26	8.21	0.757	0.003
C18:3n−3	0.69	0.61	0.058	0.005
C20:00	0.24	0.25	0.020	0.194
C20:1n−9	1.36	1.43	0.150	0.275
C20:2n−6	0.69	0.62	0.058	0.009
C20:3n−6	0.05	0.04	0.008	0.061
C20:3n−3	0.13	0.11	0.013	0.006
C22:1n−9	0.00	0.00	.	.
C20:4n−6	0.17	0.15	0.023	0.186
C22:6n−3	0.00	0.00	.	.
Indoor rearing system				
Number of pigs	14	14		
SFA	41.71	42.15	1.617	0.483
MUFA	46.52	46.82	1.661	0.342
PUFA	11.77	11.03	1.145	0.103
C10:00	0.06	0.06	0.005	0.178
C12:00	0.07	0.07	0.007	1.000
C14:00	1.30	1.26	0.080	0.130
C15:00	0.03	0.02	0.022	0.239
C16:00	25.01	24.79	0.806	0.465
C16:1n−7	2.01	1.83	0.246	0.059
C17:00	0.14	0.14	0.022	0.768
C17:1n−7	0.25	0.25	0.038	0.518
C18:00	14.89	15.60	1.181	0.123
C18:1n−9	43.09	43.48	1.533	0.509
C18:2n−6	9.99	9.36	0.994	0.108
C18:3n−3	0.84	0.73	0.076	0.001
C20:00	0.21	0.22	0.027	0.453
C20:1n−9	1.17	1.24	0.952	0.292
C20:2n−6	0.60	0.61	0.065	0.804
C20:3n−6	0.04	0.04	0.008	0.749
C20:3n−3	0.11	0.10	0.024	0.235
C22:1n−9	0.00	0.01	0.038	0.327
C20:4n−6	0.18	0.19	0.027	0.810

HP = control high-protein diet; LP = low-protein diet; RMSE = root mean square error; SFA = saturated fatty acid; MUFA = monounsaturated fatty acid; PUFA = polyunsaturated fatty acid.

2.4. RNA-Sequencing

Sequencing yielded, on average, 66.8 million total reads. The average number of read counts was 69.0, 96.1, 51.7, and 49.7 million for the outdoor LP, outdoor HP, indoor LP, and indoor HP groups, respectively. Approximately 98.4% of the reads were mapped to the annotated *Sus scrofa* 11.1 genome, with 94.5% of the mapped clean reads having unique matches. General mapping statistics can be found in Supplementary File S1.

2.4.1. Diet Effect on Differential Gene Expression in the Outdoor Rearing System

A comparison of LP and HP diets in the outdoor system revealed 65 differentially expressed genes (Figure 1A, Supplementary File S2). Among them, 28 genes were upregulated and 37 genes were downregulated, with $\log_2$ (fold change) values ranging from −3.2 (*LSMEM1*) to 4.9 (*ACSM5*). To annotate the differentially expressed genes, an over-representation analysis was performed, which revealed enrichment of genes involved in lipid metabolism (Figure 2A). These include terms related to lipid biosynthesis (GO:0008610), acyl-CoA metabolism (GO:0006637, GO:0071616), thioester metabolism (GO:0035383, GO:0035384), carboxylic acid (GO:0046394) and organic acid biosynthesis processes (GO:0016053) (Supplementary File S3). The KEGG over-representation analysis identified a significantly enriched pathway, namely propanoate metabolism (ssc00640) (Supplementary File S4).

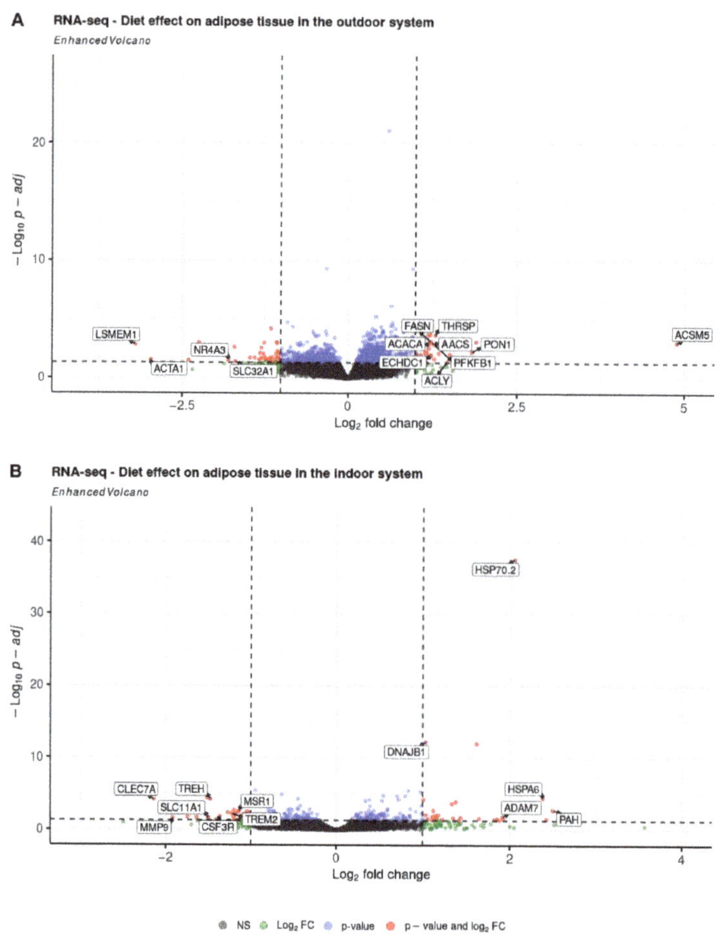

Figure 1. Volcano plots depicting genes expressed in the subcutaneous adipose tissue from pigs receiving low-protein (LP) vs. control high-protein diet (HP) in the (**A**) outdoor and (**B**) indoor

systems. The horizontal lines indicate the significance threshold of differentially expressed genes at a p-value of 0.05. The vertical lines represent the threshold of $|\log_2$ (fold change)$| > 1$. Genes upregulated in the LP compared to HP group have $\log_2$(fold change) > 1, while downregulated genes have $\log_2$ (fold change) < -1.

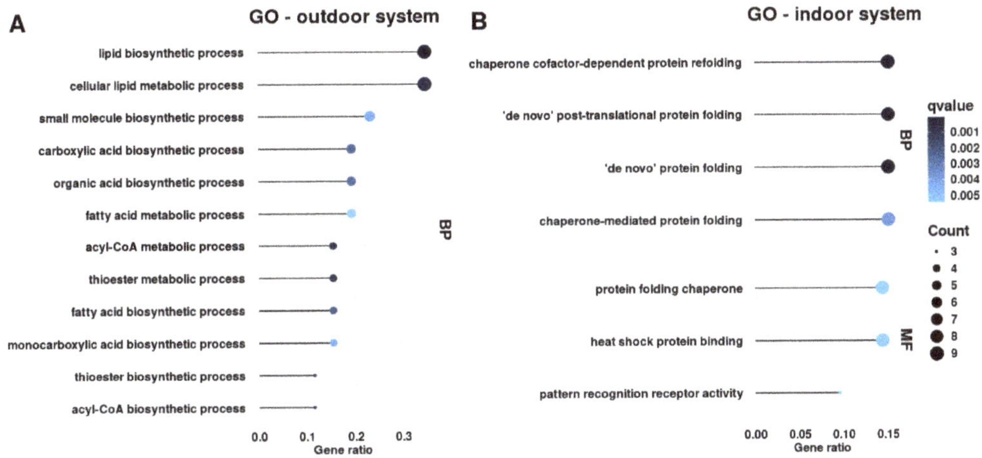

Figure 2. The gene ontology (GO) over-representation analysis of differentially expressed genes in the subcutaneous adipose tissue from pigs fed a low-protein (LP) versus control high-protein (HP) diet in the (**A**) outdoor and (**B**) indoor systems. GO analysis represents the biological process category (BP) and molecular function category (MF). The gene ratio refers to the proportion of differentially expressed genes relative to the total number of genes associated with a given GO term.

2.4.2. Diet Effect on Differential Gene Expression in the Indoor Rearing System

Differential gene expression analysis comparing LP and HP groups identified 58 differentially expressed genes (Figure 1B, Supplementary File S5). Among them, 30 genes were upregulated and 28 genes were downregulated with $\log_2$ (fold change) ranging from −2.1 (*CLEC7A*) to 2.5 (*PAH*). Gene ontology over-representation analysis revealed an enrichment of genes involved in biological processes and molecular functions related to chaperone cofactor-dependent protein folding (GO:0051085), 'de novo' post-translational protein folding (GO:0051084), and protein folding chaperone (GO:0044183) (Figure 2B, Supplementary File S6). KEGG over-representation analysis did not detect any significant pathways.

2.5. Candidate Gene Expression Analysis Using NanoString nCounter Gene Expression Assay

2.5.1. Effect of Diet on the Candidate Gene Expression in the Outdoor System

In addition to the lipogenic genes already detected with RNA-seq (e.g., *FASN*, *ACACA*, and *ACLY*), the expression of several other candidate genes related to lipid metabolism was measured. In the LP group compared to the HP group (Figure 3A, Supplementary Files S7 and S8), *G6PD*, *ME1*, *SCD*, *ELOVL6*, and *LPL* genes ($p < 0.05$) were upregulated. On the other hand, *PCK1* and *PPARG* showed a tendency to be downregulated ($p > 0.10$) in LP group.

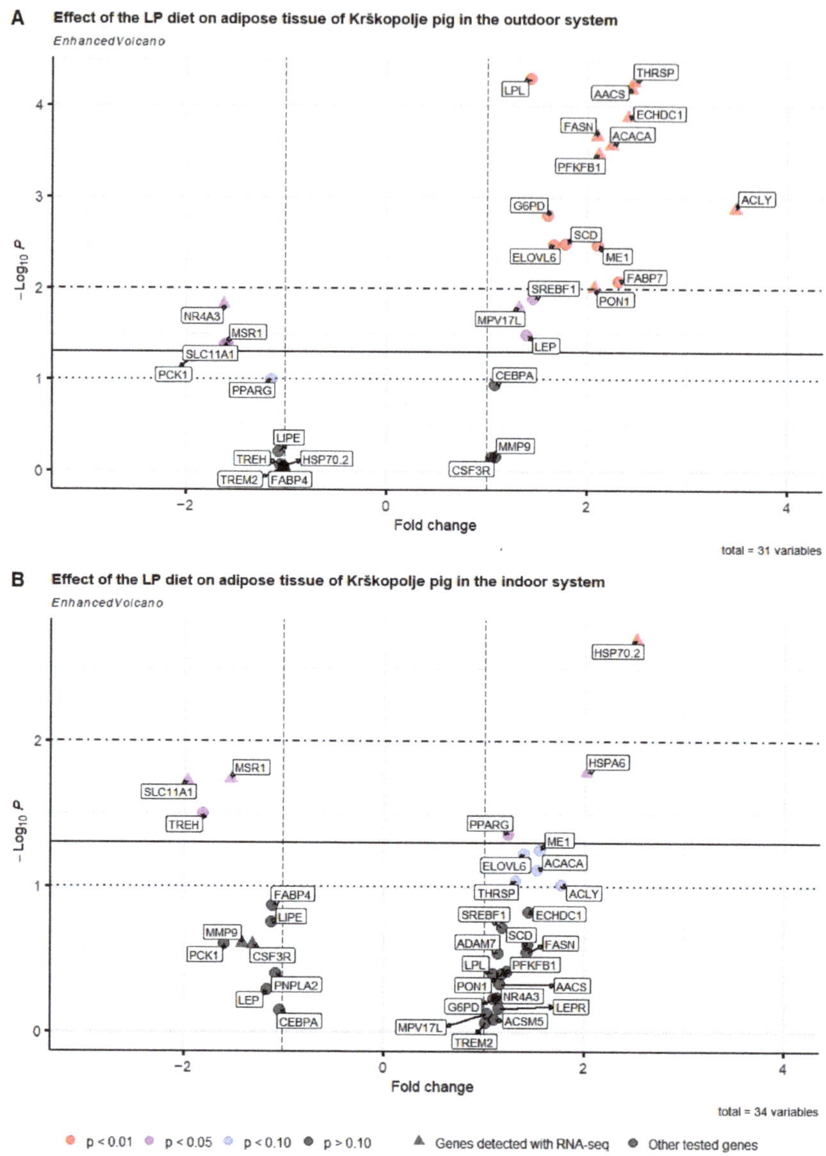

Figure 3. Volcano plot showing each gene's $-\log_{10}(p\text{-value})$ and fold change in the (**A**) outdoor and (**B**) indoor systems comparing low-protein (LP) versus control high-protein (HP) diets in Krškopolje pig breed using the NanoString nCounter gene expression assay. The x-axis indicates fold change. Horizontal lines indicate p-value thresholds. Genes are colored in red if the resulting p-value is below 0.01, in purple if the resulting p-value is between 0.01 and 0.05, in blue if the resulting p-value is between 0.05 and 0.10, and in black if the resulting p-value is higher than 0.10. The triangles show genes that were also differentially expressed using the RNA sequencing method, while the circle denotes other tested genes. Genes upregulated in the LP compared to HP group have fold change > 1, while downregulated genes have fold change < -1.

2.5.2. Effect of Diet on the Candidate Gene Expression in the Indoor System

A comparison of the LP and HP groups (Figure 3B, Supplementary Files S7 and S8) showed that the *PPARG* gene was significantly upregulated ($p < 0.05$) in the LP group, while the expression of *ELOVL6*, *ME1*, *ACACA*, and *ACLY* tended to be increased ($p > 0.10$). Conversely, the expression of *TREH* was significantly downregulated in the LP group compared to the HP group ($p < 0.05$).

2.6. RNA-Seq Validation Using Nanostring nCounter Gene Expression Assay

To validate the RNA-seq results in the outdoor system, the expression of 10 genes was selected based on expression fold difference and due to the importance in lipid and energy metabolism (Figure 3A; Supplementary Files S7 and S8). The *MPV17L*, *PON1*, *FASN*, *ACACA*, *ECHDC1*, *AACS*, *THRSP*, *PFKFB1* and *ACLY* genes were upregulated ($p < 0.01$), while the *NR4A3* gene was downregulated in the LP compared to the HP group.

To validate the RNA-seq results in the indoor system, the expression of eight genes was measured (i.e., *ADAM7*, *CSF3R*, *HSPA6*, *HSP70.2*, *MMP9*, *MSR1*, *SLC11A1*, and *TREM1*) (Figure 3B, Supplementary Files S7 and S8). Among them, *HSPA70.2* and *HSPA6* were significantly upregulated ($p < 0.05$), while the *SLC11A1* and *MSR1* genes were downregulated ($p < 0.05$) in the LP compared to the HP group. Although the relative expression of the other four genes exhibited the same trend (i.e., *ADAM7*, *TREM2*, *MMP9*, and *CSF3R*), the difference was not significant ($p > 0.10$).

3. Discussion

3.1. The Effect of the Diet on Adipose Tissue of Krškopolje Pigs Reared in the Outdoor System

In the outdoor system, gene ontology enrichment analysis of RNA-seq data highlighted genes involved in lipid metabolism (*ACLY*, *FASN*, *ACACA*, *ACSM5*, *MOGAT2*, *ACSS2*, *RDH5*, and *CH25H*). Many of the upregulated genes in the LP compared to the HP group were directly related to fatty acid synthesis (*FASN*, *ACLY*, *ACACA*, *ACSM5*, and *THRSP*) or triglyceride synthesis (*MOGAT2*). This aligns with increased lipogenic enzyme activity and greater backfat thickness observed in the present study. Higher lipid deposition in pigs receiving LP diet could be due to lysine deficiency during fattening. In addition to higher *FASN* and *ACLY* gene expression, RNA-seq also revealed upregulation of the *ACACA* gene, which catalyzes the conversion of acetyl-CoA to malonyl-CoA. Malonyl-CoA is further catalyzed by *FASN* to synthesize palmitate. Both the *ACACA* and *FASN* genes have previously been identified as candidate genes for fatness and performance in pigs [15,16] and have been linked to fatty acid composition [17]. In the present study, upregulation of the *THRSP*, encoding thyroid hormone responsive spot 14, was detected, being a direct modulator of lipogenesis [18]. Polymorphisms and the expression of this gene have previously been associated with lipid synthesis and lipid concentration [19,20] and its expression has been linked to fatty acid composition [20]. The present study also detected the upregulation of the *MOGAT2* gene, which catalyzes the esterification of glycerol monoesters to diglycerides. This gene has previously been associated with larger fat deposition [21]. The targeted gene expression approach used in the present study additionally demonstrated the upregulation of *ME1* and *G6PD* genes, which provide nicotinamide adenine dinucleotide phosphate for fatty acid synthesis. Regarding the effects of the LP diet on lipogenic gene expression or enzyme activity, the relevant literature is generally very contradictory. For example, in fatty Wujin pigs [11], the LP diet led to increased expression of *ACACA*, *ME1*, and *G6PD* genes. On the contrary, the expression of the *ACACA* gene was downregulated in fatty Alentejano pigs receiving the LP diet [13]. In studies on the Iberian and modern breeds, no apparent effect of lysine-deficiency on lipogenic gene expression was detected [5,6]. In the present study, the targeted gene expression approach additionally revealed the upregulation of *SCD* and *ELOVL6*, genes responsible for the desaturation of SFA to MUFA. This is consistent with the higher MUFA content (especially oleic acid), and lower PUFA content in LP pigs reared outdoors. In contrast to our results, in a study on the Cinta Senese pig breed reared also outdoors [8], the MUFA content of subcutaneous adipose tissue remained unaffected by the LP diet, while lower SFA and higher PUFA levels were observed. These inconsistencies highlight the complexity of dietary effects on fat composition, which can be influenced by multiple factors. Variations in experimental conditions, such as the age of the animals, the extent of protein restriction, and environmental factors, likely contribute to the variability in reported results, making direct comparisons between studies challenging.

In the present study, RNA-seq analysis of the animals reared in the outdoor system additionally detected upregulation of genes involved in oxidative stress, such as paraoxonase 1 (*PON1*) and xanthine dehydrogenase (*XDH*), and mitochondrial functions (*MPV17L*, *ECHDC1*, *ACSM5*). Paraoxonase-1 is an antioxidant enzyme that maintains normal levels of lipid metabolism intermediates [22]. In addition, xanthine dehydrogenase is involved in adipocyte reactive oxygen species (ROS) generation under oxidative stress conditions [23]. ROS production is increased during adipocyte differentiation and the development of obesity [24], which also coincides with greater subcutaneous adipose tissue gain observed in LP pigs. The upregulation of genes involved in oxidative stress could also be a consequence of the environmental conditions, since animals in the outdoor system had increased energy demands for thermoregulation and physical activity [25,26]. Other upregulated genes (*ACSM5*, *ECHDC1*) detected in the present study were implicated in mitochondrial functions, more specifically with mitochondrial fatty acid β-oxidation [27]. Consistent with this, the transcriptional regulator *NR4A3*, which plays a role in lipid and carbohydrate metabolism including fatty acid β-oxidation, was downregulated in the LP group. Notably, studies have shown that knock-out of the *NR4A3* gene also impairs glucose uptake in adipocytes [28].

3.2. The Effect of the Diet on Adipose Tissue of Krškopolje Pigs Reared in the Indoor System

In the indoor system, Krškopolje pigs receiving the LP diet demonstrated enrichment of heat shock proteins (HSPs) (i.e., *HSP70.2*, *HSPA6*, and *DNAJB1*). HSPs are a family of proteins with protective functions under various stress conditions, such as nutritional stress, tissue hypoxia, or inflammatory response [29,30]. Elevated levels of intracellular HSP70 (i.e., *HSP70.2* and *HSPA6* genes) play a cytoprotective and anti-inflammatory role in adipose tissue, while extracellular *HSP70* exhibits proinflammatory activity when released into the extracellular space [31]. In the present study, the upregulation of HSPs in pigs fed the LP diet could be associated with anti-inflammatory and cytoprotective response, which is also supported by the downregulation of genes related to the innate immune system (*CLEC7A*, *TLR8*, *MSR1*, *TREM2*, *CPM*, *CSF3R*, and *SLC11A1*). Most of the immunity-related genes discovered in the present study are associated with macrophages, which are the most abundant immune cell type in adipose tissue and crucial for regulating adipose tissue homeostasis [32]. Macrophages perform a phagocytic function that involves the clearance of apoptotic and necrotic cells, and contributes to tissue remodeling [33]. For example, the LP diet in the present study affected the downregulation of the receptors *TREM2*, *MSR1*, and *CLEC7A*, which are expressed on immune cells (including macrophages) and act as receptors for different ligands [34–36]. In addition, a candidate gene expression approach used in this study showed the upregulation of the *PPARG* gene in the LP compared to HP group. This gene regulates adipogenesis, and supports the observed differences in histomorphological traits (i.e., a tendency toward a higher number of adipocytes per ROI in the inner layer) observed in LP compared to HP groups. Besides adipogenesis,

PPARG can regulate key macrophage activities, such as differentiation, inflammation and polarization. The activation of *PPARG* has been reported to suppress the immune response of macrophages [33,37]. Accordingly, a downregulated metalloproteinase *MMP9* (encoding gelatinase B), which is involved in extracellular matrix degradation, was detected in the LP group by RNA-seq. *MMP9* can be expressed on fibrolytic macrophages, thus promoting phagocytosis (in HP compared to LP group) [32]. Furthermore, a study in mice has shown that LP diet decreases the number and functionality of macrophages [38] and that lysine deficiency can limit the synthesis of immunity-related proteins necessary for immune function [39]. The targeted candidate gene expression approach of the present study did not detect differences in the expression of genes involved in lipid metabolism or energy homeostasis. This aligns with the absence of phenotypic differences in adiposity-related traits (i.e., lipogenic enzyme activity and fatty acid composition), as well as no differences in backfat thickness. The most likely explanation is a much lower lysine content in the LP group of pigs raised indoors compared to those reared outdoors [10].

3.3. Conclusions

The present study on the local fatty Krškopolje pigs revealed that lowering dietary protein affected adipose tissue differently, depending on the environment. In the indoor system, the LP diet reduced the adipocyte size in the outer layer, while it did not affect lipogenic enzyme activities and fatty acid composition (measured in the inner layer). These findings were supported by both targeted and untargeted gene expression analyses, which showed no differences in lipid metabolism genes but indicated the suppression of genes involved in the immune system, particularly macrophages with phagocytic functions. In contrast, feeding an LP diet in the outdoor system resulted in increased lipogenic enzyme activities, higher MUFA, and lower PUFA content compared to the HP group. Gene expression analysis further confirmed the upregulation of genes associated with lipid deposition, fatty acid saturation, oxidative stress, and mitochondrial function.

4. Materials and Methods

4.1. Animals and Sample Collection

Two separate field trials were performed to test the effects of reduced dietary protein in a local pig breed. Krškopolje pigs were reared either in indoor (n = 2 × 14; 14 surgically castrated males and 14 females evenly distributed among the groups) or in outdoor conditions (fenced open area with trees and shelter) (n = 2 × 12; 16 surgically castrated males and 8 females evenly distributed among the groups). Animals from the outdoor system were reared according to organic standards and received organic feed ad libitum, while the animals reared in the indoor system received the same amount of feed based on the same ingredients, but of conventional origin. In both systems, animals received either standard (HP group: fed 15% crude protein from 20 to 80 kg, 12.5% crude protein from 80 to 100 kg, and 10% crude protein from 100 kg to slaughter) or reduced protein diet (LP group: fed 15% crude protein from 20 to 60 kg and 10% crude protein from 60 kg to slaughter). Diets were isoenergetic and isoproteic but differed in lysine content, with the organic diet containing less lysine. The trial began at an average age (mean ± SE) of 102 ± 6, 97 ± 4, 101 ± 2, and 103 ± 3 days for the outdoor HP, outdoor LP, indoor HP, and indoor LP groups, respectively. The corresponding initial body weights were 23.9 ± 0.8, 24.8 ± 0.7, 25.0 ± 1.1, and 24.9 ± 1.3 kg (mean ± SE). No significant differences were observed in the starting age or body weight between dietary regimes (p-values > 0.10). The animals were slaughtered at an average age of 331 ± 6, 326 ± 4, 329 ± 2, and 331 ± 3 days (mean ± SE), and body weights of 165.2 ± 4.5, 165.3 ± 2.7, 184.0 ± 3.9, and 182.5 ± 3.3 kg (mean ± SE) for the outdoor HP, outdoor LP, indoor HP, and indoor LP groups, respectively.

No significant differences were observed in the final age or body weight within dietary regimens (p-values > 0.10).

Within the first 30 min after slaughter, carcasses were split apart and backfat thickness was measured at three anatomical sites (withers, last rib, and above *gluteus medius* muscle). The inner subcutaneous adipose tissue layer was collected at the level of the last rib and frozen in liquid nitrogen. The samples were stored at −80 °C for subsequent RNA extraction. For RNA-seq and NanoString nCounter gene expression assay, 24 subcutaneous adipose tissue samples from the indoor group (n = 2 × 12) were selected, along with all samples from the outdoor system group (n = 2 × 12). For the determination of histomorphological traits, lipogenic enzyme activities, and fatty acid composition, the subcutaneous adipose tissue of all animals was collected (n = 2 × 14 for the indoor and 2 × 12 for the outdoor comparisons).

4.2. Adipose Tissue Histomorphological Traits

The cellularity of adipose tissue was measured as described in Škrlep et al. [4]. Briefly, the inner and the outer subcutaneous adipose tissue layers sampled at the level of the last rib (a cuboid of 2 × 1 × 1 cm) were dissected and fixed in 10% buffered formalin. Tissue sections of 5 μm (microtome Leica SM2000R, Nussloch, Germany) were cut in a vertical plane through the dermis and both layers of subcutaneous adipose tissue. The sections were stained with hematoxylin and eosin (HE) and cover slipped (Gemini AS slide stainer and cover slipper ClearVue, Thermo Fisher Scientific, Runcorn, UK). For histomorphometric analysis, images were obtained by a light microscope (Eclipse Ni-U, Nikon, Tokyo, Japan) and a digital camera (DS -Fi1, Nikon, Tokyo, Japan). For measuring the histomorphological parameters of the inner and outer subcutaneous adipose tissue layer, 3 images per sample (each measuring 1.1 mm^2) were obtained for each layer at 10× objective magnification, and the number and cross-sectional area (CSA) of adipocytes were determined using macro protocol in Fiji 1.53 software [40].

4.3. Determination of Lipogenic Enzyme Activity

Fatty acid synthase (FAS), citrate cleavage enzyme (CCL), glucose-6-phosphate dehydrogenase (G6PD), and malic enzyme (ME) were determined in subcutaneous adipose tissue according to the original methods of Bazin and Ferré [41]. Approximately 1 g of frozen pulverized sample was homogenized in 10 mL of ice-cold sucrose buffer (0.25 M sucrose, 1 mM ethylenedinitrilotetraacetic acid, 1 mM dithiothreitol, and 1 mM phenylmethylsulphonyl fluoride) and the homogenate centrifuged at 12,000× g at 0 °C for 1 h. The activities of lipogenic enzymes were assessed spectrophotometrically by measuring the absorbance at 340 nm against the blank for 10 min at 37 °C.

4.4. Fatty Acid Composition Analysis

Fatty acid composition of subcutaneous adipose tissue was assessed as previously described in Škrlep et al. [42]. After pulverizing the samples in liquid nitrogen, 100 mg of subcutaneous adipose tissue were weighed and transmethylated in situ in 0.5 M NaOH solution in methanol and dichloromethane (incubated for 50 min at 90 °C) and afterwards by the addition of 14% BF3 in methanol (incubated for 10 min at 90 °C). The fatty acid methyl esters (FAMEs), resulting from the reaction, were extracted using hexane. The separation of FAMEs was performed by Hewlett–Packard 6890 Gas Cromatograph (Agilent, Santa Clara, CA, USA) fitted with Supelco SP-2560 Capillary column (100 m × 0.25 mm i.d. × 0.20 μm; Supelco, Bellafonte, PA, USA) using nitrogen as a carrying gas and flame-ionization detector (FID). The initial temperature set at 80 °C was increased in the following order: to 160 °C at 20 °C/min, then to 198 °C at 1 °C/min for a total running time of 94.4 min. The temperature of the injector was set at 220 °C and of the FID to 200 °C.

Specific FAMEs were identified using mixtures of standards (Supelco 37 Component FAME mix), while nonadecanoic acid was used as an internal standard to determine tissue fatty acid concentrations.

4.5. RNA-Sequencing Analysis

4.5.1. RNA Extraction, Library Preparation, and Sequencing

Total RNA was extracted from approximately 90 mg of subcutaneous adipose tissue using the RNeasy Lipid Tissue Mini Kit (Qiagen GmbH, Hilden, Germany) according to the manufacturer's instructions. The quality of the extracted RNA was assessed by measuring the ratio in optical density between 260 and 280 nm using Eppendorf BioSpectrometer (Eppendorf AG, Hamburg, Germany). Samples were quantified using Qubit 4.0 Fluorometer (Life Technologies, Carlsband, CA, USA). The integrity of RNA was assessed with Agilent 2100 Bioanalyzer (Agilent, Santa Clara, CA, USA). The extracted RNA samples had an RNA integrity number larger than 6.2 and OD260/OD280 ratio between 1.7 and 2.0. Prior to library preparation, equal amounts of total RNA from two individuals were pooled together, generating 6 pools within each dietary group.

RNA sequencing libraries were prepared using the NEBNext Ultra II RNA Library Prep Kit for Illumina (NEB, Ipswich, MA, USA) according to the manufacturer's instructions. The libraries were validated using an NGS Kit on an Agilent 5300 Fragment Analyzer (Agilent Technologies, Palo Alto, CA, USA) and quantified by a Qubit 4.0 Fluorometer (Life Technologies, Carlsband, CA, USA).

The multiplexed sequencing libraries were loaded onto the Illumina NovaSeq 6000 generating 2×150 paired-end reads. Raw sequence data were converted into fastq files and de-multiplexed using Illumina bcl2fastq program (version 2.20). The RNA-seq experiment was submitted to the NCBI Gene Expression Omnibus (GEO) (Bethseda, MD, USA) and was assigned GEO accession number GSE292746.

4.5.2. Bioinformatic Analysis of RNA-Seq Data

The quality control of the raw data was performed using FastQC v1.0.0 software. The sequence reads were trimmed using Trimmomatic v.0.36 [43] to remove adapter sequences and poor-quality nucleotides. The trimmed reads were mapped to *Sus scrofa* 11.1 reference genome (release 108) using STAR aligner v.2.5.2b [44]. Unique gene hit counts were calculated using featureCounts (Subread package v.1.5.2.) [45] and summarized using the gene_id feature in the annotation file. Unique reads that fell within the exon region were counted.

Gene hit counts were further used for differential gene expression analysis using DESeq2 [46]. In both production systems (i.e., outdoor and indoor systems), differentially expressed genes in the LP compared to HP diets were searched. The Wald test was used to generate and calculate p-values and $\log_2$ fold changes. The resulting p-values were adjusted using the Benjamini–Hochberg method. Significant differential expression threshold was set at the adjusted p-value of 0.05 and $|\log_2(\text{fold change})| > 1$. Volcano plots were created using EnhancedVolcano R package.

4.5.3. Gene Ontology and KEGG Pathway Over-Representation Analyses

Gene ontology (GO) over-representation analysis of differentially expressed genes was performed by the R package ClusterProfiler v.3.17 [47]. Gene ontology terms with the adjusted q value (adjusted p-value by Benjamini and Hochberg method) below 0.05 were considered significantly enriched.

Additionally, R package ClusterProfiler v.3.17 [48] was used to test the statistical enrichment of differentially expressed genes in the Kyoto Encyclopedia of Genes and

Genomes (KEGG) pathways using Fisher's exact test. Test results were subjected to multiple testing corrections of the p-values by the Benjamini and Hochberg method.

4.6. Candidate Gene Expression Analysis and Validation of RNA-Seq Data Using NanoString

A total of 48 test genes with 5 housekeeping genes were selected for gene expression quantification in individual animals on the nCounter Sprint (NanoString Technologies, Seattle, WA, USA) using Elements TagSets. Test genes were selected from the list of differentially expressed genes detected by RNA-seq based on expression fold difference and due to their importance in lipid and energy metabolism. Additional genes were selected from the literature survey.

For the outdoor group, the following candidate genes were selected from the literature: *CEBPA, ELOVL6, G6PD, LEP, LIPE, LPL, ME1, PCK1, PNPLA2, PPARG,* and *SCD*. For RNA-seq validation, the following genes were selected from the list of differentially expressed genes: *AACS, ACACA, ACLY, ACSM5, ECHDC1, FASN, MPV17L, PFKFB1, PON1, THRSP, NR4A3, MOGAT2, TRPV4,* and *XDH*. For the indoor group, the following candidate genes were selected from the literature: *ACACA, ACLY, CEBPA, ELOVL6, FABP4, FASN, G6PD, LEP, LEPR, LIPE, LPL, ME1, PCK1, PNPLA2, PPARG, SCD, SREBF1,* and *TREH*. For RNA-seq validation, the following genes were selected: *CSF3R, HSP70.2, MMP9, MSR1, SLC11A1, TREM2, ADAM7, ATP6V0D2, CLEC7, SLC27A5,* and *TAS2R40*. More details are provided in Supplementary File S8.

The oligonucleotide probes used were designed by NanoString Technologies (Seattle, WA, USA) and synthesized by IDT Technologies (Coralville, IA, USA) (Supplementary File S8). Prior to analysis, RNA samples were diluted to 40 ng/μL in RNAse-free water. The assay was performed according to the manufacturer's protocol.

The nSolver analysis software v.4.0 was used for the raw data quality control (i.e., reporter code count files or RCC files) analysis. The RCC files were imported and underwent software quality control routine. For all samples, fields of view registration (imaging) was more than 99%; binding density was between 0.2 and 0.71; positive control linearity was 1.0; and the limit of detection was more than 0.5 fM positive control and $\leq$2 standard deviations above the mean of the negative controls.

The background threshold count value was defined as the average value plus two standard deviations of negative control probes. Testing genes having more than 40% of samples above background threshold count value were further considered for the analysis. Counts of eight genes (*MOGAT2, TAS2R40, TRPV4, XDH, SLC27A5, ATP6V0D2, CLEC7A,* and *PAH*) were below threshold; therefore, they were excluded from the analysis. Gene *FABP7* was excluded from the indoor comparison, and genes *LEPR, ACSM5, HSPA6,* and *ADAM7* were excluded from the outdoor comparison due to a low number of samples being above background threshold (Supplementary Files S8). Positive control normalization factor was computed using geometric mean. Afterwards, the housekeeping normalization factor was calculated using geometric mean of 4 housekeeping genes, i.e., *B2M, PPIA, TOP2B,* and *YWHAZ* with the percentage of coefficient of variation ranging from 13.6 to 18.9% for the indoor comparison, and 22.5 to 26.7% for the outdoor comparison. Housekeeping gene *GAPDH* was excluded from the normalization due to a high coefficient of variation. A Welch t-test was performed to calculate the significance of the fold change. EnhancedVolcano R package was used to present the results.

4.7. Statistical Analysis

Data from histomorphological analyses, lipogenic enzyme activities, and fatty acid composition were analyzed using SAS statistical software version 9.1 (SAS Institute Inc., Cary, NC, USA) with the PROC MIXED procedure. The effect of diet was evaluated within conventional indoor and organic outdoor rearing systems. In both cases, the model included fixed effects of the diet and sex. Least square means were calculated and compared using the LSMEANS procedure with the PDIFF option.

Supplementary Materials: The following supporting information can be downloaded at: https://www.mdpi.com/article/10.3390/ijms26094440/s1.

Author Contributions: Conceptualization, M.Š. and M.Č.-P.; methodology, K.P., M.Č.-P., M.U., M.V., J.B. and M.Š.; validation, K.P.; formal analysis, K.P. and M.Š.; investigation, K.P., M.Č.-P., M.U., M.V., J.B. and M.Š.; data curation, K.P.; writing—original draft preparation, K.P. and M.Š.; writing—review and editing, K.P., M.Č.-P., M.V., J.B. and M.Š.; visualization, K.P.; supervision, M.Š.; project administration, M.Š. and M.Č.-P.; funding acquisition, M.Š. and M.Č.-P. All authors have read and agreed to the published version of the manuscript.

Funding: This study has received funding from the European Union's Horizon 2020 research and innovation program (GEroNIMO project, grant agreement No 101000236) and the Slovenian Research and Innovation Agency (grants Z4-60178, J4-3094, P4-0133, P4-0053). The research was performed using the Research infrastructure ELIXIR-SI (https:/elixir-slovenia.org), funded by the European Regional Development Fund, the Ministry of Science, Education and Sports, and by the Slovenian Research and Innovation Agency.

Institutional Review Board Statement: The study was conducted in accordance with Slovenian animal protection law (Zakon o zaščiti živali, 2013). All work was performed as part of the routine operations of the farms with full owner consent. No procedures requiring ethical approval under Directive 2010/63/EU (2010) were carried out on the animals. Adipose tissue samples were collected after slaughter.

Informed Consent Statement: Not applicable.

Data Availability Statement: RNA-sequencing data are available under accession GSE292746. Other data are available upon direct request to the authors.

Acknowledgments: The authors wish to thank Tomaž Sket, Špela Velikonja Bolta, Jani Debeljak, and Petra Štolfa from Agricultural Institute of Slovenia for their help with laboratory analyses, and Polonca Ferk, Brane Leskošek, Jan Gojznikar, and Katja Črnec from the University of Ljubljana, Medical faculty for performing gene expression laboratory analysis on the Nanostring nCounter platform.

Conflicts of Interest: The authors declare no conflicts of interest.

References

1. Batorek Lukač, N.; Tomažin, U.; Škrlep, M.; Kastelic, A.; Poklukar, K.; Čandek-Potokar, M. Krškopoljski Prašič. In *European Local Pig Breeds—Diversity and Performance. A Study of Project TREASURE*; Intech Open: Rijeka, Croatia, 2019.
2. Poklukar, K.; Čandek-Potokar, M.; Lukač, N.B.; Škrlep, M. Biochemical and Gene Expression Differences Associated with Higher Fat Deposition in Krškopolje Pigs in Comparison with Lean Hybrid Pigs. *Livest. Sci.* **2023**, *272*, 105247. [CrossRef]
3. Brossard, L.; Nieto, R.; Charneca, R.; Araujo, J.P.; Pugliese, C.; Radović, Č.; Čandek-Potokar, M. Modelling Nutritional Requirements of Growing Pigs from Local Breeds Using InraPorc. *Animals* **2019**, *9*, 169. [CrossRef] [PubMed]
4. Škrlep, M.; Poklukar, K.; Vrecl, M.; Brankovič, J.; Čandek-Potokar, M. Growth Performance, Carcass Quality, and Lipid Metabolism in Krškopolje Pigs and Modern Hybrid Pigs: Comparison of Genotypes and Evaluation of Dietary Protein Reduction. *Animals* **2024**, *14*, 3331. [CrossRef] [PubMed]
5. Palma-Granados, P.; Seiquer, I.; Benítez, R.; Óvilo, C.; Nieto, R. Effects of Lysine Deficiency on Carcass Composition and Activity and Gene Expression of Lipogenic Enzymes in Muscles and Backfat Adipose Tissue of Fatty and Lean Piglets. *Anim. Int. J. Anim. Biosci.* **2019**, *13*, 2406–2418. [CrossRef]
6. Tejeda, J.F.; Hernández-Matamoros, A.; Paniagua, M.; González, E. Effect of Free-Range and Low-Protein Concentrated Diets on Growth Performance, Carcass Traits, and Meat Composition of Iberian Pig. *Animals* **2020**, *10*, 273. [CrossRef]
7. Madeira, M.S.; Rolo, E.A.; Alfaia, C.M.; Pires, V.R.; Luxton, R.; Doran, O.; Bessa, R.J.B.; Prates, J.A.M. Influence of Betaine and Arginine Supplementation of Reduced Protein Diets on Fatty Acid Composition and Gene Expression in the Muscle and Subcutaneous Adipose Tissue of Cross-Bred Pigs. *Br. J. Nutr.* **2016**, *115*, 937–950. [CrossRef]

8. Aquilani, C.; Sirtori, F.; Franci, O.; Acciaioli, A.; Bozzi, R.; Pezzati, A.; Pugliese, C. Effects of Protein Restriction on Performances and Meat Quality of Cinta Senese Pig Reared in an Organic System. *Animals* **2019**, *9*, 310. [CrossRef]
9. Muñoz, M.; Fernández-Barroso, M.A.; López-García, A.; Caraballo, C.; Nuñez, Y.; Óvilo, C.; González, E.; García-Casco, J.M. Consequences of a Low Protein Diet on the Liver and *Longissimus Dorsi* Transcriptome of Duroc × Iberian Crossbred Pigs. *Animal* **2021**, *15*, 100408. [CrossRef]
10. Škrlep, M.; Poklukar, K.; Millet, S.; Čandek-Potokar, M. Reducing Protein Levels in Diets for Local Pig Breeds: A Case Study on Fat-type Krškopolje Pig. *Anim. Sci. J.* 2025; *under review*.
11. Zhao, S.; Wang, J.; Song, X.; Zhang, X.; Ge, C.; Gao, S. Impact of Dietary Protein on Lipid Metabolism-Related Gene Expression in Porcine Adipose Tissue. *Nutr. Metab.* **2010**, *7*, 6. [CrossRef]
12. Wang, J.; Zhao, S.M.; Song, X.L.; Pan, H.B.; Li, W.Z.; Zhang, Y.Y.; Gao, S.Z.; Chen, D.W. Low Protein Diet Up-Regulate Intramuscular Lipogenic Gene Expression and down-Regulate Lipolytic Gene Expression in Growth—Finishing Pigs. *Livest. Sci.* **2012**, *148*, 119–128. [CrossRef]
13. Madeira, M.S.; Pires, V.M.R.; Alfaia, C.M.; Costa, A.S.H.; Luxton, R.; Doran, O.; Bessa, R.J.B.; Prates, J.A.M. Differential Effects of Reduced Protein Diets on Fatty Acid Composition and Gene Expression in Muscle and Subcutaneous Adipose Tissue of Alentejana Purebred and Large White × Landrace × Pietrain Crossbred Pigs. *Br. J. Nutr.* **2013**, *110*, 216–229. [CrossRef] [PubMed]
14. Hasan, M.S.; Wang, Y.; Feugang, J.M.; Zhou, H.; Liao, S.F. RNA Sequencing Analysis Revealed Differentially Expressed Genes and Their Functional Annotation in Porcine Longissimus Dorsi Muscle Affected by Dietary Lysine Restriction. *Front. Vet. Sci.* **2023**, *10*, 1233292. [CrossRef] [PubMed]
15. Malgwi, I.H.; Halas, V.; Grünvald, P.; Schiavon, S.; Jócsák, I. Genes Related to Fat Metabolism in Pigs and Intramuscular Fat Content of Pork: A Focus on Nutrigenetics and Nutrigenomics. *Animals* **2022**, *12*, 150. [CrossRef]
16. Stachowiak, M.; Nowacka-Woszuk, J.; Szydlowski, M.; Switonski, M. The *ACACA* and *SREBF1* Genes Are Promising Markers for Pig Carcass and Performance Traits, but Not for Fatty Acid Content in the Longissimus Dorsi Muscle and Adipose Tissue. *Meat Sci.* **2013**, *95*, 64–71. [CrossRef]
17. Muñoz, G.; Alves, E.; Fernández, A.; Óvilo, C.; Barragán, C.; Estellé, J.; Quintanilla, R.; Folch, J.M.; Silió, L.; Rodríguez, M.C.; et al. QTL Detection on Porcine Chromosome 12 for Fatty-Acid Composition and Association Analyses of the Fatty Acid Synthase, Gastric Inhibitory Polypeptide and Acetyl-Coenzyme A Carboxylase Alpha Genes. *Anim. Genet.* **2007**, *38*, 639–646. [CrossRef]
18. Colbert, C.L.; Kim, C.-W.; Moon, Y.-A.; Henry, L.; Palnitkar, M.; McKean, W.B.; Fitzgerald, K.; Deisenhofer, J.; Horton, J.D.; Kwon, H.J. Crystal Structure of Spot 14, a Modulator of Fatty Acid Synthesis. *Proc. Natl. Acad. Sci. USA* **2010**, *107*, 18820–18825. [CrossRef]
19. Wang, X.; Cheng, J.; Qin, W.; Chen, H.; Chen, G.; Shang, X.; Zhang, M.; Balsai, N.; Chen, H. Polymorphisms in 5′ Proximal Regulating Region of THRSP Gene Are Associated with Fat Production in Pigs. *3 Biotech* **2020**, *10*, 267. [CrossRef]
20. Corominas, J.; Ramayo-Caldas, Y.; Puig-Oliveras, A.; Estellé, J.; Castelló, A.; Alves, E.; Pena, R.N.; Ballester, M.; Folch, J.M. Analysis of Porcine Adipose Tissue Transcriptome Reveals Differences in de Novo Fatty Acid Synthesis in Pigs with Divergent Muscle Fatty Acid Composition. *BMC Genom.* **2013**, *14*, 843. [CrossRef]
21. Xing, K.; Wang, K.; Ao, H.; Chen, S.; Tan, Z.; Wang, Y.; Xitong, Z.; Yang, T.; Zhang, F.; Liu, Y.; et al. Comparative Adipose Transcriptome Analysis Digs out Genes Related to Fat Deposition in Two Pig Breeds. *Sci. Rep.* **2019**, *9*, 12925. [CrossRef]
22. Meneses, M.J.; Silvestre, R.; Sousa-Lima, I.; Macedo, M.P. Paraoxonase-1 as a Regulator of Glucose and Lipid Homeostasis: Impact on the Onset and Progression of Metabolic Disorders. *Int. J. Mol. Sci.* **2019**, *20*, 4049. [CrossRef] [PubMed]
23. Zhou, Y.; Li, H.; Xia, N. The Interplay Between Adipose Tissue and Vasculature: Role of Oxidative Stress in Obesity. *Front. Cardiovasc. Med.* **2021**, *8*, 650214. [CrossRef] [PubMed]
24. Manna, P.; Jain, S.K. Obesity, Oxidative Stress, Adipose Tissue Dysfunction, and the Associated Health Risks: Causes and Therapeutic Strategies. *Metab. Syndr. Relat. Disord.* **2015**, *13*, 423–444. [CrossRef]
25. Pugliese, C.; Madonia, G.; Chiofalo, V.; Margiotta, S.; Acciaioli, A.; Gandini, G. Comparison of the Performances of Nero Siciliano Pigs Reared Indoors and Outdoors. 1. Growth and Carcass Composition. *Meat Sci.* **2003**, *65*, 825–831. [CrossRef]
26. Bee, G.; Guex, G.; Herzog, W. Free-Range Rearing of Pigs during the Winter: Adaptations in Muscle Fiber Characteristics and Effects on Adipose Tissue Composition and Meat Quality Traits. *J. Anim. Sci.* **2004**, *82*, 1206–1218. [CrossRef]
27. Adeva-Andany, M.M.; Carneiro-Freire, N.; Seco-Filgueira, M.; Fernández-Fernández, C.; Mouriño-Bayolo, D. Mitochondrial β-Oxidation of Saturated Fatty Acids in Humans. *Mitochondrion* **2019**, *46*, 73–90. [CrossRef]
28. Zhang, C.; Zhang, B.; Zhang, X.; Sun, G.; Sun, X. Targeting Orphan Nuclear Receptors NR4As for Energy Homeostasis and Diabetes. *Front. Pharmacol.* **2020**, *11*, 587457. [CrossRef]
29. Timofeev, Y.S.; Kiselev, A.R.; Dzhioeva, O.N.; Drapkina, O.M. Heat Shock Proteins (HSPs) and Cardiovascular Complications of Obesity: Searching for Potential Biomarkers. *Curr. Issues Mol. Biol.* **2023**, *45*, 9378–9389. [CrossRef]
30. Moura, C.S.; Lollo, P.C.B.; Morato, P.N.; Amaya-Farfan, J. Dietary Nutrients and Bioactive Substances Modulate Heat Shock Protein (HSP) Expression: A Review. *Nutrients* **2018**, *10*, 683. [CrossRef]

31. Krause, M.; Heck, T.G.; Bittencourt, A.; Scomazzon, S.P.; Newsholme, P.; Curi, R.; de Bittencourt, P.I.H. The Chaperone Balance Hypothesis: The Importance of the Extracellular to Intracellular HSP70 Ratio to Inflammation-Driven Type 2 Diabetes, the Effect of Exercise, and the Implications for Clinical Management. *Mediat. Inflamm.* **2015**, *2015*, 249205. [CrossRef]
32. Dahdah, N.; Tercero-Alcázar, C.; Malagón, M.M.; Garcia-Roves, P.M.; Guzmán-Ruiz, R. Interrelation of Adipose Tissue Macrophages and Fibrosis in Obesity. *Biochem. Pharmacol.* **2024**, *225*, 116324. [CrossRef] [PubMed]
33. Yu, L.; Gao, Y.; Aaron, N.; Qiang, L. A Glimpse of the Connection between PPARγ and Macrophage. *Front. Pharmacol.* **2023**, *14*, 1254317. [CrossRef] [PubMed]
34. Colmenares, M.T.C.; Matos, A.d.O.; Dantas, P.H.d.S.; Neto, J.R.D.C.; Silva-Sales, M.; Sales-Campos, H. Unveiling the Impact of TREM-2+ Macrophages in Metabolic Disorders. *Cell. Immunol.* **2024**, *405–406*, 104882. [CrossRef] [PubMed]
35. Vogel, A.; Brunner, J.S.; Hajto, A.; Sharif, O.; Schabbauer, G. Lipid Scavenging Macrophages and Inflammation. *Biochim. Biophys. Acta (BBA) Mol. Cell Biol. Lipids* **2022**, *1867*, 159066. [CrossRef]
36. Odegaard, J.I.; Ricardo-Gonzalez, R.R.; Eagle, A.R.; Vats, D.; Morel, C.R.; Goforth, M.H.; Subramanian, V.; Mukundan, L.; Ferrante, A.W.; Chawla, A. Alternative M2 Activation of Kupffer Cells by PPARδ Ameliorates Obesity-Induced Insulin Resistance. *Cell Metab.* **2008**, *7*, 496–507. [CrossRef]
37. Heming, M.; Gran, S.; Jauch, S.-L.; Fischer-Riepe, L.; Russo, A.; Klotz, L.; Hermann, S.; Schäfers, M.; Roth, J.; Barczyk-Kahlert, K. Peroxisome Proliferator-Activated Receptor-γ Modulates the Response of Macrophages to Lipopolysaccharide and Glucocorticoids. *Front. Immunol.* **2018**, *9*, 893. [CrossRef]
38. Santos, E.W.; de Oliveira, D.C.; Hastreiter, A.; Beltran, J.S.d.O.; Rogero, M.M.; Fock, R.A.; Borelli, P. High-Fat Diet or Low-Protein Diet Changes Peritoneal Macrophages Function in Mice. *Nutrire* **2016**, *41*, 6. [CrossRef]
39. Liao, S.F.; Wang, T.; Regmi, N. Lysine Nutrition in Swine and the Related Monogastric Animals: Muscle Protein Biosynthesis and Beyond. *SpringerPlus* **2015**, *4*, 147. [CrossRef]
40. Schneider, C.A.; Rasbad, W.S.; Eliceiri, K.W. NIH Image to ImageJ: 25 years of image analysis. *Nat. Methods* **2012**, *9*, 671–675. [CrossRef]
41. Bazin, R.; Ferré, P. Assays of Lipogenic Enzymes. In *Adipose Tissue Protocols*; Ailhaud, G., Ed.; Springer: New York, NY, USA, 2001; pp. 121–127.
42. Škrlep, M.; Poklukar, K.; Kress, K.; Vrecl, M.; Fazarinc, G.; Lukač, N.B.; Weiler, U.; Stefanski, V.; Čandek-Potokar, M. Effect of Immunocastration and Housing Conditions on Pig Carcass and Meat Quality Traits1. *Transl. Anim. Sci.* **2020**, *4*, 1224–1237. [CrossRef]
43. Bolger, A.M.; Lohse, M.; Usadel, B. Trimmomatic: A Flexible Trimmer for Illumina Sequence Data. *Bioinformatics* **2014**, *30*, 2114–2120. [CrossRef] [PubMed]
44. Dobin, A.; Davis, C.A.; Schlesinger, F.; Drenkow, J.; Zaleski, C.; Jha, S.; Batut, P.; Chaisson, M.; Gingeras, T.R. STAR: Ultrafast Universal RNA-Seq Aligner. *Bioinformatics* **2013**, *29*, 15–21. [CrossRef] [PubMed]
45. Liao, Y.; Smyth, G.K.; Shi, W. The R Package Rsubread Is Easier, Faster, Cheaper and Better for Alignment and Quantification of RNA Sequencing Reads. *Nucleic Acids Res.* **2019**, *47*, e47. [CrossRef] [PubMed]
46. Love, M.I.; Huber, W.; Anders, S. Moderated Estimation of Fold Change and Dispersion for RNA-Seq Data with DESeq2. *Genome Biol.* **2014**, *15*, 550. [CrossRef]
47. Wu, T.; Hu, E.; Xu, S.; Chen, M.; Guo, P.; Dai, Z.; Feng, T.; Zhou, L.; Tang, W.; Zhan, L.; et al. clusterProfiler 4.0: A Universal Enrichment Tool for Interpreting Omics Data. *Innovation* **2021**, *2*, 100141. [CrossRef]
48. Yu, G. Thirteen Years of clusterProfiler. *Innovation* **2024**, *5*, 100722. [CrossRef]

International Journal of *Molecular Sciences*

MDPI

Article

Role of Acyl-CoA Thioesterase 7 in Regulating Fatty Acid Metabolism and Its Contribution to the Onset and Progression of Bovine Clinical Mastitis

Bin Zhou [1,2], Bohao Zhang [2,3], Jiangyuan Han [1], Junjun Zhang [1,2], Jianfu Li [1,2], Weitao Dong [2,3], Xingxu Zhao [1,2,3], Yong Zhang [1,2,3] and Quanwei Zhang [1,2,3,*]

1 College of Life Sciences and Biotechnology, Gansu Agricultural University, Lanzhou 730030, China; zhoub@st.gsau.edu.cn (B.Z.); hanjiangy@gsau.edu.cn (J.H.); jun2272136040@126.com (J.Z.); li18419220108@163.com (J.L.); zhaoxx@gsau.edu.cn (X.Z.); zhychy@126.com (Y.Z.)
2 Gansu Key Laboratory of Animal Generational Physiology and Reproductive Regulation, Lanzhou 730070, China; zhangbhgs@163.com (B.Z.); d.wt2008@163.com (W.D.)
3 College of Veterinary Medicine, Gansu Agricultural University, Lanzhou 730070, China
* Correspondence: zhangqw@gsau.edu.cn

Abstract: Clinical mastitis (CM) is a prevalent and severe inflammatory disease in dairy cows affecting the mammary glands. Fatty acid (FA) metabolism and associated enzymes are crucial for many physiological and pathological processes in dairy cows. However, the relationships among FA metabolism, FA-associated enzymes, and CM, as well as the mechanisms underlying their interactions, in dairy cows are not fully understood. The aim of this study was to characterize biological process (BP) terms, pathways, and differentially expressed proteins (DEPs) related to FA metabolism from our previous data-independent acquisition proteomic study. Six BPs involving 14 downregulated and 20 upregulated DEPs, and four pathways involving 10 downregulated and 11 upregulated DEPs related to FA synthesis and metabolism were systematically identified. Associated analysis suggested that 12 candidate DEPs obtained from BPs and pathways, especially acyl-CoA thioesterase 7 (ACOT7), regulate long-chain FA (LCFA) elongation and the biosynthesis of unsaturated FAs. Immunohistochemical and immunofluorescence staining results showed that ACOT7 was present mainly in the cytoplasm of mammary epithelial cells. The qRT-PCR and Western blotting results showed that *ACOT7* mRNA and protein levels in the mammary glands of the CM group were significantly upregulated compared to those in the healthy group. This evidence indicates that ACOT7 is positively correlated with CM onset and progression in Holstein cows. These findings offer novel insights into the role of FA metabolism and related enzymes in CM and offer potential targets for the development of therapeutic strategies and biomarkers for the prevention and treatment of CM in dairy cows.

Keywords: clinical mastitis; long-chain fatty acids; ACOT7; bioinformatics

1. Introduction

Clinical mastitis (CM), a prevalent and severe disease in dairy cows, is triggered by physical, chemical, and microbial factors that lead to localized inflammation in the bovine mammary glands. The mammary gland, as the organ for milk production, functions normally in lactation, producing various fatty acids (FAs), proteins, and other minor components of milk. FAs are crucial components of milk fat and are a key energy source for mammary epithelial cells (MECs) [1,2]. MECs either take up free FAs from the plasma or synthesize them internally, esterify them into various FAs, and secrete them into milk [2]. The endogenous accumulation or unbalanced production of FAs, occurring both naturally in the mammary gland or upon exogenous microbial infection with Gram-negative bacteria, results in the production of endotoxins, such as lipopolysaccharide (LPS), which translocate into the bloodstream and lead to an inflammatory response [3], particularly in the mammary gland. Consequently, pathogen infections prompt the immune system to initiate inflammatory responses in the mammary gland, resulting in tissue damage and disturbances in FA metabolism [4,5], which manifest as typical CM-associated characteristics of bovine mammary glands and milk. This severely affects milk yield and quality, resulting in decreased milk production, higher veterinary expenses, and prolonged herd management challenges, which collectively have a substantial impact on the economic sustainability of the dairy industry [6,7]. However, the pathological molecular mechanisms associated with FAs during the regulation of CM development in dairy cows are not completely understood.

FAs, including short-chain FAs (SCFAs), long-chain FAs (LCFAs), and medium-chain FAs (MCFAs), depending on their type, intake, or consumption patterns and specific health conditions, can affect multiple aspects of inflammation, such as leukocyte chemotaxis, adhesion molecule expression, and leukocyte–endothelial adhesive interactions [6]. This is especially apparent with chronic inflammatory diseases, such as mastitis, ketoacidosis, and endometritis in mammals [8,9]. For example, SCFAs can modulate the production of inflammatory mediators, such as interferon-gamma (IFN-γ), interleukin (IL)-2, and IL-10 in inflammatory cells [10]. SCFAs are the potential immunomodulatory metabolites in controlling *Staphylococcus-aureus*-mediated mastitis [11]. Further, saturated FAs (SFAs), a subset of LCFAs, are closely associated with chronic inflammation [12]. Elevated SFA levels can activate the Toll-like receptor (TLR) 4 signaling pathway, leading to the increased production of pro-inflammatory cytokines, such as tumor necrosis factor-alpha (TNF-α) and IL-6, thereby promoting the onset of mastitis [13,14]. Levels of these pro-inflammatory cytokines are upregulated during the pathogenesis of inflammation-based mastitis in dairy cows [15]. Moreover, excessive SFA intake can aggravate mastitis in dairy cows, which is dependent on alterations to the fluidity of cell membranes, disruptions in signal transduction, and enhanced lipid synthesis, intensifying the inflammatory response in MECs [16]. Considering the role of fatty acids in promoting the upregulation of inflammatory cytokines, it is plausible to hypothesize that this process may be driven by dysregulation of fatty acid metabolism, thereby contributing to the development of mastitis. Interestingly, owing to the beneficial effects of MCFAs on anti-pathogen activity, immune-improving reactions, inflammation prevention, and the suppression of damage caused by inflammation, they have been used as feed supplements for monogastric animals [17]. However, their application in ruminants remains controversial. Collectively, FA metabolism has crucial roles in various physiological and pathological processes in animals, both as an important source of energy and a key component of cell membranes, while occupying an important position in signal transduction and metabolic regulation.

Enzymes associated with FA metabolism can influence inflammatory responses via different pathways. Fatty acid synthase (FASN), a key metabolic enzyme, is significantly correlated with the expression of inflammatory cytokines and enzymes in bovine MECs [7]. Moreover, acyl-CoA thioesterases (ACOTs), comprising a key class of lipases responsible for the β-oxidation of LCFAs [18,19], can affect various inflammatory signaling pathways via LCFA accumulation [20]. Some enzymes involved in FA synthesis, such as cyclooxygenases (COXs), lipoxygenases (LOXs), FASN, acyl-CoA synthetase long-chain family member (ACSL)-1, and peroxisome proliferator-activated receptor gamma (PPARG), alter oxidative stress or antioxidant activity in LPS-treated bovine MECs [21]. This evidence illustrates that enzymes associated with FA metabolism are closely associated with inflammatory diseases caused by abnormal or unbalanced FA metabolism.

Overall, exploring the roles of enzymes associated with FA metabolism and the inflammatory response will provide an innovative direction for understanding physiological and pathological processes in animals, particularly in dairy cows with mastitis. However, in dairy cows, the relationships among FA metabolism, FA-associated enzymes, and CM, in addition to the underlying mechanisms, are not fully understood. The aim of this study was to screen differentially expressed proteins (DEPs) associated with FA metabolism and investigate their functions in healthy dairy cows and those with CM. The expression patterns and distributions of core DEPs were evaluated in the mammary glands of healthy and CM-affected dairy cows, which could provide new strategies for the prevention and treatment of bovine mastitis.

2. Results

2.1. Identification of Candidate DEPs Related to FA Synthesis or Metabolism Based on GO Terms

Gene Ontology (GO) terms related to FA synthesis or metabolism and associated candidate DEPs were screened based on 3739 DEPs and 819 biological process (BP) terms from data-independent acquisition (DIA) proteomics (Figure 1). Six BPs ($p < 0.05$ and p.adjust < 0.05; Table S1), including the FA metabolic process, fatty-acyl-CoA metabolic process, LCFA metabolic process and biosynthetic process, and LCFA and MCFA fatty-acyl-CoA metabolic processes, and 34 DEPs (Table S2) were identified (Figure 1A). The heat map and volcano plot showed that these proteins, including 14 DEPs for which expression was downregulated and 20 for which expression was upregulated, were differentially expressed between the Con/C and CM groups (Figure 1B,C). The Venn diagram further indicated that only acyl-CoA thioesterase 7 (ACOT7) was shared among the six BPs (Figure 1D). ClueGO results indicated that 13 of the 34 DEPs were directly related to five BPs, and ACOT7, an important candidate DEP, participates in five BPs related to FA biosynthesis or metabolism.

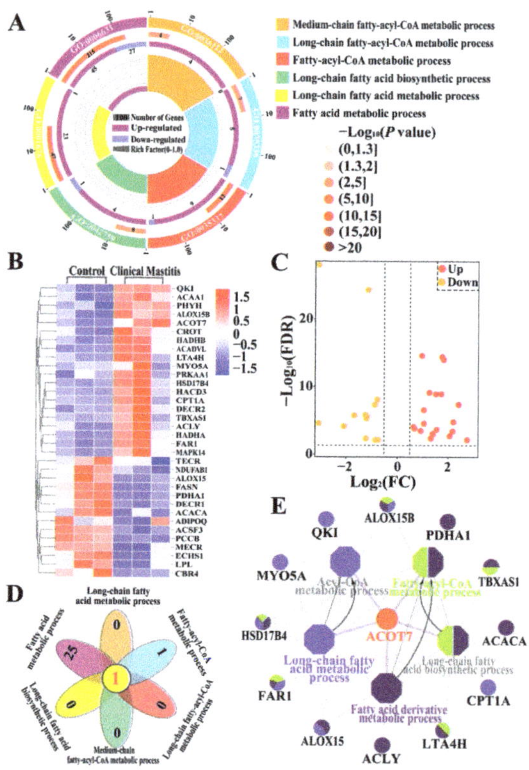

Figure 1. Identification of candidate DEPs and BPs related to FA synthesis or metabolism based on the GO enrichment data. (**A**) Candidate DEPs and BPs related to FA synthesis or metabolism. (**B**) Heat map of 34 candidate DEPs. (**C**) Volcano plot of 34 DEPs, including 14 downregulated and 20 upregulated DEPs. (**D**) Venn diagram of candidate DEPs in six BPs. (**E**) Interactive relationship analysis of candidate DEPs and BPs related to FA synthesis or metabolism using ClueGo.

2.2. Identification of Candidate DEPs Related to FA Synthesis or Metabolism in the Pathways

Pathways related to FA synthesis or metabolism and their associated DEPs were identified according to a Kyoto Encyclopedia of Genes and Genomes (KEGG) enrichment analysis of the DIA proteomic data (Figure 2). Four pathways (Table S3) fulfilling the $p < 0.05$ and p.adjust < 0.05 criteria were obtained (Figure 2A). The heat map and volcano plot results showed that 21 DEPs (Table S4), including 11 for which expression was upregulated and 10 for which expression was downregulated, were screened in the CM group after overlapping them with recurrent DEPs (Figure 2B,C). The protein–protein interaction (PPI) network suggested that four, six, twelve, and seven DEPs directly participate in the four respective pathways (Figure 2D). Additionally, ACOT7 is involved in regulating three identified pathways, namely the biosynthesis of unsaturated fatty acids (UFAs), FA degradation, and FA elongation.

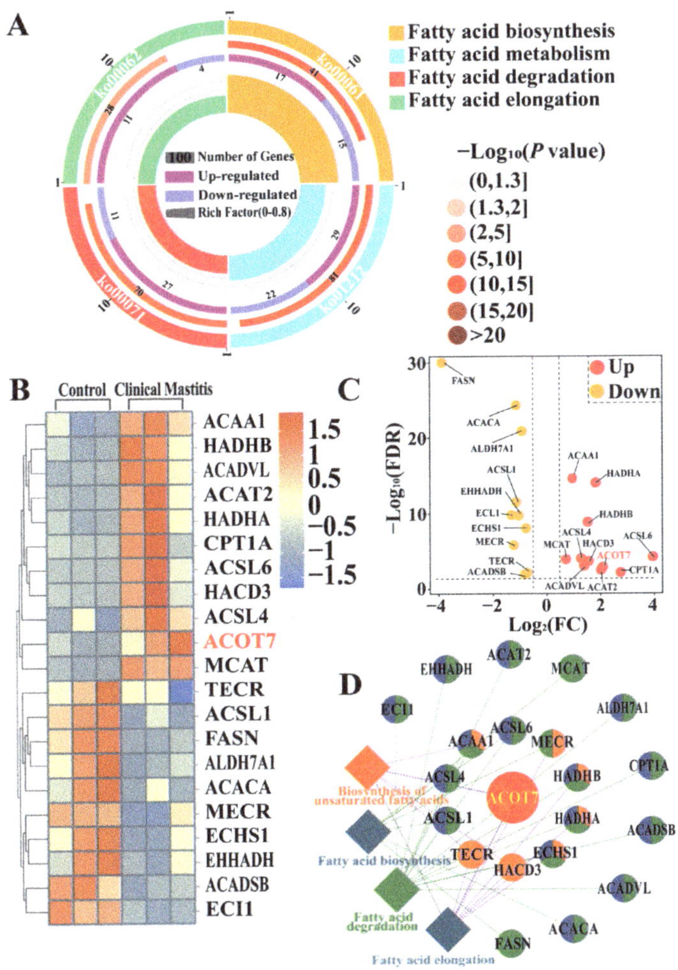

Figure 2. Identification of candidate DEPs and pathways related to FA synthesis or metabolism based on pathway enrichment. (**A**) Candidate DEPs and pathways related to FA synthesis or metabolism. (**B**) Heat map of 21 candidate DEPs. (**C**) Volcano plot of 21 DEPs, including 10 downregulated and 11 upregulated DEPs. (**D**) Interactive relationship analysis of candidate DEPs and pathways related to FA synthesis or metabolism using ClueGo.

2.3. Identification of Core DEPs Related to FA Synthesis or Metabolism Based on Conjoint Analysis of BPs and Pathways

The core DEPs related to FA synthesis or metabolism in dairy cows with CM were identified based on the selected BPs and pathways using the conjoint analysis method (Figure 3). A Venn diagram showed that 12 of the 43 candidate DEPs were co-expressed in six BPs and four pathways involved in FA synthesis or metabolism (Figure 3A). Among the twelve shared DEPs, expression levels of seven were upregulated and five were downregulated; in particular, expression levels of carnitine O-palmitoyltransferase 1 (CPT1A), ACOT7, and FASN exhibited changes that exceeded two fold (Figure 3B). The Sankey diagram showed that ACOT7, as a core DEP, was associated with six BPs and the FA elongation pathway (Figure 3C). The PPI network further indicated that ACOT7 was a core DEP that interacts with other candidate DEPs and is involved in FA and mitochondrial energy metabolism (Figure 3D).

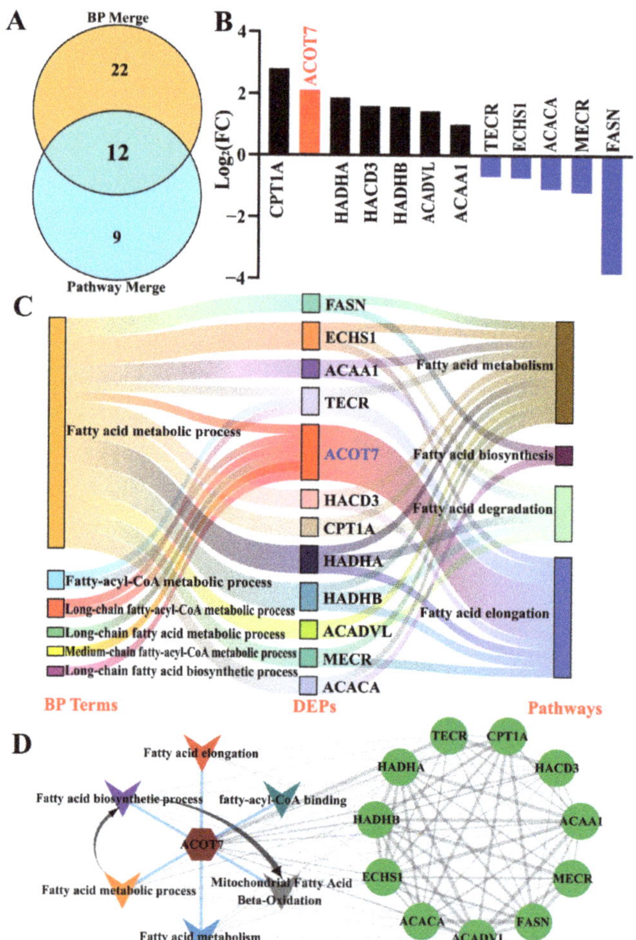

Figure 3. Identification of core DEPs related to FA synthesis or metabolism based on BPs and pathways. (**A**) Venn diagram of candidate DEPs among the merged BPs and pathways. (**B**) Relative expression levels of 12 DEPs quantified by DIA proteomics; the *y*-axis represents the $\log_2$ (FC) values. (**C**) Sankey diagram of the BPs, shared DEPs, and pathways related to FA synthesis or metabolism. (**D**) Constructed PPI network of the shared DEPs using ClueGO.

2.4. Morphological Observation and Distribution Analysis of ACOT7 in Bovine Mammary Tissue

H&E staining revealed an orderly arrangement of either columnar or cuboidal MECs in the intact alveoli of the bovine mammary gland tissues of the Con group. In contrast, epithelial cells that were detached from mammary tissue, along with predominant neutrophil infiltration in atrophied and deformed alveoli, were observed in the mammary gland tissues of the CM group (Figure 4A). Immunohistochemistry staining (IHC) indicated that ACOT7 was primarily expressed in the cytoplasm of MECs, with increased positive staining observed in the CM group (Figure 4B). No positive ACOT7 staining was observed in the negative control group (Figure 4C). The integrated optical density (IOD) results indicated significantly upregulated expression of ACOT7 in the CM group ($p < 0.01$) compared to that in the Con group (Figure 4D).

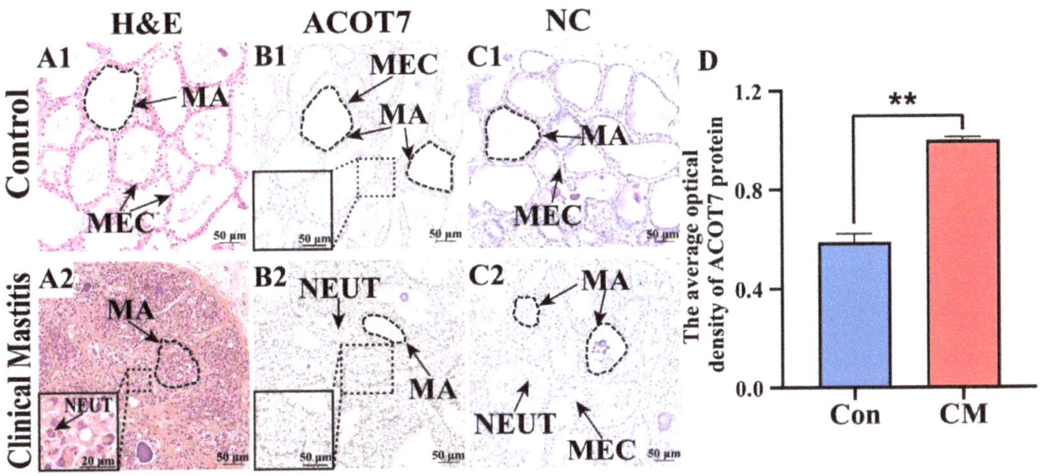

Figure 4. Distribution analysis of ACOT7 in mammary glands of the two groups. (**A**) Pathological variation in bovine mammary glands of the healthy control group (Con/C; **A1**) and clinical mastitis groups (CM; **A2**) based on H&E staining. (**B**) ACOT7 distribution in bovine mammary glands of the Con/C (**B1**) and CM (**B2**) groups based on immunohistochemical staining. (**C**) Negative control (NC) for Con/C (**C1**) and CM (**C2**) groups. (**D**) Gray values of positive expression of ACOT7 protein quantified using ImageJ 1.44p software. Con/C, control group. CM, clinical mastitis group. NC, negative control; MA: mammary alveoli; MECs, mammary epithelial cells; NEUT, neutrophil. Scale bars, 50 μm (200× magnification). ** represents $p < 0.01$.

2.5. Co-Localization Analysis of ACOT7 Protein in Bovine Mammary Tissue

IF results revealed that nuclei labeled with 4′,6-Diamidino-2-phenylindole (DAPI) appeared in alveoli with an orderly arrangement in the Con group, whereas irregular alveoli were observed in the CM group (Figure 5A). The expression of Cytokeratin (CK)-18 protein, a marker of epithelial cells, showed a similar trend, especially with necrotic and detached MECs present in the incomplete alveoli of the CM group (Figure 5B). Positive fluorescence ACOT7 signals were also present in the MECs of the alveoli in both the Con and CM groups, with different degrees of staining (Figure 5C). The co-localization results suggested that CK-18 and ACOT7 were co-expressed in the cytoplasm of MECs (Figure 5D). The IOD results of IF staining indicated significantly upregulated expression of ACOT7 in the CM group ($p < 0.01$) compared to that in the Con group (Figure 5E). These findings suggest that co-localization of ACOT7 and CK-18 proteins are correlated with the MECs' function and the occurrence of CM in dairy cows.

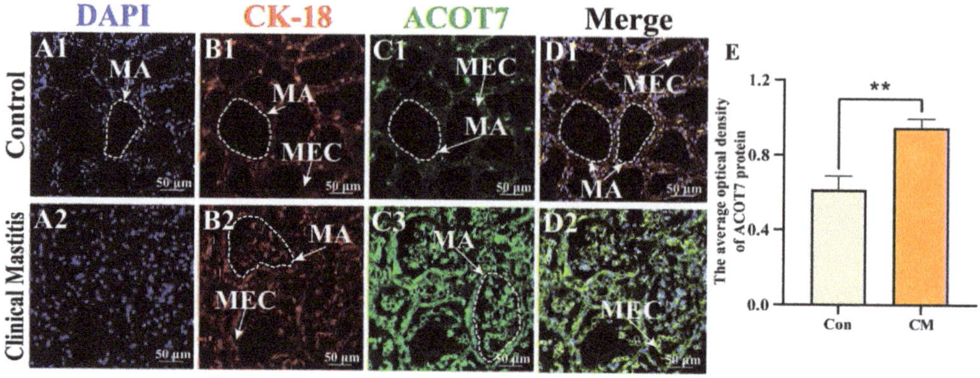

Figure 5. Co-localization analysis of ACOT7 in mammary glands of the two groups. (**A–D**) Co-localization analysis of ACOT7 in bovine mammary glands of the Con/C and CM groups: nuclei (blue, **A1**,**A2**), CK-18 (red, **B1**,**B2**), ACOT7 (green, **C1**,**C2**), merged with CK18 and ACOT7 (**D1**,**D2**) staining, respectively. (**E**) Positive IF signals of ACOT7 protein quantified using ImageJ software. Con/C, control group. CM, clinical mastitis group. MA, mammary alveoli; MECs, mammary epithelial cells; Scale bars, 50 μm (200× magnification). ** represents $p < 0.01$.

2.6. *ACOT7* mRNA and Protein Expression Patterns in Bovine Mammary Tissue

The relative expression levels of *ACOT7* mRNA and protein were evaluated in bovine mammary tissues using qRT-PCR and Western blotting (Figure 6). Compared with that in the Con group, the relative expression level of *ACOT7* mRNA was significantly upregulated in the mammary gland tissues of the CM group (Figure 6A). Moreover, ACOT7 protein expression was present but different in each sample of the Con and CM groups (Figure 6B). Results of the average IOD of each band showed that the relative expression level of ACOT7 in the CM group was significantly upregulated compared to that in the Con group ($p < 0.01$). These findings suggest that the upregulation of ACOT7 expression is positively correlated with the occurrence of CM in dairy cows.

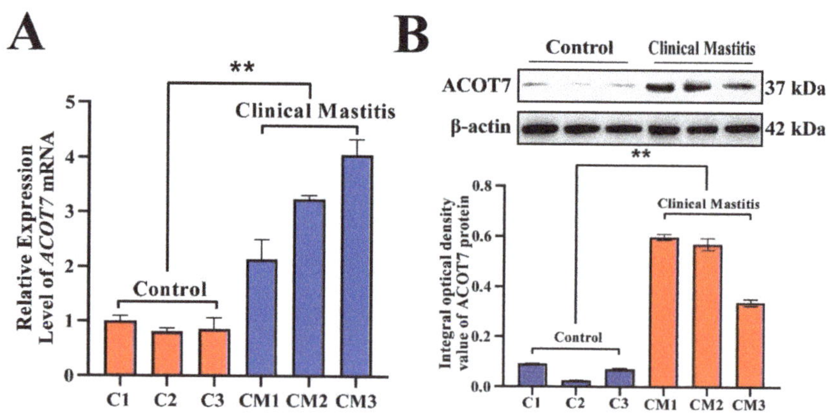

Figure 6. Expression patterns of *ACOT7* mRNA and protein in mammary glands of the two groups. (**A**) Relative expression level of *ACOT7* mRNA, monitored via qRT-PCR assays. (**B**) Protein bands and relative expression level of ACOT7, monitored via Western blot assays and using ImageJ software, respectively. Con/C, control group. CM, clinical mastitis group. Data are presented as means ± SEM. ** represents $p < 0.01$.

3. Discussion

Studies have indicated rapid energy metabolism in dairy cows in peak lactation or high-milk-yield stages, which prioritizes the mobilization of body fat [22,23]. The catabolism of FAs (i.e., beta-oxidation) not only provides energy to the body but also generates a series of metabolites with important roles in transcriptional control and cell signaling [24]. Some specific FA metabolites (i.e., SCFAs and n-3 polyunsaturated FAs) have anti-inflammatory effects, whereas others (i.e., LCFAs) promote inflammatory responses [25,26]. The unbalanced production of various FAs or overburdened lipid mobilization in FA metabolism causes a negative energy balance and/or disproportional energy metabolism. These metabolic disturbances impair immune function, reduce fertility and milk production, and ultimately decrease the economic benefits associated with dairy cows affected by mastitis [9]. Enzymes associated with FA metabolism and their metabolites are closely associated with FA lipolysis and synthesis in cow mastitis [2,4]. This study offers novel insights by identifying ACOT7 as a key enzyme that connects fatty acid metabolism to inflammation, thereby addressing a significant gap in the literature concerning its role in dairy cows with mastitis. However, the systematic exploration of using enzymes or proteins to elucidate this relationship requires further refinement.

In the present study, candidate DEPs related to FA synthesis or metabolism were screened based on GO terms and pathways in the DIA proteomic data generated from the mammary gland tissues of lactating Holstein cows with or without CM. Six BPs involving 34 DEPs and four pathways involving 21 DEPs related to FA synthesis and metabolism were identified. Among the DEPs, eight were enriched in three long-chain FA metabolic processes, and seven of the eight were involved in the FA metabolic process, especially ACOT7, indicating that LCFA metabolic processes are associated with the onset and progression of CM in dairy cows. LCFAs cause significant changes in the blood and milk samples of cows with CM, which could be related to the inflammatory response, metabolic status, and milk fat composition [27,28]. LCFAs are organic compounds that support various biological processes and maintain homeostasis [29]. For example, lauric acid, palmitic acid, and stearic acid are LCFAs that can promote persistent inflammation and induce Cyclooxygenase-2 (COX-2) expression through a nuclear factor kappa-B (NF-κB)-dependent mechanism in macrophages [14]. Further, SFAs are crucial components of bacterial endotoxins and can indirectly stimulate TLR-dependent signaling [30–32], which is also the main mechanism through which pathogenic bacteria infect the mammary tissue during the occurrence of cow mastitis. In addition, the identified pathways, including FA biosynthesis, metabolism, degradation, and elongation, are indispensable for FA metabolism. Previous studies have suggested that FASN in FA biosynthesis [33], CPT1A in FA metabolism [34], ACSLs in FA degradation [35], and very-long-chain enoyl-CoA reductase (TECR) in FA elongation are associated with the development of cow mastitis [36]. A further bioinformatic analysis indicated that LCFA metabolic pathways have a direct relationship with ACOT7 expression, suggesting that ACOT7 might influence the development of CM through its role in FA synthesis or metabolism.

ACOT7, a predominant member of the ACOT family, is an extensively studied protein that regulates the intracellular levels of FAs, acyl-CoA, and coenzyme A by catalyzing the hydrolysis of acyl-CoA to produce free fatty acids (FFAs) and coenzyme A [18]. By regulating the cell cycle, ACOT7 drives the development of mammary gland diseases; moreover, ACOT7 specifically hydrolyzes arachidonoyl-CoA to release arachidonic acid [18,19], underscoring its close association with LCFAs. In mouse macrophages, LPS-induced *ACOT7* mRNA expression is mediated by the TLR4-Myeloid Differentiation Primary Response 88 (MyD88) pathway, underscoring its role in inflammation. Similarly, ACOT7 plays a critical role in neurotoxicity prevention by modulating neuronal fatty acid metabolism, thus providing protection to the nervous system [37]. These studies highlight the broad physiological functions of the ACOT7 family in fatty acid metabolism and inflammation, suggesting their potential involvement in various inflammatory conditions. These findings bear similarities to our results; however, reports on the involvement of ACOT7 in CM development through fatty acid pathways in Holstein cows are limited. In this study, DIA proteomic results indicated high ACOT7 expression in the CM group and suggested that it participates in the regulation of FA elongation and UFA biosynthesis, which is consistent with the mRNA and protein expression findings. The elevated expression of ACOT7 may represent an adaptive response in dairy cows during lactation, a period of negative energy balance, where it increases the release of free fatty acids for β-oxidation to meet the energy demands. The biosynthesis of UFAs is closely related to LCFA elongation. In addition, these processes function together in multiple organisms to produce FAs of varying lengths and saturation levels, such as polyunsaturated fatty acids [38], which are involved in the regulation of inflammatory responses [27] and antibacterial activity [39], and influence gene expression and the gut microbiota [40] in dairy cows with mastitis. Co-localization results also confirmed that ACOT7 was predominantly expressed in the cytoplasm of MECs. These findings suggest that high ACOT7 expression in CM affects FA metabolism, leading to the decomposition of long-chain acyl-CoA and the release of FFAs that stimulate TLR4 receptor signaling, thereby influencing the onset of CM in dairy cows [14].

Based on these experimental results and previous research, we propose a potential mechanism underlying the role of ACOT7 in CM (Figure 7). Specifically, ACOT7 hydrolyzes long-chain acyl-CoA to acetyl-CoA in the endoplasmic reticulum, thereby promoting LCFA synthesis [41]. These FAs are transported to the extracellular space and bloodstream by FA transport proteins (FATPs) and FA-binding proteins (FABPs) [42,43]. This process activates the TLR4 pathway, triggers NF-κB signaling cascades, and ultimately promotes the expression and release of inflammatory factors, such as TNF-α and IL-1β, thereby enhancing the inflammatory response. However, this study still has certain limitations, and validated assays based on results at the cellular and animal level need to confirm this. Addressing these aspects would enhance the credibility of the interpretations. Nevertheless, we systematically identified candidate DEPs related to FA synthesis or metabolism that contribute to the onset and progression of CM in dairy cows, providing new insights into the prevention and treatment of mastitis in dairy cows. In addition to the fundamental findings on the role of ACOT7 and fatty acid metabolism in CM, this study holds significant practical implications for dairy farming. Moreover, the findings on ACOT7 provide a theoretical basis for the identification of new drug targets and potential biomarkers for CM.

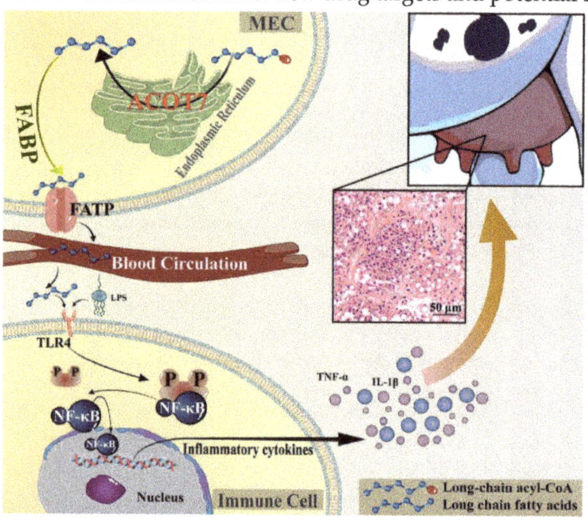

Figure 7. Proposed molecular mechanism underlying the effects of ACOT7 in the mammary glands of dairy cows with clinical mastitis (CM). MECs, mammary epithelial cells; FABP, fatty-acid-binding protein; FATP, fatty acid transport protein; TLR4, Toll-like receptor 4; TNF-α, tumor necrosis factor-α; IL-1β, interleukin-1β; LPS, lipopolysaccharide. Scale bar, 50 μm (200× magnification).

4. Materials and Methods

4.1. Tissue Sample Preparation

The lactating Holstein cows utilized in the present study were randomly selected from a commercially managed dairy farm in Wuzhong City (Ningxia Hui Autonomous Region, China). Veterinary clinical diagnoses and milk tests, including somatic cell count (SCC) and the Lanzhou Mastitis Test (LMT), were performed as previously described [44,45]. Holstein cows without typical clinical changes, with an SCC of 7×10^4–1×10^5 cells/mL and a negative LMT, were classified as the 3 healthy control group (Con/C). In contrast, Holstein cows with typical disease signs, an SCC of 1×10^5–15×10^5 cells/mL, and a positive LMT were assigned to the CM group. To ensure similarity in age, parity, body condition (aged 5–7 years, 3rd to 5th lactation, 60–250 days in milk), and the absence of other diseases, cows in the Con/C and CM groups (n = 3 per group) were transported to the slaughterhouse, where mammary gland samples were collected after slaughter. The tissues were immediately fixed in 4% paraformaldehyde, or promptly frozen in liquid nitrogen, and stored at −80 °C in the laboratory. This study was approved by the Ethics Committee of Gansu Agricultural University (No. GSAU-AEW-2018-0128).

4.2. Bioinformatic Analysis

GO and KEGG annotations were performed based on DIA proteomic sequencing data from mammary gland samples from the Con/C and CM groups of Holstein cows (accession number: IPX0003382000/PXD028100). A bioinformatic analysis was conducted to identify GO categories, KEGG pathways, and DEPs related to FA synthesis or metabolism, with significance set at $p < 0.05$ and p.adjust < 0.05. Venn diagrams, enrichment circle plots, heat maps, and volcano plots were created using R language and the OmicShare online platform (http://www.omicshare.com/tools, accessed on 20 February 2024) [45]. PPI networks were constructed to screen the potential candidate functional DEPs related to FA synthesis or metabolism using STRING v12.0 (EMBL, Heidelberg, Germany), Cytoscape 3.10.0 (Cytoscape Consortium, La Jolla, CA, USA), and ClueGO 2.5.10 software (Cytoscape Consortium, La Jolla, CA, USA). The potential mechanisms associated with the candidate DEPs were predicted through KEGG analysis and a literature review, and mechanistic diagrams were created using Adobe Illustrator 2022 (Adobe, San Jose, MA, USA) and MedPeer 2.4.241127 (MedPeer®, Beijing, China).

4.3. Histochemistry Staining

Fixed mammary gland tissues were processed using standard paraffin (Solarbio, Beijing, China)-embedding techniques and sectioned into 5 μm thick slices using a microtome (Leica, Wetzlar, Germany), as described previously [45]. H&E staining was performed as previously described [45,46]. Briefly, after deparaffinization in xylene and dehydration in a gradient of ethanol, the slices were stained using an H&E staining solution kit (Servicebio, Wuhan, China). The slices were examined and imaged using an optical microscope and imaging system (Nikon, Tokyo, Japan). Six fields from each view were randomly selected for each sample.

4.4. Immunohistochemistry Staining (IHC)

After deparaffinization and hydration, slices were subjected to antigen retrieval using sodium citrate (Servicebio, Wuhan, China). Endogenous catalase was deactivated and blocked using 3% H_2O_2 and 10% donkey serum, respectively (Solarbio, Beijing, China). The slices were incubated with a rabbit anti-ACOT7 primary antibody at a 1:200 dilution (Proteintech, Wuhan, China). Secondary antibody incubation and chromogenic reactions were performed using an StreptAvidin–Biotin Complex (SABC) staining kit (Boster, Wuhan, China) and a 3-diaminobenzidine (DAB) substrate kit (Solarbio, Beijing, China), as described previously [46]. Phosphate-Buffered Saline (PBS) was used instead of primary antibodies in the NC group. The sections were examined and imaged using a Zeiss LSM 800 confocal microscope (Carl Zeiss, Oberkochen, Germany). Ten random fields of view were selected for each section. ImageJ 1.44p software (National Institutes of Health, Rockville, MD, USA) was used to measure the average positive expression in each field, and the relative expression levels were analyzed. All immunostaining assays and measurements were performed in triplicate.

4.5. Immunofluorescence Staining (IF)

The blocked slices were incubated with rabbit anti-CK-18 and rabbit anti-ACOT7 primary antibodies (1:100; Proteintech, Wuhan, China), as described previously [46]. The slices were incubated with fluorescent secondary antibodies with different labels (1:350 dilution; Alexa Fluor® 488 for ACOT7 and Alexa Fluor® 647 for CK-18, Proteintech, Wuhan, China), as described previously [46]. The nuclei were stained with 0.5 μg/mL DAPI (Solarbio, Beijing, China). Finally, the sections were observed and imaged using an Echo-Labs upright/inverted fluorescence microscope and imaging system (Olympus, Tokyo, Japan). The positive IF signals were measured using ImageJ software, as described in IHC staining. All staining assays and measurements were performed in triplicate.

4.6. RNA Extraction, cDNA Synthesis, and qRT-PCR Detection

Total RNA was extracted from the tissues (20 mg of each tissue sample frozen at −80 °C) of the Con/C and CM groups using the FastPure RNA isolation kit (Vazyme, Nanjing, China), RNA quantification was performed using a NanoDrop-8000 spectrophotometer (Thermo Fisher Scientific, Waltham, MA, USA), and RNA integrity was evaluated through 1% denaturing formaldehyde agarose gel electrophoresis. And, a 500 ng RNA sample was used for complementary DNA (cDNA) synthesis using the BioTeke Thermo RT Kit (BioTeke, Beijing, China) according to the manufacturer's instructions, as previously described [45,47]. Primers for bovine β-actin and ACOT7 (Table S5) were designed using Primer Premier 5.0 software (PREMIER Biosoft, San Francisco, CA, USA) and synthesized by Qingke Biological Technology Co., Ltd. (Shanxi, China). A qRT-PCR was performed using cDNA as the template, as previously described [46]. The relative expression levels of β-actin were taken as the internal control. The results were calculated using the $2^{-\Delta\Delta Ct}$ method. All qRT-PCR assays were performed at least in triplicate.

4.7. Western Blotting

Total protein was isolated from the tissues (100 mg per sample stored at −80 °C) of the Con and CM groups using a Radioimmunoprecipitation Assay (RIPA) kit (Solarbio, Beijing, China) and quantified using a Bicinchoninic Acid Protein Assay (BCA) kit (Solarbio), according to the manufacturer's instructions. Total proteins (50 μg) were used to detect the expression levels of ACOT7 with a 1:800 dilution ratio, and β-actin with a 1:4500 dilution ratio (Bioss, Beijing, China), in the Con/C and CM groups, as described previously [45,47]. The bands were visualized using Enhanced Chemiluminescence (ECL) (Solarbio, Beijing, China) and analyzed using ImageJ. All immunoblot assays were performed at least in triplicate.

4.8. Statistical Analysis

Statistical analysis was performed using SPSS 23.0 (SPSS Inc., Chicago, IL, USA). Data are expressed as means ± standard deviations (X ± SD), unless otherwise stated. Comparisons between groups were performed using the least significant difference test, whereas multiple group comparisons were performed via one-way ANOVA. Statistical significance was set at $p < 0.05$. Graphs were created using GraphPad Prism 8.0.2 (GraphPad Software Inc., San Diego, CA, USA).

5. Conclusions

Based on DIA proteomics data, we identified 34 DEPs associated with six BPs, and four pathways involving 21 DEPs related to fatty acid metabolism and synthesis. Some DEPs such as CPT1A, FASN, and TECR regulate fatty acid metabolism and CM development by modulating pathways such as fatty acid β-oxidation, the fatty acid metabolic process, and the long-chain fatty-acyl-CoA metabolic process. Crucially, comprehensive analysis indicates that ACOT7 functions as a key enzyme potentially involved in the pathogenesis of CM in dairy cows via fatty acid metabolic pathways. Furthermore, IHC and IF staining demonstrated that ACOT7 was primarily localized in the MECs of dairy cows. Notably, compared to the Con/C group, the *ACOT7* mRNA and protein levels were significantly upregulated in the mammary glands of the CM group. Therefore, this study proposes a mechanism of ACOT7, which may contribute to the development of CM. Combining proteomics could help define the function of ACOT7 in CM, shedding light on its potential as a predictive biomarker or even a therapeutic target for novel CM treatment strategies using knockout models, and providing a theoretical foundation for further investigation into the physiological roles of fatty acids in the mammary tissues of dairy cows.

Supplementary Materials: The following supporting information can be downloaded at: https://www.mdpi.com/article/10.3390/ijms252313046/s1.

Author Contributions: Conceptualization, Q.Z., W.D. and J.H.; methodology, Q.Z., W.D. and J.H.; software, B.Z. (Bohao Zhang) and J.L.; validation, B.Z. (Bin Zhou), J.Z. and J.L.; formal analysis, Q.Z. and W.D.; investigation, J.H., B.Z. (Bin Zhou), B.Z. (Bohao Zhang) and J.L.; resources, Y.Z. and X.Z.; data curation, B.Z. (Bohao Zhang) and B.Z. (Bin Zhou); writing—original draft preparation, Q.Z. and W.D.; writing—review and editing, Q.Z. and Y.Z.; visualization, B.Z. (Bin Zhou), J.Z. and J.L.; supervision, Q.Z., X.Z. and Y.Z.; project administration, Q.Z., Y.Z. and X.Z.; funding acquisition, Q.Z., X.Z. and Y.Z. All authors have read and agreed to the published version of the manuscript.

Funding: This study was supported by the Education-industrial support plan of the Gansu Provincial Department (2024CYZC-30), Innovation-Driven Assistance Project of Gansu Province (GXH20240328-4), the National Natural Science Foundation of China (U21A20262), the Excellent Doctoral Project of Science and Technology Plan Funding of Gansu Province (24JRRA676), and the Young Scholars Science Foundation of Gansu Agricultural University (GAU-KYQD-2021-30).

Institutional Review Board Statement: The study was conducted according to the guidelines of the Declaration of Helsinki and was approved by the Institutional Animal Care and Use Committee of Gansu Agricultural University (approval number GSAU-AEW-2018-0128).

Informed Consent Statement: Informed consent was obtained from the commercial farm involved in the study.

Data Availability Statement: The data that support the findings of this study are available from the corresponding author upon reasonable request.

Acknowledgments: We thank Gansu Key Laboratory of Animal Generational Physiology and Reproductive Regulation for providing instruments, equipment, and related materials.

Conflicts of Interest: The authors declare no conflict of interest.

References

1. Sordillo, L.M.; Contreras, G.A.; Aitken, S.L. Metabolic Factors Affecting the Inflammatory Response of Periparturient Dairy Cows. *Anim. Health Res. Rev.* **2009**, *10*, 53–63. [CrossRef] [PubMed]
2. Zhang, H.; Shen, Z.; Yang, Z.; Jiang, H.; Chu, S.; Mao, Y.; Li, M.; Chen, Z.; Aboragah, A.; Loor, J.J.; et al. Abundance of Solute Carrier Family 27 Member 6 (*SLC27A6*) in the Bovine Mammary Gland Alters Fatty Acid Metabolism. *Food Funct.* **2021**, *12*, 4909–4920. [CrossRef] [PubMed]
3. Marshall, J.C. Lipopolysaccharide: An Endotoxin or an Exogenous Hormone? *Clin. Infect. Dis.* **2005**, *41*, S470–S480. [CrossRef]
4. Luo, Y.; Kong, Z.; Yang, B.; He, F.; Huan, C.; Li, J.; Yi, K. Relationship between Microflora Changes and Mammary Lipid Metabolism in Dairy Cows with Mastitis. *Animals* **2023**, *13*, 2773. [CrossRef]
5. Wang, Y.; Nan, X.; Zhao, Y.; Jiang, L.; Wang, H.; Zhang, F.; Hua, D.; Liu, J.; Yang, L.; Yao, J.; et al. Discrepancies among Healthy, Subclinical Mastitic, and Clinical Mastitic Cows in Fecal Microbiome and Metabolome and Serum Metabolome. *J. Dairy Sci.* **2022**, *105*, 7668–7688. [CrossRef]
6. Calder, P.C. Marine Omega-3 Fatty Acids and Inflammatory Processes: Effects, Mechanisms and Clinical Relevance. *Biochim. Biophys. Acta (BBA) Mol. Cell Biol. Lipids* **2015**, *1851*, 469–484. [CrossRef] [PubMed]
7. Li, L.; Tang, W.; Zhao, M.; Gong, B.; Cao, M.; Li, J. Study on the Regulation Mechanism of Lipopolysaccharide on Oxidative Stress and Lipid Metabolism of Bovine Mammary Epithelial Cells. *Physiol. Res.* **2021**, *70*, 777–785. [CrossRef]
8. Calder, P.C. Long-Chain Fatty Acids and Inflammation. *Proc. Nutr. Soc.* **2012**, *71*, 284–289. [CrossRef]

9. Esposito, G.; Irons, P.C.; Webb, E.C.; Chapwanya, A. Interactions between Negative Energy Balance, Metabolic Diseases, Uterine Health and Immune Response in Transition Dairy Cows. *Anim. Reprod. Sci.* **2014**, *144*, 60–71. [CrossRef]
10. Vinolo, M.A.R.; Rodrigues, H.G.; Nachbar, R.T.; Curi, R. Regulation of Inflammation by Short Chain Fatty Acids. *Nutrients* **2011**, *3*, 858–876. [CrossRef]
11. Akhtar, M.; Naqvi, S.U.-A.-S.; Liu, Q.; Pan, H.; Ma, Z.; Kong, N.; Chen, Y.; Shi, D.; Kulyar, M.F.-E.-A.; Khan, J.A.; et al. Short Chain Fatty Acids (SCFAs) Are the Potential Immunomodulatory Metabolites in Controlling *Staphylococcus aureus*-Mediated Mastitis. *Nutrients* **2022**, *14*, 3687. [CrossRef]
12. Chait, A.; Kim, F. Saturated Fatty Acids and Inflammation: Who Pays the Toll? *Arterioscler. Thromb. Vasc. Biol.* **2010**, *30*, 692–693. [CrossRef] [PubMed]
13. Rocha, D.M.; Caldas, A.P.; Oliveira, L.L.; Bressan, J.; Hermsdorff, H.H. Saturated Fatty Acids Trigger TLR4-Mediated Inflammatory Response. *Atherosclerosis* **2016**, *244*, 211–215. [CrossRef] [PubMed]
14. Rogero, M.; Calder, P. Obesity, Inflammation, Toll-Like Receptor 4 and Fatty Acids. *Nutrients* **2018**, *10*, 432. [CrossRef] [PubMed]
15. Aitken, S.L.; Corl, C.M.; Sordillo, L.M. Immunopathology of Mastitis: Insights into Disease Recognition and Resolution. *J. Mammary Gland. Biol. Neoplasia* **2011**, *16*, 291–304. [CrossRef] [PubMed]
16. Ingvartsen, K.L.; Moyes, K. Nutrition, Immune Function and Health of Dairy Cattle. *Animal* **2013**, *7*, 112–122. [CrossRef]
17. Wang, Z.; Wang, Q.; Tang, C.; Yuan, J.; Luo, C.; Li, D.; Xie, T.; Sun, X.; Zhang, Y.; Yang, Z.; et al. Medium Chain Fatty Acid Supplementation Improves Animal Metabolic and Immune Status during the Transition Period: A Study on Dairy Cattle. *Front. Immunol.* **2023**, *14*, 1018867. [CrossRef]
18. Forwood, J.K.; Thakur, A.S.; Guncar, G.; Marfori, M.; Mouradov, D.; Meng, W.; Robinson, J.; Huber, T.; Kellie, S.; Martin, J.L.; et al. Structural Basis for Recruitment of Tandem Hotdog Domains in Acyl-CoA Thioesterase 7 and Its Role in Inflammation. *Proc. Natl. Acad. Sci. USA* **2007**, *104*, 10382–10387. [CrossRef]
19. Jung, S.H.; Lee, H.C.; Hwang, H.J.; Park, H.A.; Moon, Y.-A.; Kim, B.C.; Lee, H.M.; Kim, K.P.; Kim, Y.-N.; Lee, B.L.; et al. Acyl-CoA Thioesterase 7 Is Involved in Cell Cycle Progression via Regulation of PKCζ–P53–P21 Signaling Pathway. *Cell Death Dis.* **2017**, *8*, e2793. [CrossRef]
20. Wall, V.Z.; Barnhart, S.; Kramer, F.; Kanter, J.E.; Vivekanandan-Giri, A.; Pennathur, S.; Bolego, C.; Ellis, J.M.; Gijón, M.A.; Wolfgang, M.J.; et al. Inflammatory Stimuli Induce Acyl-CoA Thioesterase 7 and Remodeling of Phospholipids Containing Unsaturated Long (≥C20)-Acyl Chains in Macrophages [S]. *J. Lipid Res.* **2017**, *58*, 1174–1185. [CrossRef]
21. Ma, N.; Wei, G.; Zhang, H.; Dai, H.; Roy, A.C.; Shi, X.; Chang, G.; Shen, X. Cis-9, Trans-11 CLA Alleviates Lipopolysaccharide-Induced Depression of Fatty Acid Synthesis by Inhibiting Oxidative Stress and Autophagy in Bovine Mammary Epithelial Cells. *Antioxidants* **2021**, *11*, 55. [CrossRef] [PubMed]
22. Kuhla, B.; Metges, C.C. Proteomic Tools Help Understanding the Metabolic Adaptation to Negative Energy Balance in Dairy Cows. In *Energy and Protein Metabolism and Nutrition in Sustainable Animal Production, Proceedings of the 4th International Symposium on Energy and Protein Metabolism and Nutrition, Sacramento, CA, USA, 9–12 September 2013*; Oltjen, J.W., Kebreab, E., Lapierre, H., Eds.; Academic Publishers: Wageningen, The Netherlands, 2013; pp. 213–221. ISBN 978-90-8686-781-3.
23. Abiso, M.; Yohannes, B.; Alemu, B. Combating Negative Effect of Negative Energy Balance in Dairy Cows: Comprehensive Review. *Approaches Poult. Dairy Vet. Sci.* **2019**, *6*, 1–6. [CrossRef]
24. Wang, Z.; Hou, X.; Shang, G.; Deng, G.; Luo, K.; Peng, M. Exploring Fatty Acid β-Oxidation Pathways in Bacteria: From General Mechanisms to DSF Signaling and Pathogenicity in Xanthomonas. *Curr. Microbiol.* **2024**, *81*, 336. [CrossRef] [PubMed]
25. Corrêa, R.O.; Vieira, A.; Sernaglia, E.M.; Lancellotti, M.; Vieira, A.T.; Avila-Campos, M.J.; Rodrigues, H.G.; Vinolo, M.A.R. Bacterial Short-Chain Fatty Acid Metabolites Modulate the Inflammatory Response against Infectious Bacteria. *Cell. Microbiol.* **2017**, *19*, e12720. [CrossRef] [PubMed]
26. Jeong, H.Y.; Moon, Y.S.; Cho, K.K. ω-6 and ω-3 Polyunsaturated Fatty Acids: Inflammation, Obesity and Foods of Animal Resources. *Food Sci. Anim. Resour.* **2024**, *44*, 988–1010. [CrossRef]
27. Mavangira, V.; Gandy, J.C.; Zhang, C.; Ryman, V.E.; Jones, A.D.; Sordillo, L.M. Polyunsaturated Fatty Acids Influence Differential Biosynthesis of Oxylipids and Other Lipid Mediators during Bovine Coliform Mastitis. *J. Dairy Sci.* **2015**, *98*, 6202–6215. [CrossRef]
28. Atroshi, F.; Rizzo, A.; Kangasniemi, R.; Sankari, S.; Työppönen, T.; Österman, T.; Parantainen, J. Role of Plasma Fatty Acids, Prostaglandins and Antioxidant Balance in Bovine Mastitis. *J. Vet. Med. Ser. A* **1989**, *36*, 702–711. [CrossRef]
29. Jannas-Vela, S.; Espinosa, A.; Candia, A.A.; Flores-Opazo, M.; Peñailillo, L.; Valenzuela, R. The Role of Omega-3 Polyunsaturated Fatty Acids and Their Lipid Mediators on Skeletal Muscle Regeneration: A Narrative Review. *Nutrients* **2023**, *15*, 871. [CrossRef]
30. Erridge, C.; Samani, N.J. Saturated Fatty Acids Do Not Directly Stimulate Toll-Like Receptor Signaling. *Arterioscler. Thromb. Vasc. Biol.* **2009**, *29*, 1944–1949. [CrossRef]
31. Anderson, E.K.; Hill, A.A.; Hasty, A.H. Stearic Acid Accumulation in Macrophages Induces Toll-like Receptor 4/2-Independent Inflammation Leading to Endoplasmic Reticulum Stress-Mediated Apoptosis. *Arterioscler. Thromb. Vasc. Biol.* **2012**, *32*, 1687–1695. [CrossRef]
32. Lancaster, G.I.; Langley, K.G.; Berglund, N.A.; Kammoun, H.L.; Reibe, S.; Estevez, E.; Weir, J.; Mellett, N.A.; Pernes, G.; Conway, J.R.W.; et al. Evidence That TLR4 Is Not a Receptor for Saturated Fatty Acids but Mediates Lipid-Induced Inflammation by Reprogramming Macrophage Metabolism. *Cell Metab.* **2018**, *27*, 1096–1110.e5. [CrossRef] [PubMed]
33. Huma, Z.I.; Sharma, N.; Kour, S.; Tandon, S.; Guttula, P.K.; Kour, S.; Singh, A.K.; Singh, R.; Gupta, M.K. Putative Biomarkers for Early Detection of Mastitis in Cattle. *Anim. Prod. Sci.* **2020**, *60*, 1721–1736. [CrossRef]

34. Ceciliani, F.; Lecchi, C.; Urh, C.; Sauerwein, H. Proteomics and Metabolomics Characterizing the Pathophysiology of Adaptive Reactions to the Metabolic Challenges during the Transition from Late Pregnancy to Early Lactation in Dairy Cows. *J. Proteom.* **2018**, *178*, 92–106. [CrossRef]
35. Delosière, M.; Bernard, L.; Viala, D.; Fougère, H.; Bonnet, M. Milk and Plasma Proteomes from Cows Facing Diet-Induced Milk Fat Depression Are Related to Immunity, Lipid Metabolism and Inflammation. *Animal* **2023**, *17*, 100822. [CrossRef] [PubMed]
36. Cecchinato, A.; Macciotta, N.P.P.; Mele, M.; Tagliapietra, F.; Schiavon, S.; Bittante, G.; Pegolo, S. Genetic and Genomic Analyses of Latent Variables Related to the Milk Fatty Acid Profile, Milk Composition, and Udder Health in Dairy Cattle. *J. Dairy Sci.* **2019**, *102*, 5254–5265. [CrossRef] [PubMed]
37. Xie, X.; Chen, C.; Feng, S.; Zuo, S.; Zhao, X.; Li, H. Acyl-CoA Thioesterase 7 Is Transcriptionally Activated by Krüppel-Like Factor 13 and Promotes the Progression of Hepatocellular Carcinoma. *J. Hepatocell. Carcinoma* **2021**, *8*, 1623–1641. [CrossRef]
38. Cook, H.W. Chapter 5—Fatty Acid Desaturation and Chain Elongation in Eukaryotes. In *New Comprehensive Biochemistry*; Vance, D.E., Vance, J.E., Eds.; Biochemistry of Lipids, Lipoproteins and Membranes; Elsevier: Amsterdam, The Netherlands, 1996; Volume 31, pp. 129–152.
39. Hogan, J.S.; Smith, K.L.; Todhunter, D.A.; Schoenberger, P.S. Growth Responses of Environmental Mastitis Pathogens to Long-Chain Fatty Acids1. *J. Dairy Sci.* **1988**, *71*, 245–249. [CrossRef] [PubMed]
40. Feng, J.; Wang, Q.; Yang, W.; Liu, J.; Gao, M.-Q. Omega-3 Polyunsaturated Fatty Acids Ameliorated Inflammatory Response of Mammary Epithelial Cells and Mammary Gland Induced by Lipopolysaccharide. *Acta Biochim. Biophys. Sin.* **2021**, *53*, 1142–1153. [CrossRef]
41. Yamada, J.; Kurata, A.; Hirata, M.; Taniguchi, T.; Takama, H.; Furihata, T.; Shiratori, K.; Iida, N.; Takagi-Sakuma, M.; Watanabe, T.; et al. Purification, Molecular Cloning, and Genomic Organization of Human Brain Long-Chain Acyl-CoA Hydrolase. *J. Biochem.* **1999**, *126*, 1013–1019. [CrossRef]
42. Lipke, K.; Kubis-Kubiak, A.; Piwowar, A. Molecular Mechanism of Lipotoxicity as an Interesting Aspect in the Development of Pathological States-Current View of Knowledge. *Cells* **2022**, *11*, 844. [CrossRef]
43. Stahl, A. Fatty Acid Transport Proteins: A Current View of a Growing Family. *Trends Endocrinol. Metab.* **2001**, *12*, 266–273. [CrossRef]
44. Zhang, Q.; Bai, X.; Shi, J.; Wang, X.; Zhang, B.; Dai, L.; Lin, T.; Gao, Y.; Zhang, Y.; Zhao, X. DIA Proteomics Identified the Potential Targets Associated with Angiogenesis in the Mammary Glands of Dairy Cows with Hemorrhagic Mastitis. *Front. Vet. Sci.* **2022**, *9*, 980963. [CrossRef] [PubMed]
45. Zhang, Q.; Bai, X.; Lin, T.; Wang, X.; Zhang, B.; Dai, L.; Shi, J.; Zhang, Y.; Zhao, X. HMOX1 Promotes Ferroptosis in Mammary Epithelial Cells via FTH1 and Is Involved in the Development of Clinical Mastitis in Dairy Cows. *Antioxidants* **2022**, *11*, 2221. [CrossRef] [PubMed]
46. Zhang, Q.; Wang, Q.; Zhang, Y.; Cheng, S.; Hu, J.; Ma, Y.; Zhao, X. Comprehensive Analysis of MicroRNA⁻Messenger RNA from White Yak Testis Reveals the Differentially Expressed Molecules Involved in Development and Reproduction. *Int. J. Mol. Sci.* **2018**, *19*, 3083. [CrossRef] [PubMed]
47. Zhang, B.; Lin, T.; Bai, X.; An, X.; Dai, L.; Shi, J.; Zhang, Y.; Zhao, X.; Zhang, Q. Sulfur Amino Acid Metabolism and the Role of Endogenous Cystathionine-γ-Lyase/H2S in Holstein Cows with Clinical Mastitis. *Animals* **2022**, *12*, 1451. [CrossRef]

International Journal of
Molecular Sciences

MDPI

Article

Transcriptional Dynamics and Key Regulators of Adipogenesis in Mouse Embryonic Stem Cells: Insights from Robust Rank Aggregation Analysis

Mouza Alzaabi [1], Mariam Khalili [1], Mehar Sultana [2] and Mohamed Al-Sayegh [1,2,*]

1 Division of Biology, New York University Abu Dhabi, Saadiyaat Island, Abu Dhabi P.O. Box 129188, United Arab Emirates; maa9608@nyu.edu (M.A.); mk8962@nyu.edu (M.K.)

2 Center for Genomics & Systems Biology, New York University Abu Dhabi, Saadiyaat Island, Abu Dhabi P.O. Box 129188, United Arab Emirates; ms8783@nyu.edu

* Correspondence: ma3803@nyu.edu; Tel.: +971-26284560

Abstract: Embryonic stem cells are crucial for studying developmental biology due to their self-renewal and pluripotency capabilities. This research investigates the differentiation of mouse ESCs into adipocytes, offering insights into obesity and metabolic disorders. Using a monolayer differentiation approach over 30 days, lipid accumulation and adipogenic markers, such as *Cebpb*, *Pparg*, and *Fabp4*, confirmed successful differentiation. RNA sequencing revealed extensive transcriptional changes, with over 15,000 differentially expressed genes linked to transcription regulation, cell cycle, and DNA repair. This study utilized Robust Rank Aggregation to identify critical regulatory genes like PPARG, CEBPA, and EP300. Network analysis further highlighted *Atf5*, *Ccnd1*, and *Nr4a1* as potential key players in adipogenesis and its mature state, validated through RT-PCR. While key adipogenic factors showed plateaued expression levels, suggesting early differentiation events, this study underscores the value of ESCs in modeling adipogenesis. These findings contribute to our understanding of adipocyte differentiation and have significant implications for therapeutic strategies targeting metabolic diseases.

Keywords: adipogenesis; embryonic stem cells; transcriptomics; robust ranking aggregation

1. Introduction

Embryonic stem cells (ESCs) possess remarkable self-renewal and pluripotency capabilities, making them invaluable tools for developmental biology and regenerative medicine [1]. ESCs can differentiate into various cell types, offering insights into lineage commitment and the molecular underpinnings of development. One crucial differentiation pathway is adipogenesis, the process through which preadipocytes mature into adipocytes, which is essential for understanding obesity, metabolic disorders, and potential therapeutic avenues for adipose-tissue-related conditions. Adipogenesis has been extensively studied in various cell models, including preadipocyte cell lines (e.g., 3T3-L1) [2], mesenchymal stem cells (MSCs) [3], and embryonic fibroblasts [4,5]. However, using ESCs to study adipogenesis offers distinct advantages. ESCs provide a more physiologically relevant model for early development and differentiation processes, enabling researchers to investigate the initial stages of adipocyte lineage commitment and the factors influencing this process from a pluripotent state [6]. This can lead to a deeper understanding of the molecular mechanisms underlying adipogenesis and the potential discovery of novel regulatory pathways [6].

Mouse ESCs, derived from the inner cell mass of blastocysts, have the unique ability to proliferate indefinitely while maintaining the potential to differentiate into all three germ layers [1,7,8]. This pluripotency is regulated by a network of transcription factors, including Oct4, Sox2, and Nanog, which maintain the undifferentiated state and prevent lineage-specific gene expression [9,10]. The differentiation of ESCs into specific cell types involves complex signaling pathways and epigenetic modifications, making them an ideal model for studying early developmental processes [11,12]. Recent advancements in RNA sequencing have further elucidated the heterogeneity and plasticity within ESC populations, highlighting subpopulations with distinct differentiation potentials. These findings underscore the dynamic nature of ESCs and the intricate regulatory networks governing their pluripotency and differentiation [13]. ESC has been utilized in various studies due to its capability to differentiate in vitro into various generic specialized cell types, including neurons [14], blood vessels [15], hepatocytes [16], and osteocytes [17]. In addition to cell types, several studies have utilized ESC differentiation to understand adipogenesis [6,18–20].

Adipogenesis is a highly regulated process that plays a critical role in energy homeostasis and metabolism. It involves the conversion of mesenchymal stem cells or preadipocytes into mature adipocytes characterized by lipid accumulation and the expression of adipogenic markers, such as PPARγ and C/EBPα [21,22]. Key signaling pathways involved in adipogenesis include the Wnt/β-catenin pathway, the insulin signaling pathway, and the BMP signaling pathway, all of which contribute to the regulation of adipocyte differentiation and function [22–24]. Recent studies have identified additional molecular regulators of adipogenesis, such as miRNAs and long non-coding RNAs (lncRNAs), which modulate gene expression and cellular function during adipocyte differentiation [25,26]. Dysregulation of adipogenesis is associated with metabolic disorders, including obesity, insulin resistance, and type 2 diabetes [22,27,28]. Therefore, understanding the molecular mechanisms governing adipogenesis is of great importance for developing therapeutic strategies to combat these conditions. Given the complexities of these metabolic conditions as well as the dynamical nature of adipogenesis, global genome, specifically transcriptional, changes can occur. Therefore, investigating the transcriptomic changes during the differentiation of ESCs into adipocytes offers valuable insights into the fundamental processes of cell fate determination and tissue development. From a clinical perspective, this knowledge can inform the development of regenerative therapies and interventions for metabolic diseases. By elucidating the key regulatory networks and signaling pathways involved in adipogenesis, researchers can identify potential targets for therapeutic intervention and enhance our understanding of adipose tissue biology [22,29,30]. For instance, the advent of CRISPR/Cas9 genome editing technology has revolutionized functional genomics, allowing for the precise manipulation of genes involved in adipogenesis. This technology enables the identification and validation of novel adipogenic regulators, providing deeper insights into the genetic and epigenetic control of adipocyte differentiation [31,32]. However, current approaches address computational/algorithmic analysis within gene expression changes, such robust ranking aggregation (RRA) [33].

This study employs RRA to analyze the transcriptomic data obtained from mouse ESCs undergoing adipogenesis. RRA is a powerful statistical method designed to identify differentially expressed genes with high confidence, even in the presence of noise and variability [33]. By ranking genes based on their expression changes across multiple conditions, RRA provides a robust framework for identifying key regulatory elements and pathways involved in adipocyte differentiation. This approach is particularly beneficial for understanding the dynamic and complex nature of gene expression during cellular differentiation [34]. RRA has been successfully applied in various fields of biomedical research, demonstrating its utility in integrating heterogeneous datasets and identifying robust biomarkers and therapeutic targets [35,36]. By leveraging the power of RRA, this study aims to provide a comprehensive analysis of the transcriptomic changes during ESC-to-adipocyte differentiation, revealing either critical regulatory genes and/or novel insights. This knowledge can have significant implications for both basic biology and clinical applications, offering potential avenues for therapeutic interventions in metabolic disorders and advancing our understanding of adipose tissue development and fat differentiation.

2. Results

2.1. Capacity of Mouse ESC (mESC::HB9::GFP) to Differentiate into Adipogenic-like Features

To comprehend the capability of ESC to differentiate into adipogenic-like features, the mESC::*hb9*::GFP cell line, a well-established motor neuron progenitor cell line [37], was used in this study. Based on previous studies [19,20], a monolayer experimental layout designed was performed, encompassing a period of 25 days with or without adipogenic differentiating factors followed by 5 days with serum-free medium (Figure 1A). On day 30, non-differentiating media did not show signs of lipid formation, whereas in the differentiation media condition there was more lipid formation present, as evidenced by O Red Oil staining (Figure 1B). Furthermore, quantification of lipid particles was observed to be statistically significant (Figure 1C). To make in-depth observations at the molecular level, specific gene markers were tested, including pluripotent stem cell marker *Oct4* [19], adipogenic transcription markers *Cebpb* and *Pparg*, and mature adipose marker *Fabp4* [4,5]. The pluripotent stem cell marker *Oct4* was significantly downregulated at day 30 under differentiating and non-differentiating conditions as opposed to day 0 (Figure 1D). Meanwhile, when comparing the adipogenic markers at day 30 (differentiation/non-differentiation), there was a slight increase in the transcriptional markers *Cebpb* and *Pparg* and a significant increase in the mature adipose tissue marker *Fabp4*. Collectively, these observations suggest that adipogenesis progressed throughout this experimental layout.

2.2. RNA-Seq Correlation Analysis Displays Global Differential Gene Expression between Differentiating and Non-Differentiating Conditions

To obtain insight into the global transcriptome profile changes between conditions at the beginning timepoint of day 0 and the end of the experiment at day 30, RNA-Seq was performed on RNA samples from day 0 (D0), day 30 non-differentiating condition (D30(−) or D30min), and day 30 differentiating condition (D30(+) or D30 pos).

The correlation analysis displayed variable changes between all three conditions (Figure 2A,B). As illustrated in the principal component analysis, the main factor of this was driven by time, followed by the conditions, with PC1 at 80% and PC2 at 19% (Figure 2A). This has also been reaffirmed accordingly in the distance heatmap (Figure 2B). Due to this, differentially expressed genes (DEGs) were investigated mainly according to the condition over time (i.e., D30(+)/D0 and D30(−)/D0).

To isolate relevant DEGs, a statistical cut-off of $P\text{-adj} < 0.05$ was set for both D30(−)/D0 and D30(+)/D0 (Figure 2C,D). Based on this cut-off, there were 15,180 and 15,016 DEGs identified in D30(−)/D0 and D30(+)/D0, respectively. These genes were further investigated via gene ontology enrichment analysis, specifically at the biological process's levels (Figure 2E,F; Supplementary File: Tables S1 and S2). In the differentiation condition of D30(+)/D0, there were several processes that are involved in molecular and transcription regulation as well as DNA damage response and the cell cycle (GO:0045944~positive regulation of transcription from RNA polymerase II promoter, GO:0007049~cell cycle, GO:0006281~DNA repair) (Figure 2E; Table S1). Likewise, these terms were also observed in the non-differentiation condition of D30(−)/D0 (Figure 2F; Table S2). Therefore, this suggests that the highest impact of change is focused on transcriptional mechanisms, which may display changes in gene expression between the conditions.

2.3. Isolation of Adipose Tissue, Lipid, and/or Fat-Related DEG Genes and Their Expression

To specify the previous DEG list into specific terms related to regulatory features of adipogenesis, terms were further filtered into "adipose tissue", "lipids", and/or "fat" (Figure 3). In (+)/D30 and D30(−)/D0 conditions, there were several terms related to the features of adipose tissue; however, only with the statistically significant enrichments were GoTerms further investigated. Commonly, in both conditions, these included lipid metabolic process, fat cell differentiation, and the fatty acid metabolic process (Figure 3A,B; Tables S3 and S4).

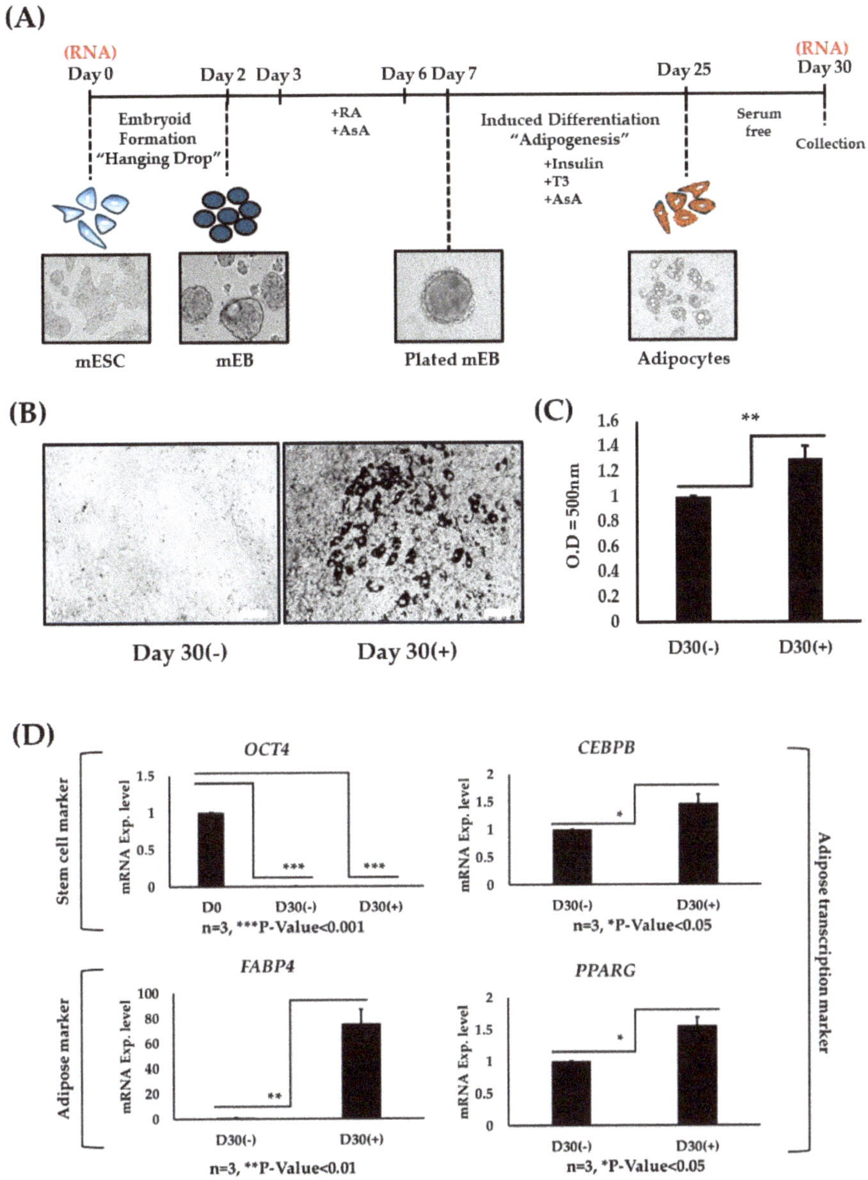

Figure 1. (**A**) Schematic experimental workflow of mouse ESC undergoing adipogenic differentiation with specified phase timepoints. Relevant brightfield images are presented at the indicated timepoints. RNA (in red) was collected at the indicated timepoint. (**B**) Oil Red O staining in non-differentiating (day 30(−)) and differentiating (day 30(+)) conditions; white bar scales to 100 um. (**C**) Spectrometry of lipid triglyceride particle isolates formed by staining at an optical density (O.D) of 500 nm; values are normalized to D30(−) (n = 3; ** p-Value < 0.01). (**D**) RT-PCR quantification of the relevant biomarkers, including *Oct4*, *Cebpb*, *Pparg*, and *Fabp4*. Apart from *Oct4*, which was normalized to the beginning timepoint (D30(−)/D0 and D30(+)/D0), the remaining genes were normalized based on the condition (D30(+)/D30(−)); all genes were run in triplicates, as indicated.

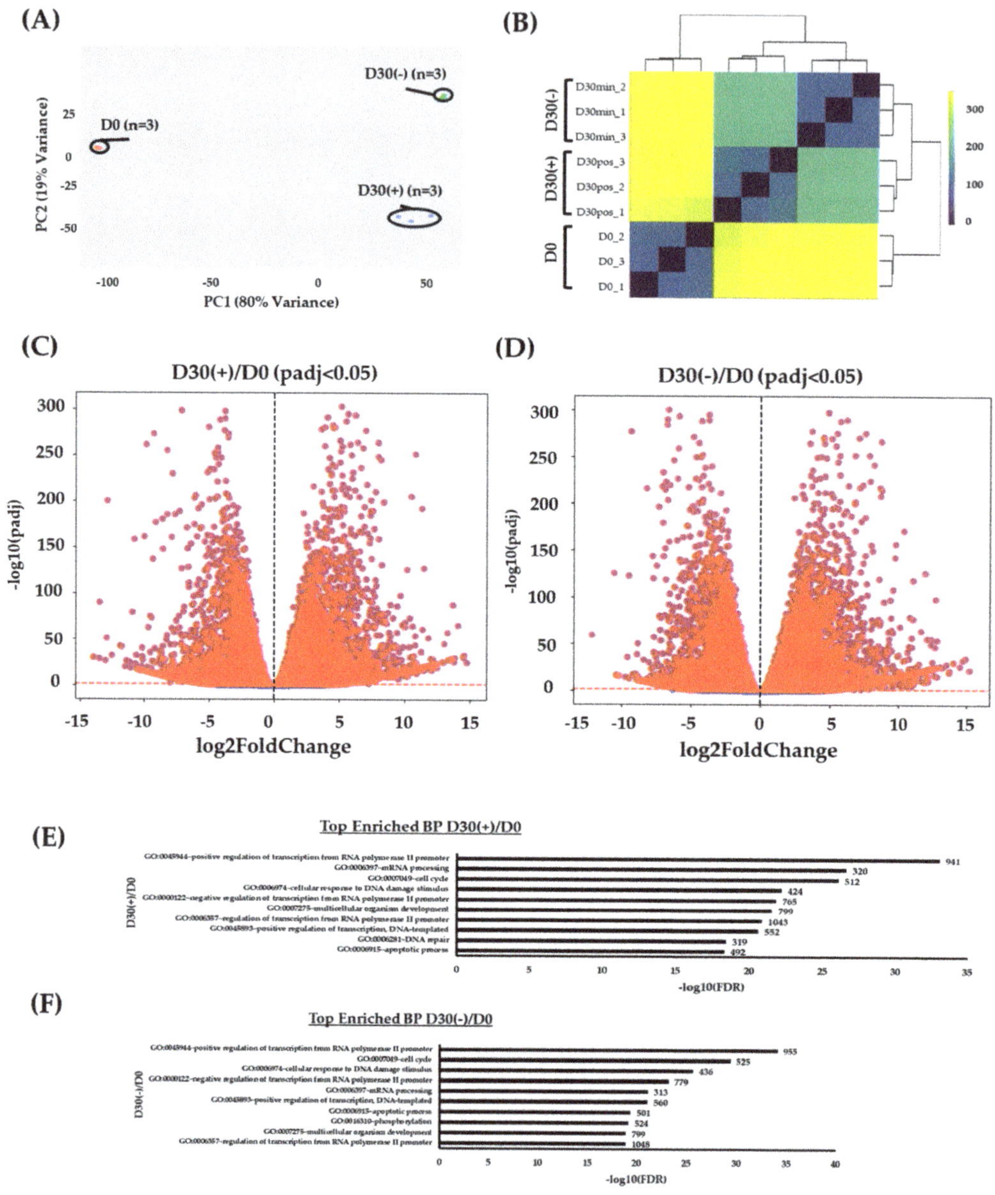

Figure 2. (**A**) Principal component analysis (PCA) of conditions D0, D30(−), and D30(+) with tight clustering of three biological replicates displaying the highest variance at 80% (PC1) followed by 19% (PC2). (**B**) Metric distance heatmap displaying the variance as observed in the previous PCA. (**C**) Volcano plot representation of DEGs in D30(+)/D0 and (**D**) D30(−)/D0 (y-axis: normalized statistical significance at −log10(P-adj); x-axis: normalized fold change of gene expression at log 2(Fold Change); the vertical line represents the cut-off of P-adj < 0.05, whereas the horizontal line represents the fold change at 0). (**E**) GoTerm enrichment analysis of top 10 biological processes in D30(+)/D0 and (**F**) D30(−)/D0. Enrichments are based on false dependent rate (FDR) values normalized with −log10(FDR).

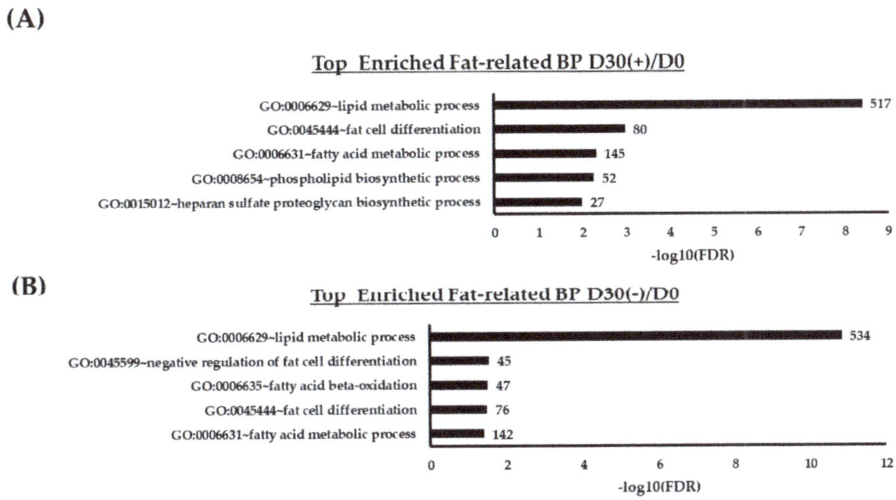

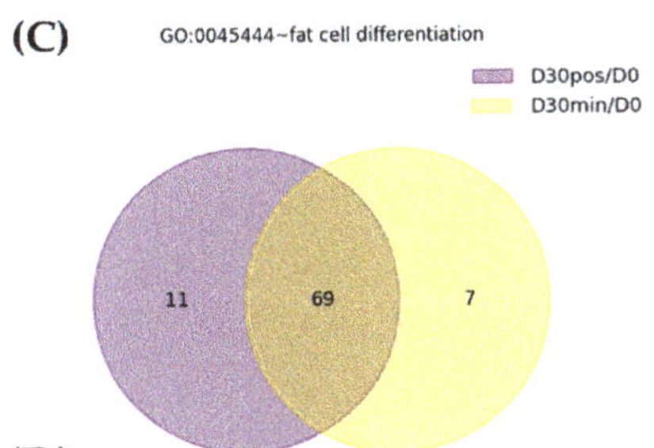

(D)

Category	Common	D30min/D0	D30pos/D0
Gene_name	Ccnd1, Psmb8, Trim32, Atf5, Egr2, Eif2ak3, Osbpl11, Creb5, Bbs12, 4932438A13Rik, Fosl2, Mettl8, Nr4a3, Zbtb7a, Lrrc8c, Adig, Cntn2, Enpp1, Tgfb1i1, Fbxo9, Grk5, Tbl1x, Zfpm2, Tcf7l2, Metrnl, Senp2, Aloxe3, Adgrf5, Steap4, Bbs7, Id4, Hmga2, Socs7, C1qtnf3, Arxes1, Ffar4, Cebpb, Pias1, Gm15290, Nr4a2, Wnt3a, Ttc8, Nipbl, Noc3l, Pparg, Gsk3b, Plcb1, Osbpl8, Inhbb, Fam120b, Cebpd, Cntnap2, Tmem120b, Kcnj8, Bbs1, Bscl2, Gdf10, Lpin1, Sdf4, Ep300, Lrp6, Cebpa, Wnt10b, Rnasel, Tmem120a, Tfap2b, Gdf6, Bbs9, Nr4a1	Cby1, Klf4, Jdp2, Smad6, Med1, Adgrf1, Ap1s2	Ankrd26, Srebf1, Arl6, Glis1, Dysf, Retn, Clip3, Arid5b, Runx1t1, Gm6484, Bbs4

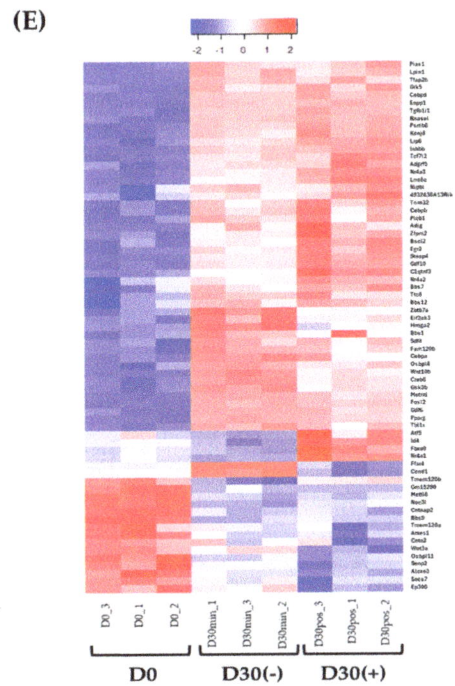

Figure 3. (**A**) GoTerm enrichment analysis of top 5 biological processes in (+)/D0 and (**B**) D30(−)/D0 filtered for "adipose", "lipid", and/or "fat" terms. Enrichments are based on the false dependent rate (FDR) values normalized with −log10(FDR). (**C**) Venn diagram comparing DEGs of the term GO:0045444~fat cell differentiation displaying common and unique genes in D30(−)/D0 and (+)/D0. (**D**) List of DEGs that are common between D30(−)/D0 and (+)/D0 or uniquely expressed in either D30(−)/D0 or (+)/D0. (**E**) A metric heatmap displaying expression patterns of each replicate in conditions D0, D30(−), and (+) of the common DEG genes identified in GO:0045444~fat cell differentiation. The heatmap is scaled to a log2 count per million [log2cpm] and a Z-Score (blue = downregulation, red = upregulation).

More specifically, the top 5 (among 36 enriched terms) enriched terms in (+)/D30 DEGs included GO:0006629~lipid metabolic process (517 genes), GO:0045444~fat cell differentiation (80 genes), GO:0006631~fatty acid metabolic process (145 genes), GO:0008654~phospholipid biosynthetic process (52 genes), and GO:0015012~heparan sulfate proteoglycan biosynthetic process (27 genes) (Figure 3A; 3S). However, the list included GO:0050873~brown fat cell differentiation (35 genes), GO:0019915~lipid storage (27 genes), GO:0045600~positive regulation of fat cell differentiation (42 genes), GO:0050872~white fat cell differentiation (21 genes), and GO:0060612~adipose tissue development (38 genes) (Table S3).

In D30(−)/D0, the top 5 (among 31 enriched terms) included GO:0006629~lipid metabolic process (534 genes), GO:0045599~negative regulation of fat cell differentiation (45 genes), GO:0006635~fatty acid beta-oxidation (47 genes), GO:0045444~fat cell differentiation (76 genes), and GO:0006631~fatty acid metabolic process (142 genes) (Figure 3B). However, the list included GO:0045165~cell fate commitment (60 genes), GO:0048665~neuron fate specification (22 genes), GO:0045600~positive regulation of fat cell differentiation (41 genes), GO:0060612~adipose tissue development (38 genes), and GO:0090336~positive regulation of brown fat cell differentiation (15 genes) (Table S4).

Even though there are variable enrichment terms between both conditions, DEGs in GO:0045444~fat cell differentiation were further investigated to delineate a common list (Figure 3C,D). There were 69 common genes that were differentially expressed between (+)/D0 and D30(−)/D0 (Figure 3C,D). Meanwhile, 11 and 7 genes were shown to be uniquely differentially expressed in (+)/D0 and D30(−)/D0, respectively (Figure 3C,D). The unique genes in (+)/D0 included *Ankrd26*, *Srebf1*, *Arl6*, *Glis1*, *Dysf*, *Retn*, *Clip3*, *Arid5b*, *Runx1t1*, *Gm6484* (*Angptl8*), and *Bns4* (Figure 3D), whereas in D30(−)/D0 the unique genes included *Cby1*, *Klf4*, *Jdp2*, *Smad6*, *Med1*, *Adgrf1*, and *Ap1s2* (Figure 3D).

The expressions of common DEGs shared between (+)/D0 and D30(−)/D0 were further investigated (Figure 3E). The heatmap revealed variable expression-pattern levels among conditions based on the scaled level. For instance, there were exclusive upregulation expressions of genes in the (+) condition, which included *Atf5*, *Id4*, *Fbxo9*, and *Nr4a1*. As opposed to this, there were exclusive upregulation expressions in D30(−), such as *Ffar4* and *Ccnd1*. In addition to these observations, there were several genes that were observed to be downregulated in (+), including *Ep300*, *Socs7*, *Aloxe3*, *Senp2*, *Osbpl11*, and *Wnt3a*. The remaining genes were observed to be differentially upregulated in both (+) and D30(−) as opposed to D0.

2.4. Network Analysis of Common DEGs and RRA

To further decipher the identified common DEGs of (+)/D0 and D30(−)/D0, a protein-to-protein interaction (PPI) was constructed (Figure 4A). The aim of this construct was to identify the centrality between the protein products of DEGs and therefore gain insight into their biological and/or molecular roles. Based on this, the construct identified 43 proteins that are interconnected with other proteins (Figure 4B). PPARG showed the highest centrality by having 18 node degrees, whereas CEBPA and CEBPB were ranked fourth (11 node degrees) and fifth (10 node degrees), respectively (Figure 4B). EP300 was ranked second with 14 node degrees, whereas GSK3B and CCND1 were ranked third with 12 node degrees. Interestingly, FFAR4 was ranked twelfth with two node degrees, whereas ATF5 was ranked eleventh (three node degrees) and NR4A1 was ranked seventh (seven node degrees), along with WNT10B. Lastly, ID4 was observed to be ranked 13 with a node degree of one.

To further identify and rank these DEGs based on changes in expression across conditions (+)/D0 and D30(−)/D0 data, RRA was applied after eliminating all proteins that had a node degree less than one (Figure 4C). The lower RRA score indicates that a gene consistently appears at top rank lists across the datasets, whereas a higher RRA score indicates that a gene less consistently appears to the top rank across the datasets (Figure 4C). The top-ranked genes identified were *Cntnap2* and *Wnt3a* with a shared score of 41. This is followed by *Bbs9* (score = 40), *Cntn2* (score = 39) and *Senp2* (score = 37). Interestingly, *Ep300* was also observed at a score of 37, followed by *Atf5* at a score of 36. For the remaining genes with variable expression patterns observed previously, *Nr4a1* scored 33, *Id4* scored 30, *Ccnd1* scored 26, and *Ffar4* scored 23. As for the known adipogenic markers, *Cebpb* was observed to have a score of 15, *Cebpa* scored 14, *Cebpd* scored 11, and *Pparg* scored 8.

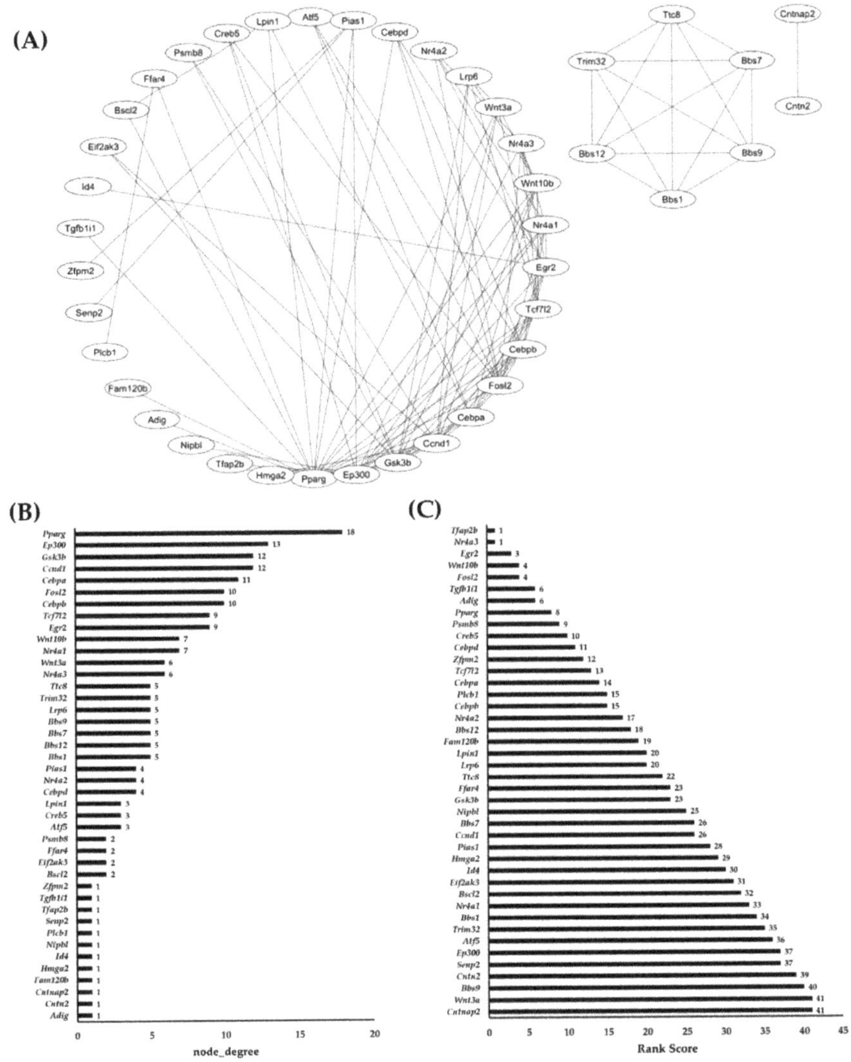

Figure 4. (**A**) A PPI interaction network (based on STRING and modified via Cytoscape) displaying interactions among the protein products of common DEGs identified previously. (**B**) Chart showing centrality by node degrees, where the higher the node the more central it is to the whole network. (**C**) Robust Rank Aggregation (RRA) score chart demonstrating the significant expression of DEGs in the (+)/D0 and D30(−)/D0 dataset. The lower the score, the higher the biological significance and effect.

2.5. Expression and RT-PCT Validation of Atf5, Ccnd1, and Nr4a1:

To validate some of the genes that showed interesting expression patterns, RT-PCR was utilized to validate *Atf5*, *Ccnd1*, and *Nr4a1* (Figure 5). Firstly, in silico data were visualized of each gene, which corresponded to the expression observed in the heatmap. It showed upregulation of genes *Atf5* (Figure 5A—left panel) and *Nr4a1* (Figure 5C—left panel) in (+) as opposed to D30(−). Conversely, *Ccnd1* was observed to be downregulated in (+) as opposed to D30(−) (Figure 5B—left panel). RT-PCR showed significant upregulation of *Atf5* (Figure 5A—right panel) and *Nr4a1* (Figure 5C—right panel) in (+) as opposed to D30(−). Meanwhile, RT-PCR showed significant downregulation of *Ccnd1* in (+) as opposed to D30(−) (Figure 5B—right panel). Collectively, RT-PCR results corresponded well with the datasets.

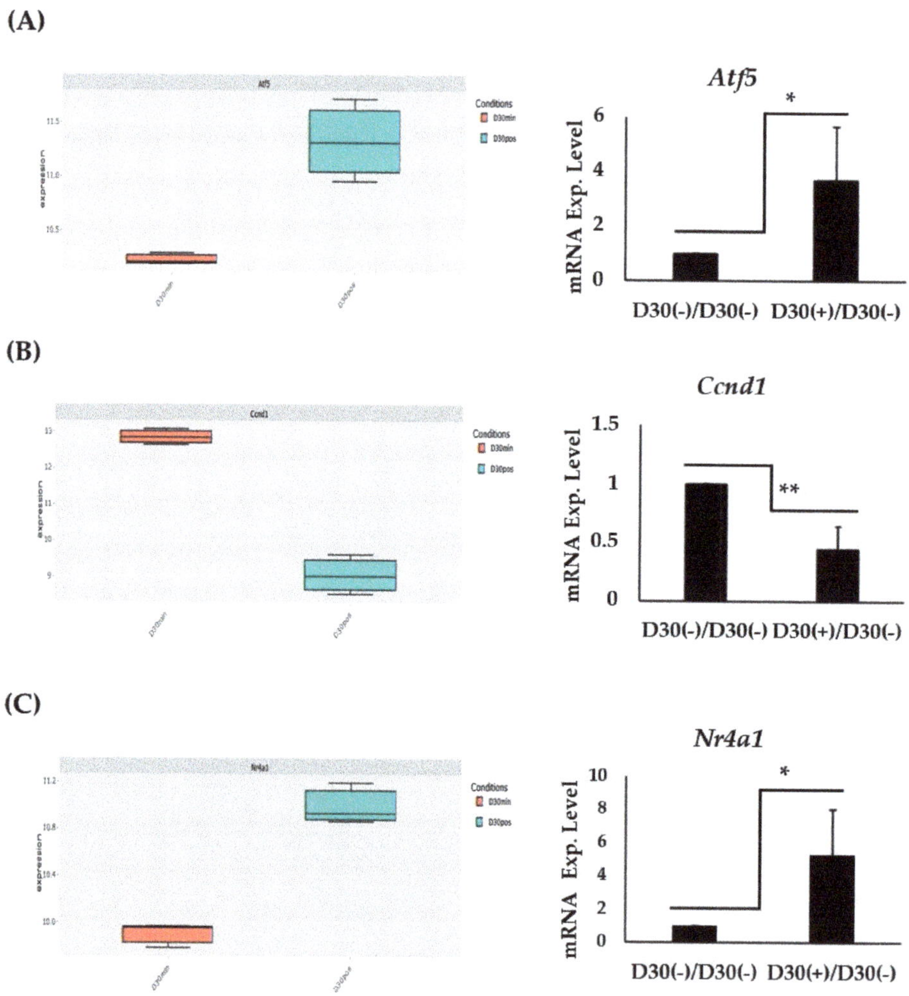

Figure 5. (**A**) Left panel: boxplot *Atf5* expression based on the RNA-Seq data. Right panel: RT-PCR of gene *Atf5* displaying mRNA expression levels (n = 3; * p-Value < 0.05). (**B**) Left panel: boxplot *Ccnd1* expression based on the RNA-Seq data. Right panel: RT-PCR of gene *Ccnd1* displaying mRNA expression levels (n = 3; ** p-Value < 0.01). (**C**) Left panel: boxplot *Nr4a1* expression based on the RNA-Seq data. Right panel: RT-PCR of gene *Nr4a1* displaying mRNA expression levels (n = 3; * p-Value < 0.05).

3. Discussion

ESCs offer a unique platform to investigate adipogenesis, providing insights into early developmental processes and regulatory mechanisms governing cell fate determination [22]. In this study, the adipogenic dynamical potential of mouse ESCs was explored using the computational approach of RRA. Applying this approach identified several key genes that would be characterized features of adipogenesis and/or mature adipose tissue.

The protocol approached in this study uses previous studies [6,18–20] that utilize ESC to understand adipogenesis. Although there are several approaches, the monolayer-directed differentiation was employed with the potential of collecting secreted material from formed adipocytes. This involved the exposure of attached EBs with soluble differentiating factors for a fixed period followed by incubation under low- to no-growth medium, such as serum-free medium. However, the main question at hand would be whether this reverses adipocyte differentiation or halts it, which remains unknown. Yet our findings demonstrate that under this approach, the mouse ESCs exhibited significant adipogenic-like features. This was characterized by lipid accumulation and upregulation of known adipogenic markers, including *Cebpb, Pparg,* and *Fabp4* in the differentiating condition as opposed to the non-differentiating condition. Indeed, the expression of this gene might be indicative of cell specification, especially as *Oct4,* one of the stem cell markers, was observed to decrease in both differentiating and non-differentiating conditions. Therefore, these observations underscore the robustness of ESCs as a model system for studying adipocyte differentiation from a pluripotent state. Comparative analysis with existing adipogenesis models, such as mesenchymal stem cells (MSCs) and preadipocyte cell lines, reveals distinct advantages of using ESCs. ESCs offer a more physiologically relevant model for early adipogenic events, enabling the study of initial lineage commitment and regulatory mechanisms not fully accessible in other systems.

Due to the dynamical transcriptional nature of adipogenesis at global genomic level, it was necessary to observe profile changes across the conditions. Bulk-RNA-Seq analysis in this study revealed substantial alteration in gene expression profiles over the period of 30 days. This is consistent with previous studies, which utilized next-generation sequencing approaches to understand adipogenesis [19,22,38–40]. However, most of these studies use mature sources of adipogenesis, whereas this study utilizes an earlier phase of cell commitment and specificity, thereby highlighting the benefit of studying adipocyte differentiation from ESCs. This study demonstrated the expression of over 15,000 genes that were involved in transcriptional regulation, cell cycle modulation, and DNA repair. Furthermore, the enrichment of GO terms related to lipid metabolism and fat cell differentiation further supports the specificity and relevance of our differentiation model.

Using, specifically, GO term GO:0045444~fat cell differentiation for RRA, we identified key regulatory genes central to adipogenesis in ESCs. Genes, such as PPARG, CEBPA, and EP300, emerged as central players in protein–protein interaction networks, underscoring their pivotal roles in adipocyte differentiation. Moreover, gene candidates, like *Atf5, Ccnd1,* and *Nr4a1,* highlighted by their consistent appearance in top-ranked gene lists across datasets, suggest potential new targets for further investigation into adipogenic regulation. These genes were further investigated for displaying distinctive expressions throughout the differentiation period. More specifically, *Atf5, Ccnd1,* and *Nra41* RT-PCR expressions were shown to correspond with the RNA-Seq data.

Studies have shown the roles of these genes and other early regulators (identified in the RNA-Seq data) in adipogenesis. These were observed to be uniquely differentially expressed in the non-differentiating conditions (D30(−)). For instance, *Klf4, Jdp2,* and *Smad6* are upregulated early in the differentiation process. *Klf4,* a member of the Kruppel-like factor family, is well-known for its role in maintaining pluripotency and has been shown to repress adipogenesis in preadipocytes by inhibiting the expression of adipogenic markers, such as PPARγ [41,42]. However, its role during the early stages of ESC differentiation into adipocytes suggests a dual function, potentially acting as a gatekeeper that balances self-renewal and lineage commitment. Similarly, *Jdp2,* a member of the AP-1 family, is known to interact with C/EBPβ, a critical transcription factor in adipogenesis [43,44]. Jdp2 facilitates adipocyte differentiation by modulating the activity of C/EBPβ, highlighting its importance in the transcriptional regulation of adipogenic genes. Interestingly, *Jdp2* has also been shown to directly regulate activating transcription factor (ATF) proteins, specifically ATF4 [45]. Furthermore, Smad6, an inhibitory Smad, plays a role in the negative regulation of BMP signaling, which is crucial for adipogenesis [46]. The upregulation of Smad6 early in differentiation might indicate a fine-tuning mechanism that controls the extent of BMP signaling, thereby influencing the commitment to the adipogenic lineage.

Further to these genes, there were also genes that were associated as mediators of metabolic and cell cycle regulation. Among these, *Ccnd1* was shown to be downregulated in both the data and, as mentioned previously, in RT-PCR. *Ccnd1*, typically associated with cell cycle regulation, has been shown to inhibit adipogenesis by blocking the activity of PPARγ and C/EBPα, two master regulators of adipocyte differentiation [47]. *Ccnd1* downregulation has been observed to be necessary for the progression of adipogenesis [47,48]. Therefore, the observed downregulation of *Ccnd1* in the current study in the later stages of differentiation might be necessary for the full activation of these adipogenic pathways, underscoring its role in the temporal regulation of cell fate decisions. Another mediator identified included *Ffar4*, which is a receptor involved in fatty acid sensing that has been implicated in promoting adipogenesis through the enhancement of insulin sensitivity and anti-inflammatory effects [49,50].

Further to the validation of *Ccnd1* in this study, *Atf5* and *Nr4a1* have also been shown to be involved in adipocyte differentiation. *Atf5*, a member of the ATF/CREB family, has been implicated in the regulation of adipogenesis through its interaction with C/EBPβ [51]. This study found ATF5 expression to be associated with the obesity-related phenotype, suggesting its role in promoting adipocyte differentiation. On the other hand, *Nr4a1*, which is an orphan nuclear receptor, is involved in the regulation of metabolic processes and has been shown to influence adipocyte maturation and lipid metabolism [52,53]. The consistent upregulation of *Nr4a1* during differentiation, in the current study, indicates its potential role in fine-tuning the adipogenic program, possibly through interactions with other nuclear receptors or transcription factors. Another factor, which was observed in this study to be highly ranked, was the dynamic expression of EP300. EP300, a histone acetyltransferase, is known to coactivate PPARγ and C/EBPα, thereby enhancing the transcription of adipogenic genes [54,55]. Its central role in the protein–protein interaction network identified in our study further supports its importance in the regulation of adipogenesis. Interestingly, this study also identified *Angptl8* (Gm6484) to be uniquely differentially expressed in differentiating conditions ((+)/D0). *Angptl8*, also known as betatrophin, is a liver-derived protein, which has been implicated in lipid metabolism and adipocyte differentiation [56]. Studies have shown that Angptl8 plays a role in promoting lipid accumulation in adipocytes and modulating systemic lipid metabolism, making it a potential target for understanding fat cell differentiation and metabolic regulation [57–59].

Although this study leverages the robustness of RRA to provide a comprehensive dynamical view of the transcriptome, especially during adipogenesis in ESCs, it is still limited. This is mainly due to having key adipogenic factors, such as (but not limited to) *Cebpa*, *Cebpb*, *Cebpd*, and *Pparg*, already being expressed at a plateau level. This leads to the inability to observe distinctive changes in expression between the differentiating and non-differentiating conditions and might suggest that this distinctive expression occurs at an earlier timepoint.

The use of RRA in this study was particularly significant from a biological perspective. RRA is a powerful computational tool that excels in integrating data from multiple sources or experimental conditions, even in the presence of noise and variability [33]. This approach allows for the identification of key regulatory genes that may be subtle or obscured in traditional differential expression analysis [33]. By ranking genes based on their expression changes across various conditions, RRA helps to prioritize biologically relevant candidates that are consistently associated with the adipogenic process [60], providing a more reliable identification of genes that are central to cellular differentiation. This current work is the first to utilize this approach using mouse ESCs as a model for adipogenesis.

The significance of RRA lies in its ability to consolidate complex transcriptomic data into meaningful biological insights [61]. In the context of adipogenesis, this approach enabled the identification of genes that are not only differentially expressed but also functionally relevant to the adipogenic phenotype. For instance, RRA highlighted the roles of EP300, Atf5, and Nr4a1, which may not have been as prominent in a traditional analysis due to the complex and dynamic nature of gene expression during differentiation. However, the limitations observed in our study underscore the importance of considering earlier time points in future experimental designs. Investigating earlier stages of differentiation could reveal significant changes in gene expression that are critical for initiating adipogenesis. Additionally, linking these early transcriptional changes to secreted protein functionalities would provide further insights into the molecular mechanisms driving adipocyte differentiation.

Despite these limitations, the findings from this study contribute significantly to basic biology and offer potential applications in clinical settings, particularly in the context of metabolic diseases. The insights gained by RRA pave the way for future research aimed at enhancing therapeutic strategies and deepening our understanding of adipose tissue biology. By identifying key regulatory networks involved in adipogenesis, this study offers a foundation for developing targeted interventions that could modulate adipocyte function and potentially treat conditions like obesity and insulin resistance.

4. Materials and Methods

4.1. mESC Culture and Adipocyte Differentiation

Mouse ESC (mESC::*hb9*::GFP) of the strain HBG3 [37] was used throughout this study. Cells were grown and maintained under ES growth media (Knockout™ DMEM (Thermo Fisher Scientific, Carlsbad, CA, USA), 14% Fetal Bovine Serum (Corning®, Woodland, CA, USA), 1× 2-Mercaptoethanol (Thermo Fisher Scientific, Carlsbad, CA, USA), 3× Mix Nucleosides (Thermo Fisher Scientific, Carlsbad, CA, USA)], and 10^7 U/mL Leukemia inhibitory factor [LIF] (Thermo Fisher Scientific, Carlsbad, CA, USA)) under humidified conditions of 5%CO_2/37 °C. Differentiation media consisted of the same components as ES growth media but without the addition of LIF. Differentiation was carried out first through the formation of embryoid bodies (EBs) through the "hanging drop" method with a cellular concentration of 1×10^5 cells/mL for 2 days [18–20]. EBs were collected and plated into 0.1% gelatin-coated wells followed by treatment with 1 μM of Retinoic Acid and 12.5 μg/mL of Ascorbic Acid in differentiation media the following day for 3 days. From day 7, attached EBs were treated with differentiation media supplemented with 12.5 μg/mL of Ascorbic Acid, 3 nM of Thyroid T3, and 0.5 μg/mL of insulin. These media were changed every 2 to 3 days up until day 25. On day 25, differentiation media were removed, and serum-free media were added for a period of 5 days (day 30). Cells were collected for further analysis on day 30.

4.2. Total RNA Extraction and cDNA Synthesis

Total RNA was extracted from all mouse ESC conditions using two methods when needed. For RNA-Seq libraries' preparation, a combination of TriZol (Life Technologies™, Thermo Fisher Scientific, Carlsbad, CA, USA) [62] and the RNeasy Mini Kit (QIAGEN, Hilden, Germany, Cat.no. 74104) was used, with modifications from the manufacturer's protocols. Cells in each well were washed once with 1 mL of PBS prior to the addition of 1 mL of TriZol reagent. TriZol lysates were added to fresh 1.5 mL tubes. Approximately 0.2 mL of chloroform was added per ml of TriZol and centrifuged at a speed of 12,000 rpm at 4 °C for 15 min. The upper aqueous phase (~400 mL in volume) was transferred to a fresh 1.5 mL tube. An equivalent volume of 70% ethanol was mixed with the RNA solution, and downstream purification was performed using an RNeasy Mini Spin Column (QIAGEN, Hilden, Germany, Cat.no. 74104) according to the manufacturer's protocol [63].

The second method was followed accordingly and solely with TriZol manufacturer instructions [62]. RNA concentrations were determined using a Thermo Scientific™ NanoDrop 2000™ (Thermo Fisher Scientific, Carlsbad, CA, USA), whereas the integrity of RNA (RIN) was evaluated on a Tape station 4200 (Agilent Technologies, Inc., Santa Clara, CA, USA) RNA Screen Tape. RIN numbers of more than 8 were further used for library preparation.

4.3. cDNA Synthesis and RT-PCR

For complementary DNA (cDNA) synthesis, a concentration amount of (0.1 ng) total RNA, from either method, was used as a template using a First-Strand Synthesis Kit (Invitrogen™, Carlsbad, CA, USA) according to the manufacturer's instructions. cDNA synthesized was further diluted 1:50 in ddH2O prior to its use as a template for real-time PCR (RT-PCR) using SYBR Green (Thermo Scientific™). A total volume of 10 μL was prepared for the SYBR Green qPCR reaction [2.5 μL diluted cDNA product or nH2O, 2.5 μL 2 mM forward + reverse primer mix (Integrated DNA Technologies, Inc., Coralville, IA, USA) and 5 uL of SYBR Green]. The qPCR thermal cycle utilized an Applied Biosystems™ StepOnePlus™ platform (Applied Biosystems™, Waltham, MA, USA), with one cycle of 50 °C for 2 min and 95 °C for 10 min followed by 45 cycles of 95 °C for 15 s and 60 °C for 1 min. The expression of gene targets' values was normalized to housekeeping gene β-actin and estimated using the $\Delta\Delta$Ct approach. Primers included the genes *Pparg*, *Cebpb*, *Fabp4*, *Oct4*, *Ccnd1*, *Atf5*, and *Nr4a1* (Supplementary File: Table S5). The housekeeping gene *Bact* was used as reference control.

4.4. Oil Red O Staining and Quantification

Mouse ESCs, under differentiating and non-differentiating conditions, were washed with 1 mL of 1× PBS (Corning®, Woodland, CA, USA) once before fixing for 5 min with 4% paraformaldehyde in PBS at room temperature. Thereafter, cells were washed three times with 1 mL of PBS and once with 60% isopropanol and completely air-dried. ESCs were then stained with Oil Red O (ORO) [6 ORO:4 ddH2O ratio] (Sigma©, Saint Louis, MO, USA) for 10 min at room temperature. Excess stain residue was removed with four ddH2O washes. Approximately 1 mL of PBS was then added for microscopic visualization. Images were processed at 460× magnification using a 20× objective lens, with transmitted bright-field light using EVOS® FLoid® Cell Imaging Station (Thermo Fisher Scientific, Carlsbad, CA, USA). For quantification, ORO stain particles were eluted with 100% isopropanol and analysed using Thermo Scientific™ Varioskan® Flash (Thermo Fisher Scientific, Carlsbad, CA, USA) for spectrophotometry readings at 500 nm.

4.5. RNA-Sequencing Library Preparation and Sequencing

Libraries were prepared using the NEBNext Ultra II RNA Library Prep Kit (New England Biolabs®, Ipswich, MA, USA) from New England Biolabs following the manufacturer's recommendations. Total RNA samples (500 ng) were subjected to cDNA construction for Illumina sequencing, in accordance with the protocol for the mRNA-Seq sample preparation kit. Oligo(dT) (Illumina, Inc., San Diego, CA, USA) magnetic beads were used to isolate poly(A) RNA from the total RNA samples. The mRNA was fragmented, and the first-strand cDNA was reverse-transcribed using random primers, following second-strand cDNA synthesis. The resulting double-stranded cDNA was ligated to NEBNext® adaptors after being end-repaired and A-tailed. Finally, PCR amplification of 12 cycles was performed for enrichment, producing a 350–400 bp fragment including adapters. The fragment size and purity of the libraries were assessed on a Tape station 4200 (Agilent Technologies, Inc., Santa Clara, CA, USA) using high-sensitivity D5000 Screen Tape and quantified with the Qubit 4.0 Fluorometer (Thermo Fisher, Waltham, MA, USA). The quantifications of the libraries required for RNA-seq were determined through real-time qPCR using a KAPA library quantification kit for the Illumina platform (KAPABIOSYSTEMS, Cape Town, South Africa). Enriched cDNA libraries were sequenced using the Illumina NextSeq 500/550 system with High Output v2 kit (150 cycles, PE), (Illumina, Inc., San Diego, CA, USA).

4.6. Computational Analysis

Correlation analyses, including principal component analysis (PCA) and distance heatmap dendrogram analysis, were generated using RNA-Seq NASCAR 2.0 via the New York University Abu Dhabi Centre of Genomic and Systems Biology (NYUAD-CGSB) Bioinformatics Online Analysis and Visualization Portal (http://nasqar2.abudhabi.nyu.edu/) [64].

DEG lists based on comparative analysis (i.e., statistical significance [p-value < 0.05]) were identified via Python scripting and were used for further gene ontology (GO) enrichment analysis using the g:Profiler database (https://biit.cs.ut.ee/gprofiler/gost, accessed on 1 March 2024) [65].

4.7. Statistical Analysis

Statistical analysis was conducting using Student's t-tests in Microsoft Excel™ (Microsoft®, Version 2406). Results are represented as means of at least three independent experiments, and a p-value < 0.05 was considered significant.

Supplementary Materials: The following supporting information can be downloaded at: https://www.mdpi.com/article/10.3390/ijms25179154/s1.

Author Contributions: Conceptualization, M.A.-S.; methodology, M.A., M.K., M.S. and M.A.-S.; validation, M.A. and M.K.; formal analysis, M.A.-S.; investigation, M.A.-S.; resources, M.A.-S.; data curation, M.A.-S.; writing—original draft preparation, M.A.-S.; writing—review and editing, M.A.-S.; visualization, M.A.-S.; supervision, M.A.-S.; project administration, M.A.-S.; funding acquisition, M.A.-S. All authors have read and agreed to the published version of the manuscript.

Funding: This research was funded by a New York University Abu Dhabi Cooporation internal grant (76-71202-ADHPG-AD231-63210).

Institutional Review Board Statement: Not applicable.

Informed Consent Statement: Not applicable.

Data Availability Statement: RNA-Seq data associated with this study were securely saved under BioProject accession number reference PRJNA1126392 in the National Library of Medicine at the National Center of Biotechnology information (NCBI).

Acknowledgments: We would like to thank the funding body at the NYUAD cooperation for their financial support throughout the development of this project.

Conflicts of Interest: The authors declare no conflicts of interest.

References

1. Varzideh, F.; Gambardella, J.; Kansakar, U.; Jankauskas, S.S.; Santulli, G. Molecular Mechanisms Underlying Pluripotency and Self-Renewal of Embryonic Stem Cells. *Int. J. Mol. Sci.* **2023**, *24*, 8386. [CrossRef] [PubMed]
2. Cave, E.; Crowther, N.J. The Use of 3T3-L1 Murine Preadipocytes as a Model of Adipogenesis. *Methods Mol. Biol.* **2019**, *1916*, 263–272. [CrossRef] [PubMed]
3. Robert, A.W.; Marcon, B.H.; Dallagiovanna, B.; Shigunov, P. Adipogenesis, Osteogenesis, and Chondrogenesis of Human Mesenchymal Stem/Stromal Cells: A Comparative Transcriptome Approach. *Front. Cell Dev. Biol.* **2020**, *8*, 561. [CrossRef]
4. Al-Sayegh, M.; Ali, H.; Jamal, M.H.; ElGindi, M.; Chanyong, T.; Al-Awadi, K.; Abu-Farha, M. Mouse Embryonic Fibroblast Adipogenic Potential: A Comprehensive Transcriptome Analysis. *Adipocyte* **2021**, *10*, 1–20. [CrossRef]
5. Al-Sayegh, M.A.; Mahmood, S.R.; Khair, S.B.A.; Xie, X.; El Gindi, M.; Kim, T.; Almansoori, A.; Percipalle, P. beta-actin contributes to open chromatin for activation of the adipogenic pioneer factor CEBPA during transcriptional reprograming. *Mol. Biol. Cell* **2020**, *31*, 2511–2521. [CrossRef]
6. Dani, C. Embryonic stem cell-derived adipogenesis. *Cells Tissues Organs* **1999**, *165*, 173–180. [CrossRef] [PubMed]
7. Evans, M.J.; Kaufman, M.H. Establishment in culture of pluripotential cells from mouse embryos. *Nature* **1981**, *292*, 154–156. [CrossRef]
8. Martin, G.R. Isolation of a pluripotent cell line from early mouse embryos cultured in medium conditioned by teratocarcinoma stem cells. *Proc. Natl. Acad. Sci. USA* **1981**, *78*, 7634–7638. [CrossRef]
9. Nichols, J.; Smith, A. Naive and primed pluripotent states. *Cell Stem Cell* **2009**, *4*, 487–492. [CrossRef]
10. Loh, Y.H.; Wu, Q.; Chew, J.L.; Vega, V.B.; Zhang, W.; Chen, X.; Bourque, G.; George, J.; Leong, B.; Liu, J.; et al. The Oct4 and Nanog transcription network regulates pluripotency in mouse embryonic stem cells. *Nat. Genet.* **2006**, *38*, 431–440. [CrossRef]
11. Young, R.A. Control of the embryonic stem cell state. *Cell* **2011**, *144*, 940–954. [CrossRef]

12. Smith, A. Formative pluripotency: The executive phase in a developmental continuum. *Development* **2017**, *144*, 365–373. [CrossRef] [PubMed]
13. Tang, F.; Barbacioru, C.; Wang, Y.; Nordman, E.; Lee, C.; Xu, N.; Wang, X.; Bodeau, J.; Tuch, B.B.; Siddiqui, A.; et al. mRNA-Seq whole-transcriptome analysis of a single cell. *Nat. Methods* **2009**, *6*, 377–382. [CrossRef] [PubMed]
14. Germain, N.; Banda, E.; Grabel, L. Embryonic stem cell neurogenesis and neural specification. *J. Cell. Biochem.* **2010**, *111*, 535–542. [CrossRef] [PubMed]
15. Huber, T.L. Dissecting hematopoietic differentiation using the embryonic stem cell differentiation model. *Int. J. Dev. Biol.* **2010**, *54*, 991–1002. [CrossRef]
16. Baharvand, H.; Hashemi, S.M.; Kazemi Ashtiani, S.; Farrokhi, A. Differentiation of human embryonic stem cells into hepatocytes in 2D and 3D culture systems in vitro. *Int. J. Dev. Biol.* **2006**, *50*, 645–652. [CrossRef]
17. Ehnes, D.D.; Price, F.D.; Shrive, N.G.; Hart, D.A.; Rancourt, D.E.; zur Nieden, N.I. Embryonic stem cell-derived osteocytes are capable of responding to mechanical oscillatory hydrostatic pressure. *J. Biomech.* **2015**, *48*, 1915–1921. [CrossRef]
18. Dani, C.; Smith, A.G.; Dessolin, S.; Leroy, P.; Staccini, L.; Villageois, P.; Darimont, C.; Ailhaud, G. Differentiation of embryonic stem cells into adipocytes in vitro. *J. Cell Sci.* **1997**, *110 Pt 11*, 1279–1285. [CrossRef]
19. Cuaranta-Monroy, I.; Simandi, Z.; Kolostyak, Z.; Doan-Xuan, Q.M.; Poliska, S.; Horvath, A.; Nagy, G.; Bacso, Z.; Nagy, L. Highly efficient differentiation of embryonic stem cells into adipocytes by ascorbic acid. *Stem Cell Res.* **2014**, *13*, 88–97. [CrossRef]
20. Cuaranta-Monroy, I.; Simandi, Z.; Nagy, L. Differentiation of Adipocytes in Monolayer from Mouse Embryonic Stem Cells. *Methods Mol. Biol.* **2016**, *1341*, 407–415. [CrossRef]
21. Rosen, E.D.; Spiegelman, B.M. Molecular regulation of adipogenesis. *Annu. Rev. Cell Dev. Biol.* **2000**, *16*, 145–171. [CrossRef]
22. Bahmad, H.F.; Daouk, R.; Azar, J.; Sapudom, J.; Teo, J.C.M.; Abou-Kheir, W.; Al-Sayegh, M. Modeling Adipogenesis: Current and Future Perspective. *Cells* **2020**, *9*, 2326. [CrossRef]
23. Christodoulides, C.; Lagathu, C.; Sethi, J.K.; Vidal-Puig, A. Adipogenesis and WNT signalling. *Trends Endocrinol. Metab.* **2009**, *20*, 16–24. [CrossRef]
24. MacDougald, O.A.; Mandrup, S. Adipogenesis: Forces that tip the scales. *Trends Endocrinol. Metab.* **2002**, *13*, 5–11. [CrossRef]
25. Sun, L.; Xie, H.; Mori, M.A.; Alexander, R.; Yuan, B.; Hattangadi, S.M.; Liu, Q.; Kahn, C.R.; Lodish, H.F. Mir193b-365 is essential for brown fat differentiation. *Nat. Cell Biol.* **2011**, *13*, 958–965. [CrossRef] [PubMed]
26. Alvarez-Dominguez, J.R.; Bai, Z.; Xu, D.; Yuan, B.; Lo, K.A.; Yoon, M.J.; Lim, Y.C.; Knoll, M.; Slavov, N.; Chen, S.; et al. De Novo Reconstruction of Adipose Tissue Transcriptomes Reveals Long Non-coding RNA Regulators of Brown Adipocyte Development. *Cell Metab.* **2015**, *21*, 764–776. [CrossRef]
27. Rosen, E.D.; MacDougald, O.A. Adipocyte differentiation from the inside out. *Nat. Rev. Mol. Cell Biol.* **2006**, *7*, 885–896. [CrossRef]
28. Kahn, S.E.; Hull, R.L.; Utzschneider, K.M. Mechanisms linking obesity to insulin resistance and type 2 diabetes. *Nature* **2006**, *444*, 840–846. [CrossRef]
29. Farmer, S.R. Transcriptional control of adipocyte formation. *Cell Metab.* **2006**, *4*, 263–273. [CrossRef] [PubMed]
30. Gregoire, F.M.; Smas, C.M.; Sul, H.S. Understanding adipocyte differentiation. *Physiol. Rev.* **1998**, *78*, 783–809. [CrossRef]
31. Doudna, J.A.; Charpentier, E. Genome editing. The new frontier of genome engineering with CRISPR-Cas9. *Science* **2014**, *346*, 1258096. [CrossRef] [PubMed]
32. Shalem, O.; Sanjana, N.E.; Zhang, F. High-throughput functional genomics using CRISPR-Cas9. *Nat. Rev. Genet.* **2015**, *16*, 299–311. [CrossRef] [PubMed]
33. Kolde, R.; Laur, S.; Adler, P.; Vilo, J. Robust rank aggregation for gene list integration and meta-analysis. *Bioinformatics* **2012**, *28*, 573–580. [CrossRef] [PubMed]
34. Su, L.J.; Chang, C.W.; Wu, Y.C.; Chen, K.C.; Lin, C.J.; Liang, S.C.; Lin, C.H.; Whang-Peng, J.; Hsu, S.L.; Chen, C.H.; et al. Selection of DDX5 as a novel internal control for Q-RT-PCR from microarray data using a block bootstrap re-sampling scheme. *BMC Genomics* **2007**, *8*, 140. [CrossRef]
35. Pavlidis, P.; Lewis, D.P.; Noble, W.S. Exploring gene expression data with class scores. In Proceedings of the Biocomputing 2002, Kauai, HI, USA, 3–7 January 2002; pp. 474–485.
36. Wang, J.; Duncan, D.; Shi, Z.; Zhang, B. WEB-based GEne SeT AnaLysis Toolkit (WebGestalt): Update 2013. *Nucleic Acids Res.* **2013**, *41*, W77–W83. [CrossRef] [PubMed]
37. Wichterle, H.; Lieberam, I.; Porter, J.A.; Jessell, T.M. Directed differentiation of embryonic stem cells into motor neurons. *Cell* **2002**, *110*, 385–397. [CrossRef] [PubMed]
38. Deutsch, A.; Feng, D.; Pessin, J.E.; Shinoda, K. The Impact of Single-Cell Genomics on Adipose Tissue Research. *Int. J. Mol. Sci.* **2020**, *21*, 4773. [CrossRef] [PubMed]
39. Yi, X.; Wu, P.; Liu, J.; Gong, Y.; Xu, X.; Li, W. Identification of the potential key genes for adipogenesis from human mesenchymal stem cells by RNA-Seq. *J. Cell. Physiol.* **2019**, *234*, 20217–20227. [CrossRef]
40. Jaager, K.; Islam, S.; Zajac, P.; Linnarsson, S.; Neuman, T. RNA-seq analysis reveals different dynamics of differentiation of human dermis- and adipose-derived stromal stem cells. *PLoS ONE* **2012**, *7*, e38833. [CrossRef]
41. Birsoy, K.; Chen, Z.; Friedman, J. Transcriptional regulation of adipogenesis by KLF4. *Cell Metab.* **2008**, *7*, 339–347. [CrossRef]
42. Rivero, S.; Diaz-Guerra, M.J.; Monsalve, E.M.; Laborda, J.; Garcia-Ramirez, J.J. DLK2 is a transcriptional target of KLF4 in the early stages of adipogenesis. *J. Mol. Biol.* **2012**, *417*, 36–50. [CrossRef] [PubMed]

43. Huang, Y.C.; Hasegawa, H.; Wang, S.W.; Ku, C.C.; Lin, Y.C.; Chiou, S.S.; Hou, M.F.; Wu, D.C.; Tsai, E.M.; Saito, S.; et al. Jun dimerization protein 2 controls senescence and differentiation via regulating histone modification. *J. Biomed. Biotechnol.* **2011**, *2011*, 569034. [CrossRef]
44. Nakade, K.; Pan, J.; Yoshiki, A.; Ugai, H.; Kimura, M.; Liu, B.; Li, H.; Obata, Y.; Iwama, M.; Itohara, S.; et al. JDP2 suppresses adipocyte differentiation by regulating histone acetylation. *Cell Death Differ.* **2007**, *14*, 1398–1405. [CrossRef]
45. Engler, M.J.; Mimura, J.; Yamazaki, S.; Itoh, K. JDP2 is directly regulated by ATF4 and modulates TRAIL sensitivity by suppressing the ATF4-DR5 axis. *FEBS Open Bio* **2020**, *10*, 2771–2779. [CrossRef]
46. Huang, H.; Song, T.J.; Li, X.; Hu, L.; He, Q.; Liu, M.; Lane, M.D.; Tang, Q.Q. BMP signaling pathway is required for commitment of C3H10T1/2 pluripotent stem cells to the adipocyte lineage. *Proc. Natl. Acad. Sci. USA* **2009**, *106*, 12670–12675. [CrossRef]
47. Fu, M.; Rao, M.; Bouras, T.; Wang, C.; Wu, K.; Zhang, X.; Li, Z.; Yao, T.P.; Pestell, R.G. Cyclin D1 inhibits peroxisome proliferator-activated receptor gamma-mediated adipogenesis through histone deacetylase recruitment. *J. Biol. Chem.* **2005**, *280*, 16934–16941. [CrossRef]
48. Marquez, M.P.; Alencastro, F.; Madrigal, A.; Jimenez, J.L.; Blanco, G.; Gureghian, A.; Keagy, L.; Lee, C.; Liu, R.; Tan, L.; et al. The Role of Cellular Proliferation in Adipogenic Differentiation of Human Adipose Tissue-Derived Mesenchymal Stem Cells. *Stem Cells Dev.* **2017**, *26*, 1578–1595. [CrossRef]
49. Oh, D.Y.; Talukdar, S.; Bae, E.J.; Imamura, T.; Morinaga, H.; Fan, W.; Li, P.; Lu, W.J.; Watkins, S.M.; Olefsky, J.M. GPR120 is an omega-3 fatty acid receptor mediating potent anti-inflammatory and insulin-sensitizing effects. *Cell* **2010**, *142*, 687–698. [CrossRef]
50. Oh, D.Y.; Walenta, E.; Akiyama, T.E.; Lagakos, W.S.; Lackey, D.; Pessentheiner, A.R.; Sasik, R.; Hah, N.; Chi, T.J.; Cox, J.M.; et al. A Gpr120-selective agonist improves insulin resistance and chronic inflammation in obese mice. *Nat. Med.* **2014**, *20*, 942–947. [CrossRef]
51. Zhao, Y.; Zhang, Y.D.; Zhang, Y.Y.; Qian, S.W.; Zhang, Z.C.; Li, S.F.; Guo, L.; Liu, Y.; Wen, B.; Lei, Q.Y.; et al. p300-dependent acetylation of activating transcription factor 5 enhances C/EBPbeta transactivation of C/EBPalpha during 3T3-L1 differentiation. *Mol. Cell Biol.* **2014**, *34*, 315–324. [CrossRef]
52. Pearen, M.A.; Muscat, G.E. Minireview: Nuclear hormone receptor 4A signaling: Implications for metabolic disease. *Mol. Endocrinol.* **2010**, *24*, 1891–1903. [CrossRef] [PubMed]
53. Qin, D.D.; Yang, Y.F.; Pu, Z.Q.; Liu, D.; Yu, C.; Gao, P.; Chen, J.C.; Zong, C.; Zhang, Y.C.; Li, X.; et al. NR4A1 retards adipocyte differentiation or maturation via enhancing GATA2 and p53 expression. *J. Cell. Mol. Med.* **2018**, *22*, 4709–4720. [CrossRef] [PubMed]
54. Mo, X.; Kowenz-Leutz, E.; Xu, H.; Leutz, A. Ras induces mediator complex exchange on C/EBP beta. *Mol. Cell* **2004**, *13*, 241–250. [CrossRef]
55. Bhattacharya, A.; Chatterjee, S.; Bhaduri, U.; Singh, A.K.; Vasudevan, M.; Sashidhara, K.V.; Guha, R.; Nazir, A.; Rath, S.K.; Natesh, N.; et al. Butyrylation Meets Adipogenesis-Probed by a p300-Catalyzed Acylation-Specific Small Molecule Inhibitor: Implication in Anti-obesity Therapy. *J. Med. Chem.* **2022**, *65*, 12273–12291. [CrossRef]
56. Zhang, R. The ANGPTL3-4-8 model, a molecular mechanism for triglyceride trafficking. *Open Biol.* **2016**, *6*, 150272. [CrossRef]
57. Ren, G.; Kim, J.Y.; Smas, C.M. Identification of RIFL, a novel adipocyte-enriched insulin target gene with a role in lipid metabolism. *Am. J. Physiol. Endocrinol. Metab.* **2012**, *303*, E334–E351. [CrossRef]
58. Abu-Farha, M.; Al-Khairi, I.; Cherian, P.; Chandy, B.; Sriraman, D.; Alhubail, A.; Al-Refaei, F.; AlTerki, A.; Abubaker, J. Increased ANGPTL3, 4 and ANGPTL8/betatrophin expression levels in obesity and T2D. *Lipids Health Dis.* **2016**, *15*, 181. [CrossRef]
59. Ghosh, A.; Leung, Y.H.; Yu, J.; Sladek, R.; Chenier, I.; Oppong, A.K.; Peyot, M.L.; Madiraju, S.R.M.; Al-Khairi, I.; Thanaraj, T.A.; et al. Silencing ANGPTL8 reduces mouse preadipocyte differentiation and insulin signaling. *Biochim. Biophys. Acta Mol. Cell Biol. Lipids* **2024**, *1869*, 159461. [CrossRef]
60. Zhang, S.; Wang, L.; Li, S.; Zhang, W.; Ma, X.; Cheng, G.; Yang, W.; Zan, L. Identification of Potential Key Genes Associated with Adipogenesis through Integrated Analysis of Five Mouse Transcriptome Datasets. *Int. J. Mol. Sci.* **2018**, *19*, 3557. [CrossRef]
61. Wang, B.; Law, A.; Regan, T.; Parkinson, N.; Cole, J.; Russell, C.D.; Dockrell, D.H.; Gutmann, M.U.; Baillie, J.K. Systematic comparison of ranking aggregation methods for gene lists in experimental results. *Bioinformatics* **2022**, *38*, 4927–4933. [CrossRef] [PubMed]
62. Rio, D.C.; Ares, M., Jr.; Hannon, G.J.; Nilsen, T.W. Purification of RNA using TRIzol (TRI reagent). *Cold Spring Harb. Protoc.* **2010**, *2010*, pdb.prot5439. [CrossRef]
63. Velasco, S.; Ibrahim, M.M.; Kakumanu, A.; Garipler, G.; Aydin, B.; Al-Sayegh, M.A.; Hirsekorn, A.; Abdul-Rahman, F.; Satija, R.; Ohler, U.; et al. A Multi-step Transcriptional and Chromatin State Cascade Underlies Motor Neuron Programming from Embryonic Stem Cells. *Cell Stem Cell* **2017**, *20*, 205–217.e8. [CrossRef] [PubMed]
64. Yousif, A.; Drou, N.; Rowe, J.; Khalfan, M.; Gunsalus, K.C. NASQAR: A web-based platform for high-throughput sequencing data analysis and visualization. *BMC Bioinform.* **2020**, *21*, 267. [CrossRef]
65. Kolberg, L.; Raudvere, U.; Kuzmin, I.; Adler, P.; Vilo, J.; Peterson, H. g:Profiler-interoperable web service for functional enrichment analysis and gene identifier mapping (2023 update). *Nucleic Acids Res.* **2023**, *51*, W207–W212. [CrossRef]

International Journal of
Molecular Sciences

MDPI

Article

Let-7a-5p Regulates Animal Lipid Accumulation by Targeting Srebf2 and Thbs1 Signaling

Jiahao Shao †, Genglong Jiang †, Yanhong Li, Meigui Wang, Tao Tang, Jie Wang, Xianbo Jia and Songjia Lai *

College of Animal Science and Technology, Sichuan Agricultural University, Chengdu 611130, China; shaojh1997@163.com (J.S.); jianggl1196282249@163.com (G.J.); lyh81236718@163.com (Y.L.); wmg1987797495@163.com (M.W.); m18483220592@163.com (T.T.); wjie68@163.com (J.W.); jaxb369@sicau.edu.cn (X.J.)

* Correspondence: laisj5794@sicau.edu.cn

† These authors contributed equally to this work.

Abstract: Recently, the trend of obesity is becoming increasingly prevalent, and the underlying pathogenesis of obesity is complex and needs to be researched further. In this study, we report a decreased expression of let-7a-5p in the white adipose tissue (WAT) of animals with obesity. Using the RNA oligo, let-7a-5p over-expression or suppression–expression is achieved, impacting the proliferation and differentiation of preadipocytes in vitro. *Srebf2* mechanistically interacts with the metabolic effect of let-7a-5p and participates in lipid accumulation by regulating *Srebf2* downstream signaling. Moreover, let-7a-5p binds to *Thbs1* to interact with the PI3K-AKT-mTOR pathway, down-regulating the phosphorylation levels of AKT, mTOR, and S6K1 to decrease lipid accumulation. In conclusion, our study highlights the physiological significance of let-7a-5p in lipid accumulation and suggests that the let-7a-5p/Srebf2 and let-7a-5p/Thbs1/PI3K-AKT-mTOR axes may represent potential mechanisms for controlling lipid accumulation in obesity.

Keywords: lipid accumulation; high-fat diet; preadipocytes; white adipose tissue

1. Introduction

Obesity is defined as an excessive body mass index (BMI) that is associated with white adipose tissue (WAT) accumulation, as well as an increased risk of endocrine disease, type II diabetes, and cancer [1–4]. WAT is a multi-depot organ that supports endocrine and energy storage. WAT has the unique ability to coordinate metabolic changes throughout the body and integrate responses to maintain metabolic homeostasis [5,6]. Thus, the maintenance of WAT lipid metabolism is essential. However, comprehending the pathological process of fat buildup is difficult because both environment and genes influence lipid metabolic regulation [7]. Nowadays, highly palatable and fat-dense foods are becoming more and more available and cheaper, and this, along with the development of food production, processing, storage, and preparation, contributes to the rise in obesity [8]. Furthermore, extensive data have indicated that multiple genes, gene products, and signaling pathways are significant contributors to the control of WAT metabolism. Overall, 15–50% of the cells in the WAT are preadipocytes [9]. During adipogenesis, preadipocyte transcription factors such as Sox9 and WNT/β-catenine undergo suppression, while the CCAAT/enhancer-binding protein (C/EBP) family of bZIP transcription factors and PPARγ are active [6,10,11]. Furthermore, some micronutrients, such as vitamins, constitute an essential functional element of the development of WAT upon their transformation into hormone-like molecules by certain enzymes [12].

microRNAs (miRNAs) pair with the 3′ untranslated regions (UTR) to regulate the post-transcriptional suppression of target genes. miRNAs usually exhibit cell type-, tissue-, and developmental stage-specific expression and influence intracorporeal metabolic processes.

Emerging data highlight the roles of miRNAs in lipid metabolic regulation and the progression of lipid dysmetabolism-related diseases. For instance, in humans and animals, normal and obese individuals differ with respect to adipose miRNA expression [13,14]. miR-143a-3p regulates preadipocyte proliferation and differentiation by targeting MAPK7 [15]. In addition, miR-214-3p promotes the differentiation of 3T3-L1 through the Ctnnb1/Wnt/β-Catenin pathway [16]. Overall, miRNAs have attracted a lot of interest for their involvement in lipid metabolism and diseases. Although major pathways leading to lipid accumulation in adipocytes have been identified, there are still many roles for miRNAs in the context of obesity that need to be investigated.

In this study, we discovered a miRNA named let-7a-5p, which is down-regulated in obese individuals, and examined its function further in vivo and in vitro. Our findings indicate that the let-7a-5p/Srebf2 and let-7a-5p/Thbs1/PI3K-AKT-mTOR axes may be crucial in regulating animal lipid accumulation and may contribute to the development of new therapies for obesity.

2. Results

2.1. Let-7a-5p Is Associated with Lipid Metabolism

miRNAs are conserved among species with distinct temporal and geographical patterns of expression in response to many biological activities [17,18]. We sequenced the miRNA expression in the adipose tissue from the SND and HFD groups to discover miRNAs that may play a function in lipid metabolism by binding protein-coding genes in the adipose tissue [19]. Interestingly, we noted that let-7a-5p was markedly decreased and showed a similar alternation in other obese animals [13,20] and obese humans [14]. In this study, we first assessed the abundance of let-7a-5p in the organs of the SND rabbits, such as the liver, heart, etc., to verify the expression pattern of let-7a-5p in animals with a normal metabolic condition. Compared with the control animals, let-7a-5p expression was observed to be high in the adipose tissue of the obese animals, suggesting that let-7a-5p may be functionally important for preserving the balance of lipid metabolism (Figure 1A). In addition, we found that let-7a-5p was significantly decreased in the WAT of rabbits subjected to the HFD treatment, which validates the sequencing data (Figure 1B). Next, the major hormones secreted from the adipose tissue, serum LEP and ADP, were next measured in the rabbits. We discovered that let-7a-5p expression in the adipose tissue was negatively correlated with serum LEP levels but positively correlated with serum ADP levels (Figure 1C). Using the Starbase database, let-7a-5p targets were captured, including *Srebf2*, *Dagla*, *Ptgs2*, and *Thbs1* (Table S7). For the let-7a-5p target genes, lipid metabolic pathways such as the mTOR signaling pathway, PI3K-AKT signaling pathway, and insulin signaling pathway were enriched in the KEGG database (Figure 1D, Table S8). According to the results of GO analysis, the let-7a-5p target genes were significantly enriched in 787 GO terms (483 biological processes (BP), 153 cellular components (CC), and 151 molecular functions (MF)), including the regulation of cell size, cell cycle, the metabolic process, and the cellular response to insulin stimulus (Figure 1E, Table S9). Taken together, these findings suggest that let-7a-5p may be a functional molecule involved in lipid metabolism.

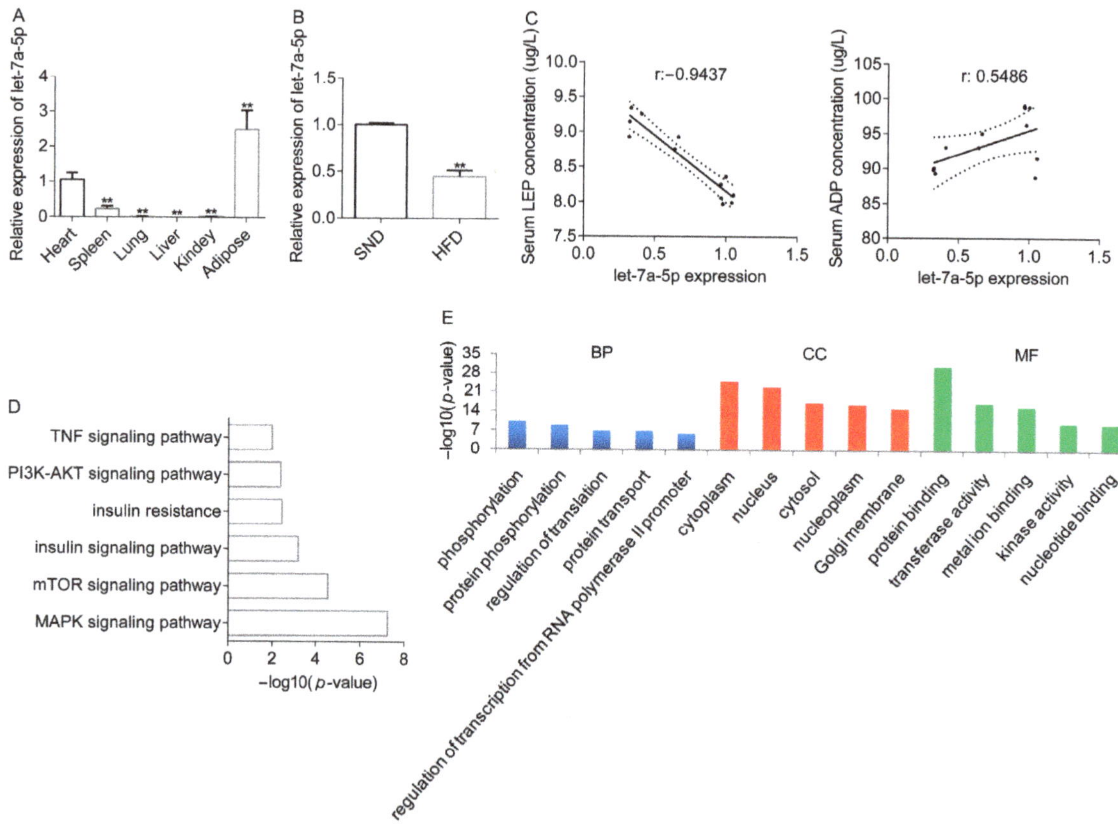

Figure 1. Let-7a-5p is associated with lipid metabolism. (**A**) Let-7a-5p expression levels in the different organs from the SND group were measured ($n = 6$). (**B**) Let-7a-5p expression was assessed in the rabbits' WAT, and the data based on the normal rabbits were used as a reference ($n = 6$). (**C**) The Pearson correlation between let-7a-5p expression in the rabbits' WAT and serum LEP (**left**) or ADP (**right**) levels ($n = 12$ in biological animals). (**D**) The pathways partly enriched in the KEGG database. (**E**) A list of the top five important GO terms in MF, CC, and BP. The data are presented as means ± SEM. ** p-value < 0.01.

2.2. Let-7a-5p Regulates Adipogenesis by Targeting the PI3K-AKT-mTOR Signaling

Measuring the effect of let-7a-5p on lipid metabolic capabilities is important since it is connected to several protein-coding genes linked to lipid metabolism and is expressed at low levels in the adipose tissue of obese rabbits. Next, by transfecting the let-7a-5p mimic, NC, let-7a-5p inhibitor, and INC into the preadipocytes, we examined the impact of let-7a-5p on the adipogenesis of rabbit preadipocytes. We verified that the transfected analogs effectively increased or lowered let-7a-5p abundance in vitro (Figure 2A). For preadipocytes proliferation, the let-7a-5p mimic decreased the absorbance to different extents at various lengths of time, but the let-7a-5p inhibitor markedly enhanced absorbance after transfection for 24 h and 72 h (Figure 2B). Following treatment with the let-7a-5p mimic, the expression of the proliferation-related protein Pcna was reduced; however, following treatment with the let-7a-5p inhibitor, it was increased (Figure 2C). Furthermore, we verified that the let-7a-5p mimic, which reversed the effects of the let-7a-5p inhibitor after 4 days of differentiation, decreased the protein levels of Fabp4 and Pparg, which are crucial transcription factors in the development and function of adipose tissue and markers of preadipocyte differentiation

(Figure 2D). Additionally, the TG accumulation in the mimic group was less than that in the NC group (Figure 2E).

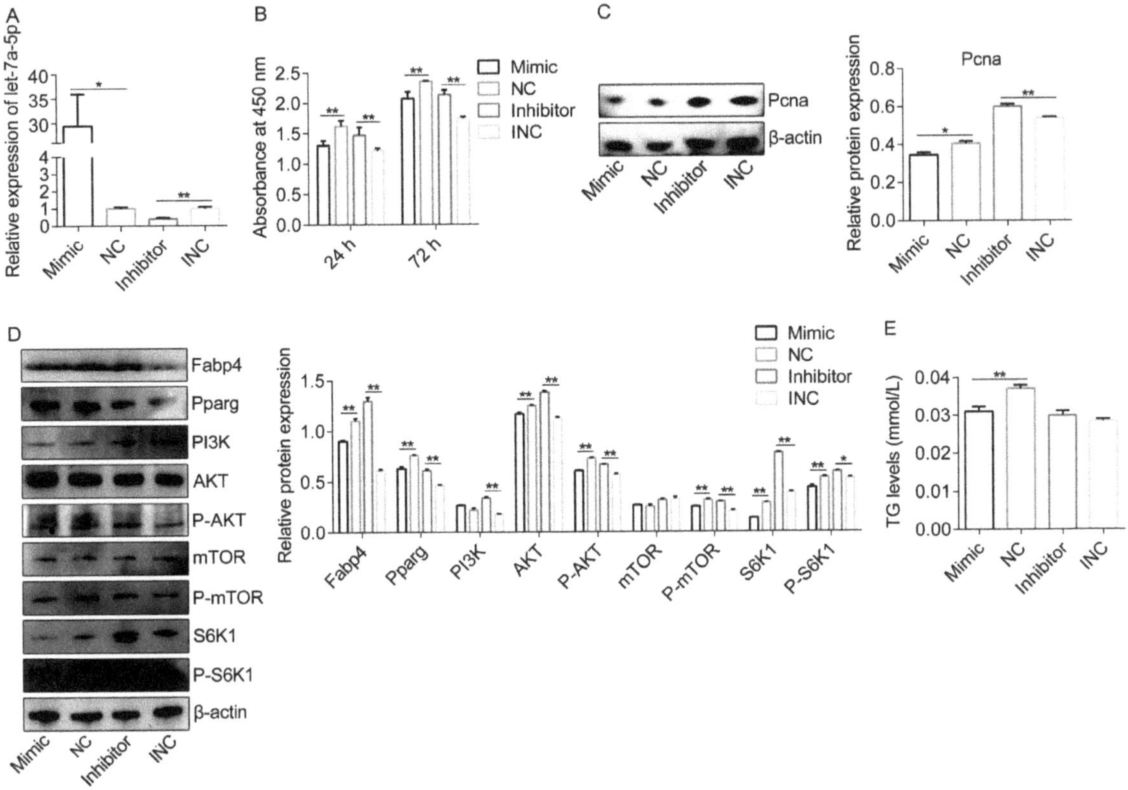

Figure 2. Let-7a-5p regulates adipogenesis by targeting the PI3K-AKT-mTOR signaling. (**A**) The miR-24-3p mimic and inhibitor's transfection efficiency were assessed (n = 6). (**B**) The absorbance of preadipocytes from rabbits transfected with let-7a-5p mimic, NC, let-7a-5p inhibitor, and INC at 24 h and 72 h (n = 6). (**C**) Representative Pcna bands in rabbit preadipocytes transfected with let-7a-5p mimic, NC, let-7a-5p inhibitor, and INC. The Pcna expression was quantitatively analyzed (n = 3). (**D**) Representative WB of Fabp4, Pparg, PI3K, AKT, P-AKT, mTOR, P-mTOR, S6K1, P-S6K1, and β-actin at 4 days after transfection with the let-7a-5p mimic, NC, let-7a-5p inhibitor, and INC in rabbits' preadipocytes in the context of differentiation. n = 3 for data analysis. (**E**) After treating the preadipocytes with the let-7a-5p mimic, NC, let-7a-5p inhibitor, and INC, the TG content was assessed (n = 8). The data are presented as means ± SEM. * p-value < 0.05; ** p-value < 0.01.

Meanwhile, we also wanted to determine whether let-7a-5p is functionally involved in the lipid metabolic regulation of the PI3K-AKT-mTOR pathway. To test this notion, we first confirmed whether let-7a-5p inhibition and over-expression could both significantly change PI3K activity (Figure 2D). Subsequently, the over-expressing of let-7a-5 resulted in the suppression of P-AKT expression, while the let-7a-5p inhibitor caused an increase in P-AKT expression (Figure 2D). The downstream phosphorylated mTOR and S6K1 showed a reduction in the let-7a-5p mimic group but drastically increased after let-7a-5p inhibition when compared with the related reference (Figure 2D). In summary, our findings show that let-7a-5p has a direct interaction with preadipocytes adipogenesis via the PI3K-AKT-mTOR pathway.

2.3. RNA-seq Revealed Potential Let-7a-5p Target Genes

In this study, we performed RNA-seq on adipose RNA collected from the SND and HFD groups to identify the potential let-7a-5p target genes. Overall, six libraries were built successfully, and sequencing for each of the libraries surpassed 45,617,148 high-quality raw reads, with an average clean read of 46,544,061 (Table S1). The Q20 (percentage of reads with a Phred quality value > 20) ranged from 97.77% to 98.07%, and the GC content of libraries ranged from 52.82% to 54.15% (Table S1). Next, the clean data were aligned to reference genomes using HISAT2, with high mapping efficiency ranging from 86.08% to 87.5% (Table S1). The gene expression was quantified using subread software based on read counts. Following the fragments per kilobase per million reads (FPKM) method, we found that the gene expression distribution was consistent between the two metabolic conditions of adipose tissue (Figure 3A).

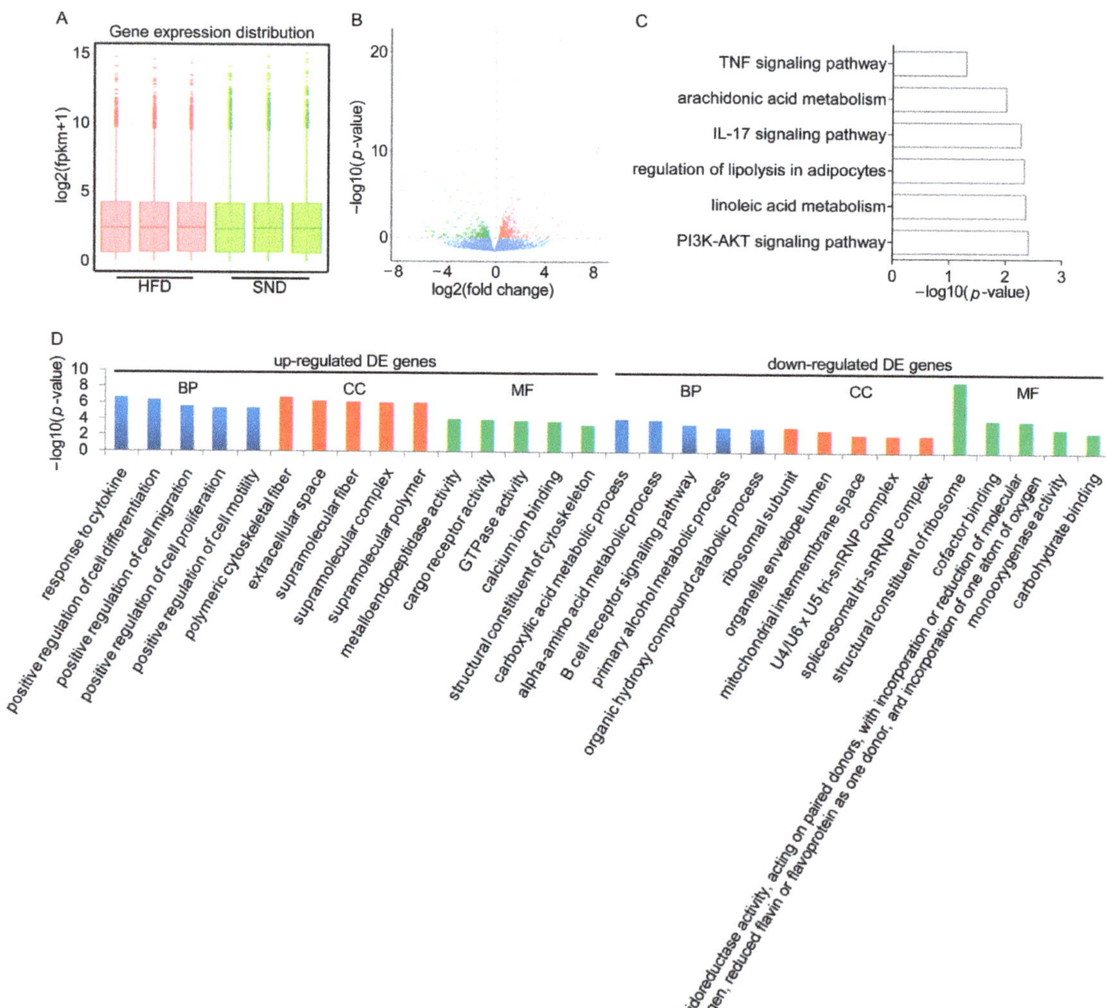

Figure 3. RNA-seq analysis revealed potential let-7a-5p target genes. (**A**) Gene expression distribution from the SND and HFD RNA-seq libraries. (**B**) Based on log2(fold change) and −log10(p-value),

a volcanic picture of the DE genes between the SND and HFD groups was constructed. The red point indicates the statistically significant up-regulated genes, whereas the green point indicates the statistically significant down-regulated genes. (**C**) The part gene sets enriched in the KEGG database are listed. (**D**) The DE genes enriched in the GO database and only the top five significant GO terms in BP, CC, and MF are listed.

We identified DE genes using the DESeq following the standard of p-value ≤ 0.05. The full list of DE genes is available in Table S2. Among them, 447 genes were down-regulated and 607 genes were up-regulated in the HFD group, respectively (Figure 3B). To better understand the functional roles of DE genes, we subsequently performed an enrichment analysis to categorize the functions of the DE genes in the KEGG database. The data indicated that the PI3K-AKT signaling pathway, IL-17 signaling pathway, arachidonic acid metabolism, linoleic acid metabolism, and TNF signaling pathway were significantly enriched (Figure 3C, Table S3). Meanwhile, based on the up-regulated DE genes, the top significant biological process GO terms are mostly responsible for signaling and cell regulatory functions, such as response to lipoprotein particle, the positive regulation of cell differentiation, the positive regulation of cell migration, the positive regulation of cell proliferation (Figure 3D, Table S4). Genes *Srebf2*, *Lep*, *Ptgs2*, and *Adam17* are important contributors and have been shown to have direct consistency with the above GO terms. Using the down-regulated DE genes, we obtained the related GO terms data (Figure 3D, Table S4). Of the significantly enriched biological process GO terms, the most down-regulated DE genes were shown to control metabolism-related GO terms. Specifically, *Sorl1*, *Lgals12*, and *Pnpla7* are prominently involved in the lipid catabolic process. *Adh2-1*, *Hacl1*, *Echs1*, and *Acad11* are involved in lipid oxidation. Thus, our data revealed a dynamic adipose gene expression under obesity condition.

2.4. Let-7a-5p Regulates Lipid Metabolism by Targeting the Srebf2 Signaling

Based on their sequence information, the Targetscan website was used to determine the possible binding site between *Srebf2* and let-7a-5p (Figure 4A). In terms of preadipocytes differentiation, *Srebf2* mRNA expression was decreased in the mimic group but increased in the inhibitor group in vitro (Figure 4B). Using the luciferase reporter assay, we observed that the luciferase activity of a luc-reporter construct containing *Srebf2* 3′ UTR (WT1) was considerably reduced by the let-7a-5p mimic (Figure 4C). Subsequently, we introduced the specific siRNA target *Srebf2* into cells in vitro to examine the essentialness of *Srebf2* in the metabolic responses of let-7a-5p toward lipid accumulation (Figure 4D). Following si-Srebf2 treatment, 3T3-L1 exhibited reduced lipid accumulation, as demonstrated via Oil Red O staining and intracellular TG analysis, which was consistent with the findings in preadipocytes with let-7a-5p mimic treatment (Figure 4E,F). Importantly, in 3T3-L1 with *Srebf2* knockdown, the further knockdown or overexpression of let-7a-5p did not affect lipid accumulation (Figure 4F). The knockdown of *Srebf2* also reduced the levels of *Fabp4* and *Pparg* genes but did not affect the expression levels of let-7a-5p, implying that *Srebf2* does not modulate the stability of let-7a-5p but might influence the stability of genes related to lipid metabolism (Figure 4G,H). Furthermore, qRT-PCR demonstrated that *Srebf2* knockdown reduced the transcription of genes linked to lipid metabolism, including *Fasn*, *Fads1*, *Elovl5*, *Scd*, *Fads2*, and *Aacs* (Figure 4I). In conclusion, our findings show that let-7a-5p binds to *Srebf2* and modulates *Srebf2*-mediated lipid metabolic signaling.

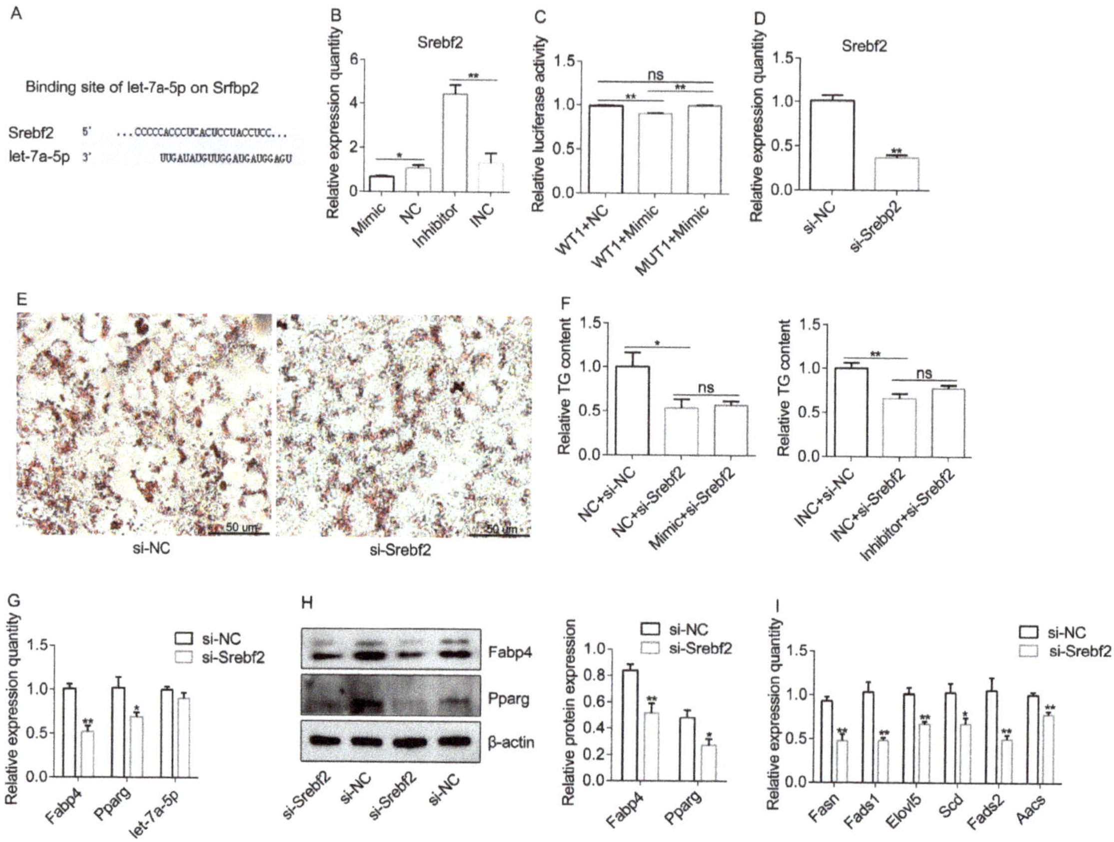

Figure 4. Let-7a-5p regulates lipid metabolism by targeting *Srebf2* signaling. (**A**) Potential binding sites between let-7a-5p and *Srebf2* were identified using the Targetscan website. The sites in the white background represent binding sites. (**B**) The mRNA of *Srebf2* at 4 days after transfection with let-7a-5p mimic, NC, let-7a-5p inhibitor, and INC in the context of differentiation. ($n = 6$). (**C**) Relative luciferase activity was measured after the co-transfection of WT1 and MUT1 with the let-7a-5p mimic and NC, respectively ($n = 3$). (**D**) After si-Srebf2 treatment, the mRNA of *Srebf2* was examined using qRT-PCR ($n = 6$). (**E**) Oil Red O staining was performed for visual observation of the lipid droplets from the si-Srebf2 and si-NC groups. (**F**) The cellular TG concentration was assessed as the indication of lipid accumulation ($n = 6$). (**G**) The expression levels of *Fabp4*, *Pparg*, and let-7a-5p in the 3T3-L1 differentiated mature adipocytes were detected using qRT-PCR ($n = 6$). (**H**) Representative WB of Fabp4 and β-actin at 4 days after transfection with si-Srebf2 and si-NC in the 3T3-L1 preadipocytes in the context of differentiation. $n = 3$ for data analysis. (**I**) qRT-PCR analysis of *Fasn*, *Fads1*, *Elovl5*, *Scd*, *Fads2*, and *Aacs* in the 3T3-L1 preadipocytes treated with si-Srebf2 and si-NC ($n = 6$). Not significant: ns. The data are presented as means ± SEM. * p-value < 0.05; ** p-value < 0.01.

2.5. A Let-7a-5p Target—Thbs1 Is Required for Adipogenesis via the PI3K-AKT-mTOR Pathway

Considering that let-7a-5p has a direct interaction with adipogenesis via the PI3K-AKT-mTOR pathway, we investigated whether a key gene acts as the mediator. We were intrigued by the identification of *Thbs1* as a putative let-7a-5p target, a functional gene linked to obesity, lipid metabolism, and adipose biology (Figure 5A). *Thbs1* is up-regulated in adipose tissue under obesity (Section 2.3), and in this study, it was negatively associated with let-7a-5p expression (Figure 5B). The luciferase activity of a luc-reporter construct incorporating *Thbs1* 3′ UTR (WT2) was shown to be greatly reduced by the let-7a-5p mimic, but this reduction was abolished by a mutation in the *Thbs1* binding site in the 3′ UTR (Figure 5C). Next, we delivered the specific siRNA target *Thbs1* into cells in vitro to investigate *Thbs1*'s essentialness in the metabolic responses of let-7a-5p toward lipid metabolism (Figure 5D). We noted that the si-Thbs1 group showed lower lipid accumulation, as determined via Oil Red O staining (Figure 5E). Proteins Fabp4 and Pparg were consistently down-regulated after si-Thbs1 treatment (Figure 5F). Meanwhile, the abundance of PI3K protein markedly decreased in the si-Thbs1 group compared with the si-NC group (Figure 5F). Decreased levels of P-AKT were observed, with no significant effect on the total AKT levels after si-Thbs1 treatment (Figure 5F). Moreover, the phosphorylation levels of the downstream mTOR and S6K1 showed a significant decrease, indicating that mTOR signaling was inhibited when *Thbs1* expression was silenced (Figure 5F). These data indicate that *Thbs1* is required for adipogenesis via the PI3K-AKT-mTOR pathway.

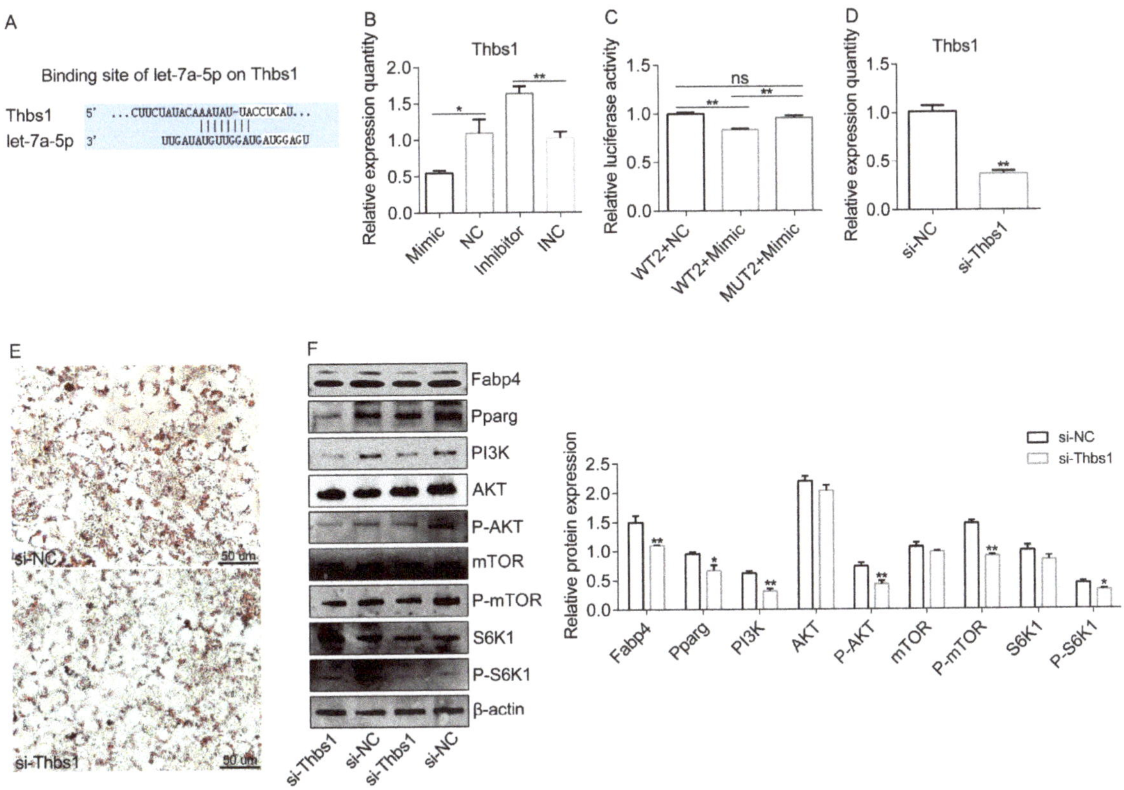

Figure 5. A let-7a-5p target—*Thbs1* is required for adipogenesis via the PI3K-AKT-mTOR pathway. (**A**) Potential binding sites between let-7a-5p and *Thbs1* were identified using the Targetscan website. The sites in the white background represent binding sites. (**B**) The mRNA of *Thbs1* at 4 days after transfection with let-7a-5p mimic, NC, let-7a-5p inhibitor, and INC in the context of differentiation. ($n = 6$). (**C**) Relative luciferase activity was measured after the co-transfection of WT2 and MUT2 with the let-7a-5p mimic and NC, respectively ($n = 3$). (**D**) After si-Thbs1 treatment, the mRNA of *Thbs1* was examined using qRT-PCR ($n = 6$). (**E**) Oil Red O staining was performed for visual observation of the lipid droplets from the si-Thbs1 and si-NC groups. (**F**) Representative bands for Fabp4, Pparg, PI3K, AKT, P-AKT, mTOR, P-mTOR, S6K1, P-S6K1, and β-actin in 3T3-L1 cells treated with si-Thbs1 and si-NC. $n = 3$ for data analysis. Not significant: ns. The data are presented as means ± SEM. * p-value < 0.05; ** p-value < 0.01.

2.6. Let-7a-5p Inhibites High-Fat-Diet-Induced Obesity in a Mice Model

At first, we discovered that let-7a-5p was strongly expressed in the adipose tissue of normal metabolic mice, but this was considerably reduced when the mice were fed a high-fat diet, which was consistent with the findings in the HFD rabbit model (Figure 6A,B). Next, we investigated whether let-7a-5p could control lipid accumulation in vivo. We subjected the mice to a high-fat diet and injected them with specific let-7a-5p agomir or agomir control in the caudal vein. After a detailed analysis, it was found that the HFD-a mice featured decreased adiposity compared with the HFD-m mice, as shown by their reduced body weight and adipose tissue weight (Figure 6C,D). Levels of serum TG, which is a marker of lipid metabolism and obesity, were significantly enhanced in the HFD-m group, but let-7a-5p agomir decreased the TG in the HFD-a group compared to the HFD-m group (Figure 6E). At the molecular levels, we examined the expression levels of lipogenesis-related protein Fabp4 in the perirenal adipose tissue of each group. As expected, its expression was decreased after consistent let-7a-5p agomir treatment in the HFD-a group compared to the HFD-m group (Figure 6F). Thus, our data indicate that let-7a-5p may play an important role in the regulation of lipid accumulation in vivo.

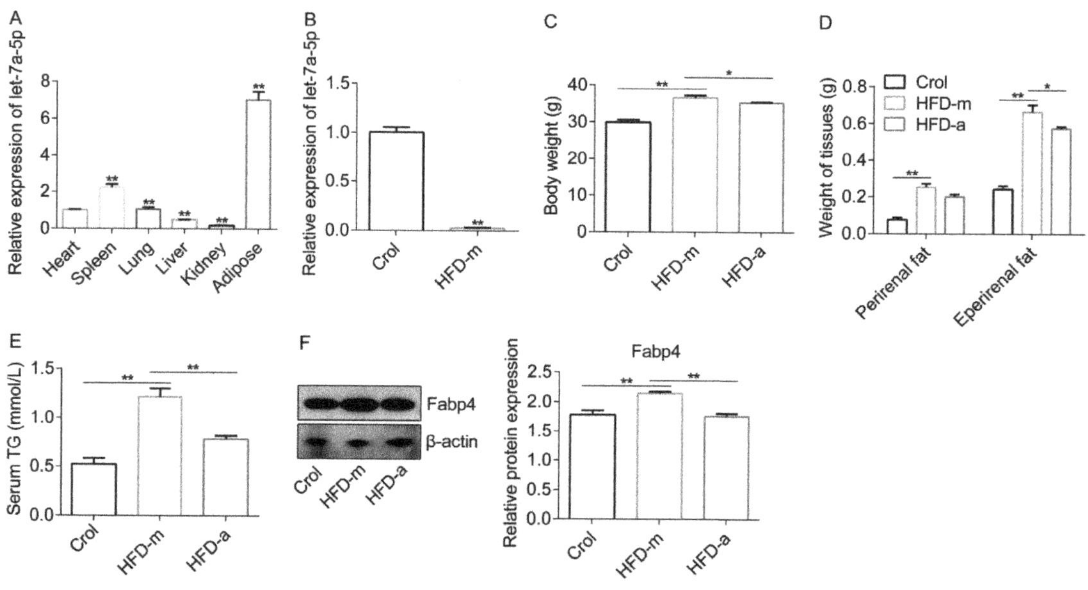

Figure 6. Let-7a-5p inhibits high-fat-diet-induced obesity in a mice model. (**A**) Let-7a-5p expression in the different organs from the Crol group was measured ($n = 6$). (**B**) Let-7a-5p expression was measured in the Crol and HFD-m groups, and the data in the Crol were used as a reference ($n = 6$). Analyses of the metabolic performances of the Crol, HFD-m, and HFD-a groups were carried out, considering body weight (**C**), fat tissue weight (**D**), serum TG abundance (**E**), and lipid metabolism-related protein in the perirenal fat (**F**). $n = 6$ in (**C**–**E**). $n = 3$ in (**F**). The data are presented as means ± SEM. * p-value < 0.05; ** p-value < 0.01.

3. Discussion

Obesity is defined by the World Health Organization (WHO) as the excessive accumulation of WAT. Therefore, it would be beneficial to better understand the molecular network and mechanism of lipid metabolism, including the roles played by miRNAs in effectively controlling lipid metabolism and preventing lipid accumulation in WAT.

Numerous studies have demonstrated that adipose miRNAs exhibit dynamic regulation in obesity and functional roles in adipogenesis [13,15]. For example, miR-24-3p inhibited preadipocytes differentiation, whereas miR-9-5p promoted preadipocytes differentiation [21,22]. According to data available online, let-7a-5p is a key miRNA with roles in bone marrow mesenchymal stem cell (BMSC) osteogenesis [23], embryo implantation [24], shoulder stiffness [25], and inflammatory response [26]. Furthermore, let-7a-5p can decrease *BZW2* expression, and it inhibits hepatoma cell proliferation, invasion, and migration and increases apoptosis [27]. Meanwhile, an oncogenic role for let-7a-5p in colorectal cancer [28] and cervical cancer [29] has been reported. However, although let-7a-5p was markedly decreased in the WAT of obese animals [13,20], its possible role in the adipogenesis and hypertrophy of adipose tissue requires further investigation. In this study, we confirmed that let-7a-5p maintained a low abundance in the WAT of obese rabbits and mice. Gain- or loss-of-function studies showed that let-7a-5p inhibited preadipocytes proliferation and differentiation in vitro. In addition, our data indicate that the over-expression of let-7a-5p altered high-fat-diet-induced body weight, tissue weight gain, and blood fat in the obese mice, suggesting that boosting the expression of let-7a-5p in the obese mice WAT was beneficial for maintaining lipid metabolic homeostasis.

Lipid accumulation is composed of many enzymatic reactions and controlled by a variety of molecules. In this study, targeting research showed that let-7a-5p is directly bound to the 3′-UTR of *Srebf2*. The Srebf2 protein is a member of the sterol regulatory element binding protein (SREBP) nuclear transcription factor family, which regulates lipid homeostasis by controlling the expression of several enzymes necessary for endogenous phospholipid, triacylglycerol, cholesterol, and fatty acid (FA) synthesis [30]. Lipid metabolism is influenced by mature forms of SREBPs that are transcriptionally active and transported to the nucleus, where they bind to the promoters of the SREBP target [31]. Srebp2 promotes the expression of target genes in the cholesterol biosynthesis pathway by binding to the sterol regulator and promoter/enhancer in the lipid synthase gene [32]. In this study, we discovered that *Srebf2* has a direct interaction with the lipid accumulation effect of let-7a-5p via downstream metabolic genes such as *Fasn*, *Fads1*, and *Scd*. Moreover, another possibility is that let-7a-5p regulates lipid accumulation via the PI3K-AKT-mTOR pathway by binding to the 3′ UTR of *Thbs1*. Thbs1 is a large adhesive extracellular matrix glycoprotein that contains multiple functional domains. In mice and humans, *Thbs1* is highly expressed in hypertrophic visceral adipose tissue [33,34]. *Thbs1*-null mice were protected against insulin resistance and the inflammation of adipose tissue [35]. Moreover, *Thbs1* deficiency prevents high-fat-diet-induced adipose tissue hypertrophy [33]. We discovered that let-7a-5p regulates the expression of *Thbs1* and that si-Thbs1 inhibits lipid accumulation by targeting the PI3K-AKT-mTOR pathway, indicating the potential connection between let-7a-5p and the metabolic network. It has been revealed that the PI3K-AKT-mTOR pathway controls lipid accumulation in animal cells by increasing the unclear accumulation of mature SREBP [36,37].

However, we cannot exclude the notion that let-7a-5p may cooperate with other actors through unidentified ways to regulate lipid accumulation. Therefore, further mechanistic studies are required to elucidate how let-7a-5p controls lipid accumulation. For instance, it is necessary to investigate the roles that let-7a-5p plays in the process of lipid metabolism in other animal models. It is also necessary to look into the other possible mechanisms of let-7a-5p and their effects on lipid metabolism, such as enhanced lipid secretion through the skin, WAT browning, and metabolomics.

4. Materials and Methods

4.1. Animals

This work is an extension of a previous study wherein an obese rabbit model was established using a high-fat diet [38]. Briefly, the rabbits in the model were subjected to a control (120 g/d, labeled as the SND group) or a high-fat diet (120 g/d, labeled as the HFD group) for 5 weeks in a thermostatic room. At the end of the trial, six rabbits from each group were selected for sampling, and three samples from each group were used for transcriptome sequencing. Next, eighteen 21-day male Kunming mice were divided into three groups. Two groups received either a normal diet (referred to as the Crol group) or a high-fat diet (referred to as the HFD-m group) for four weeks and received a total of four tail vein injections of agomir control (RiboBio, Guangzhou, China). Moreover, the remaining group was fed a high-fat diet and given four tail vein injections of let-7a-5p agomir (HFD-a group; RiboBio, Guangzhou, China) over four weeks. Every mouse had unrestricted access to food and water. All experimental protocols were performed under the direction of the Institutional Animal Care and Use Committee from the College of Animal Science and Technology, Sichuan Agricultural University, China.

4.2. Cell Culture

Cells were kept in a growth medium (GM) containing 10% fetal bovine serum (FBS; Gibco, Grand Island, NY, USA), which was stored in an incubator (Thermo Fisher Scientific, Waltham, MA, USA) at 37 °C and 5% CO_2 environment. We used a differentiation medium (DM) containing Dulbecco's modified Eagle's medium (DMEM; VivaCell, Shanghai, China) supplemented with 10% FBS, 1 µmol/L dexamethasone (DEX; Solarbio, Beijing, China), 0.5 mmol/L 3-isobutyl-1-methylxanthine (IBMX; Solarbio, Beijing, China), and 10 µg/mL insulin (Solarbio, Beijing, China). The maintenance medium (MM) consisted of DMEM with 10% FBS and 10 µg/mL insulin. According to the manufacturer's instructions, transfection was carried out using the lipofectamine 3000 reagent (Invitrogen, Carlsbad, CA, USA). The transfection reagent was mixed with RNA oligo (let-7a-5p mimic, inhibitor, negative control (NC), inhibitor negative control (INC)) and siRNA (si-Srebf2, si-Thbs1, and si-NC), and then the mixture was transfected into the cells. The concentration of transfection was 50 uM for the mimic and NC, and 100 uM for the inhibitor, INC, siRNA, and si-NC. The above RNA oligo and siRNA were purchased from Sangon Biotech Co., Ltd. (Shanghai, China) and Tsingke Biotechnology Co., Ltd. (Beijing, China), respectively, and Table S5 displays the sequence information.

4.3. Quantitative Real-Time Polymerase Chain Reaction (qRT-PCR)

Using the TRIpure reagent (Aidlab, Beijing, China), total RNA was extracted in accordance with the instructions. The quality test was conducted using an Agilent Bioanalyzer (Agilent Technologies, Santa Clara, CA, USA), and only eligible RNA was used for subsequent experiments. For the reverse-transcription of mRNAs and miRNAs, we used the RT Easy™ II (with gDNase) (FOREGENE, Chengdu, China) and the Mir-X™ miRNA First-Strand Synthesis Kit (Takara, Dalian, China), respectively. Furthermore, the NovoStart®SYBR qPCR SuperMix Plus (Novoprotein, Shanghai, China) was used in triplicate for qRT-PCR on a CFX96 instrument (Bio-Rad, Hercules, CA, USA). The relative abundance was determined using the $2^{-\Delta\Delta Ct}$ method. U6 was utilized as an internal reference, and the mRQ 3′ primer from the Mir-X™ miRNA First-Strand Synthesis Kit (Takara, Dalian, China) was utilized as the reverse primer for miRNA quantification. Furthermore, the internal reference for mRNA quantification was *β-actin* gene. Table S6 provides a detailed list of primer sequence information.

4.4. Western Blotting (WB)

Total protein extraction was carried out using the ProteinExt® Mammalian Total Protein Extraction Kit (TransGen Biotech, Beijing, China). The Bradford protein assay kit (Novoprotein, Shanghai, China) was then used to measure the protein concentration. Briefly, the denatured protein was resolved on the SurePAGE gels (GenScript Corporation, Nanjing, China) and then transferred to the PVDF membrane, which was then sealed with a sealing liquid. The membranes were initially incubated with the corresponding primary antibodies at 4 °C for 8 h. After that, they were incubated with the secondary antibodies (Goat Anti-Rabbit or Anti-Mouse IgG H&L (HRP), Zen Bioscience, Chengdu, China) for 1 h. The antibodies we used are as follows: Pcna (10205-2-AP), purchased from the Proteintech (Wuhan, China); Pparg (A0270) and Fabp4 (A0232), purchased from the Abclonal (Wuhan, China); and AKT (382804), P-AKT (310021), PI3K (251221), mTOR (380411), P-mTOR (381557), S6K1 (R25650), P-S6K1 (380880), and β-actin (380624), purchased from Zen Bioscience (Chengdu, China). Finally, the membranes were exposed to a SuperBright

Subpico ECL reagent (SUDGEN, Nanjing, China). Images were taken using a GelDoc system (Bio-Rad, Hercules, CA, USA).

4.5. Oil Red O Staining

The cells were fixed for 30 min at room temperature in 4% paraformaldehyde. Subsequently, Oil Red O (Servicebio, Wuhan, China) was mixed with deionized water (3:2) and then added to the cells for 15 min. Ultimately, PBS was used to clean the aforementioned cells until no visible impurities remained. Images were captured using an inverted microscope (Olympus, Tokyo, Japan).

4.6. Measuring Blood Markers

All mice were fasted overnight before slaughter and anesthetized via an intraperitoneal injection of chloral hydrate. Blood was collected from the heart and then centrifuged at 4 °C for 5 min to obtain the serum. The commercial enzymatic colorimetry kit (Jiancheng, Nanjing, China) was utilized to test the triglyceride (TG) concentration of the sample following established protocols. In addition, following the standard method, the rabbits' leptin (LEP) and adiponectin (ADP) levels were tested using a commercial enzyme-linked immunosorbent assay (ELISA) kit (Shanghai Hengyuan Biotechnology Co., Ltd., Shanghai, China).

4.7. Measuring Cell Proliferation

When the cell density reached 50%, transfection was carried out. Six hours later, the medium was changed to GM. A cell counting kit (CCK, Zomanbio, Beijing, China) was employed to examine the effect of let-7a-5p on cell proliferation. Briefly, after transfection for 24 h and 72 h, 10 µL CCK reagent was added to each well and incubated at 37 °C and 5% CO_2 for 2 h. The absorbance, which indicates the quantity of living cells, was then measured using a microplate reader (Thermo Fisher Scientific, Waltham, MA, USA).

4.8. Dual-Luciferase Reporter Assay

In this study, we predicted possible let-7a-5p targets using the Starbase database (https://starbase.sysu.edu.cn/, accessed on 10 July 2022). The software DAVID 6.7 (http://david.abcc.ncifcrf.gov/home.jsp, accessed on 18 May 2023) was used to conduct Gene Ontology (GO) analysis and Kyoto Encyclopedia of Genes and Genomes (KEGG) pathway enrichment analysis. To validate the proposed binding site, wild-type (WT) and mutant (MUT) pmirGLO plasmids of the target sequence were constructed by Tsingke Biotechnology Co., Ltd. (Beijing, China). Next, 24-well plates (NEST Biotechnology, Wuxi, China) were seeded with 293T cells. Using the lipofectamine 3000 reagent (Invitrogen, Carlsbad, CA, USA), the let-7a-5p mimic or NC was co-transfected with specific WT or MUT plasmids. After a day, luciferase activities were measured using the Duo-Lite TM Luciferase Assay System (Vazyme, Nanjing, China) in accordance with the manufacturer's instructions.

4.9. Transcriptome Sequencing

Sample RNA was produced using the above-described method. The M-MuLv reverse transcriptase system was used to create the first strand of cDNA, while DNA polymerase I and dNTPs were used to create the second strand. Following the manufacturer's instructions, cDNA of 250–300 bp was generated, and PCR amplification was then used to create the cDNA libraries using the NEBNext® Ultra™ RNA Library Prep Kit for Illumina® (New England BioLabs, Ipswich, MA, USA). Novogene Co., Ltd. (Beijing, China) used the Illumina NovaSeq 6000 platform to sequence the cDNA libraries. Additionally, the FASTP program (version 0.19.7) was used to clean the raw data. HISAT2 software 2.0.5 was then used to compare the clean data with the reference genome to determine the read localization information. After that, the transcript read count was determined, and differential expression (DE) analysis was performed using the R package DEseq 1.16.1. The threshold was defined to be p-value ≤ 0.05. Using the R ggplot2 tool, the DE genes between groups were displayed graphically.

4.10. Statistical Analysis

All data are presented as means ± SEM. Student's *t*-test was utilized to ascertain group differences in the statistical analysis, which was conducted using GraphPad software 6.04 (GraphPad Software Inc., La Jolla, CA, USA). Moreover, a one-way ANOVA was used for multiple comparisons. Differences were considered statistically significant at p-value < 0.05.

5. Conclusions

In conclusion, our findings suggest that the let-7a-5p/Srebf2 and let-7a-5p/Thbs1/PI3K-AKT-mTOR axes are crucial for lipid accumulation and that they may represent a potential mechanism for controlling lipid accumulation in obesity.

Supplementary Materials: The supporting information can be downloaded at: https://www.mdpi.com/article/10.3390/ijms25020894/s1.

Author Contributions: The project was planned and designed by J.S., J.W., X.J. and S.L.; Data gathering and research were carried out by J.S., M.W., G.J., Y.L. and T.T.; J.S. wrote the paper. All authors have read and agreed to the published version of the manuscript.

Funding: The National Modern Agricultural Industrial Technology System (CARS-43-A-2) and the Key Research and Development Program of Sichuan Province (2021YFYZ0033) provided funding for our research.

Institutional Review Board Statement: The experimental procedures were approved by the Animal Care and Use Committee of the College of Animal Science and Technology, Sichuan Agricultural University, China. Certification No. SYXK2019-187; Date: 29 January 2019.

Informed Consent Statement: Not applicable.

Data Availability Statement: The sequencing data are available on the NCBI database at Sequence Read Archive (SRA). The transcriptome sequencing data: SUB13840470.

Conflicts of Interest: The authors declare no conflict of interest.

References

1. Popkin, B.; Doak, C. The Obesity Epidemic Is a Worldwide Phenomenon. *Nutr. Rev.* **2010**, *56*, 106–114. [CrossRef] [PubMed]
2. Kokkoris, P.; Pi-Sunyer, F. Obesity and endocrine disease. *Endocrinol. Metab. Clin. N. Am.* **2003**, *32*, 895–914. [CrossRef] [PubMed]
3. Tahergorabi, Z.; Khazaei, M.; Moodi, M.; Chamani, E. From obesity to cancer: A review on proposed mechanisms. *Cell Biochem. Funct.* **2020**, *34*, 533–545. [CrossRef] [PubMed]
4. Goran, M.; Ball, G.; Cruz, M. Obesity and risk of type 2 diabetes and cardiovascular disease in children and adolescents. *J. Clin. Endocrinol. Metab.* **2003**, *88*, 1417–1427. [CrossRef]
5. Trayhurn, P.; Beattie, J. Physiological role of adipose tissue: White adipose tissue as an endocrine and secretory organ. *Proc. Nutr. Soc.* **2001**, *60*, 329–339. [CrossRef] [PubMed]
6. Jeyakumar, S.; Yasmeen, R.; Reichert, B.; Ziouzenkova, O. Metabolism of Vitamin A in White Adipose Tissue and Obesity. In *Carotenoids and Vitamin A in Translational Medicine*; CRC Press: Boca Raton, FL, USA, 2013; pp. 23–54.
7. Llewellyn, C.; Wardle, J. Behavioral susceptibility to obesity: Gene–environment interplay in the development of weight. *Physiol. Behav.* **2015**, *152*, 494–501. [CrossRef] [PubMed]
8. Swinburn, B.; Sacks, G.; Hall, K.; McPherson, K.; Finegood, D.; Moodie, M.; Gortmaker, S. The global obesity pandemic: Shaped by global drivers and local environments. *Lancet* **2011**, *378*, 804–814. [CrossRef]
9. Kirkland, J.; Hollenberg, C.; Kindler, S.; Gillon, W. Effects of age and anatomic site on preadipocyte number in rat fat depots. *J. Gerontol.* **1994**, *49*, B31–B35. [CrossRef]
10. Wang, Y.; Sul, H. Pref-1 regulates mesenchymal cell commitment and differentiation through Sox9. *Cell Metab.* **2009**, *9*, 287–302. [CrossRef]
11. Lagathu, C.; Christodoulides, C.; Virtue, S.; Cawthorn, W.; Franzin, C.; Kimber, W.; Nora, E.; Campbell, M.; Medina-Gomez, G.; Cheyette, B.; et al. Dact1, a nutritionally regulated preadipocyte gene, controls adipogenesis by coordinating the Wnt/β-catenin signaling network. *Diabetes* **2009**, *58*, 609–619. [CrossRef]
12. Reichert, B.; Yasmeen, R.; Jeyakumar, S.; Yang, F.; Thomou, T.; Alder, H.; Duester, G.; Maiseyeu, A.; Mihai, G.; Harrison, E.; et al. Concerted action of aldehyde dehydrogenases influences depot-specific fat formation. *Mol. Endocrinol.* **2011**, *25*, 799–809. [CrossRef] [PubMed]

13. Chartoumpekis, D.; Zaravinos, A.; Ziros, P.; Iskrenova, R.; Psyrogiannis, A.; Kyriazopoulou, V.; Habeos, I. Differential expression of microRNAs in adipose tissue after long-term high-fat diet-induced obesity in mice. *PLoS ONE* **2012**, *7*, e34872. [CrossRef] [PubMed]
14. Wang, L.; Shang, C.; Pan, H.; Yang, H.; Zhu, H.; Gong, F. MicroRNA Expression Profiles in the Subcutaneous Adipose Tissues of Morbidly Obese Chinese Women. *Obes. Facts* **2021**, *14*, 78–92. [CrossRef] [PubMed]
15. Zhang, P.; Du, J.; Wang, L.; Niu, L.; Zhao, Y.; Tang, G.; Jiang, Y.; Shuai, S.; Bai, L.; Li, X. MicroRNA-143a-3p modulates preadipocyte proliferation and differentiation by targeting MAPK7. *Biomed. Pharmacother.* **2018**, *108*, 531–539. [CrossRef] [PubMed]
16. Xi, F.; Wei, C.; Xu, Y.; Ma, L.; He, Y.; Shi, X.; Yang, G.; Yu, T. MicroRNA-214-3p Targeting Ctnnb1 Promotes 3T3-L1 Preadipocyte Differentiation by Interfering with the Wnt/β-Catenin Signaling Pathway. *Int. J. Mol. Sci.* **2019**, *20*, 1816. [CrossRef] [PubMed]
17. Fang, W.; Jia, Y.; Wang, P.; Yang, Q.; Chang, Z. Identification and profiling of *Cyprinus carpio* microRNAs during ovary differentiation by deep sequencing. *BMC Genom.* **2017**, *18*, 333.
18. Pitto, L.; Ripoli, A.; Cremisi, F.; Rainaldi, G. microRNA (interference) networks are embedded in the gene regulatory networks. *Cell Cycle* **2008**, *7*, 2458–2461. [CrossRef]
19. Wang, J.; Shao, J.; Li, Y.; Elzo, M.A.; Jia, X.; Lai, S. Genome-wide identification and characterization of perirenal adipose tissue microRNAs in rabbits fed a high-fat diet. *Biosci. Rep.* **2021**, *41*, BSR20204297. [CrossRef]
20. Youssef, E.; Elfiky, A.; BanglySoliman; Abu-Shahba, N.; Elhefnawi, M. Expression profiling and analysis of some miRNAs in subcutaneous white adipose tissue during development of obesity. *Genes. Nutr.* **2020**, *15*, 8. [CrossRef]
21. Lin, Z.; Tang, Y.; Li, Z.; Li, J.; Yu, C.; Yang, C.; Liu, L.; Wang, Y.; Liu, Y. miR-24-3p Dominates the Proliferation and Differentiation of Chicken Intramuscular Preadipocytes by Blocking ANXA6 Expression. *Genes* **2022**, *13*, 635. [CrossRef]
22. Luo, G.; Hu, S.; Lai, T.; Wang, J.; Wang, L.; Lai, S. MiR-9-5p promotes rabbit preadipocyte differentiation by suppressing leptin gene expression. *Lipids Health Dis.* **2020**, *19*, 126. [CrossRef] [PubMed]
23. Ma, W.; Dou, Q.; Ha, X. Let-7a-5p inhibits BMSCs osteogenesis in postmenopausal osteoporosis mice. *Biochem. Biophys. Res. Commun.* **2019**, *510*, 53–58. [CrossRef]
24. Shakerzadeh, J.; Movahedin, M.; Eidi, A.; Roodbari, N.; Parivar, K. Forced Suppression of let-7a-5p in Mouse Blastocysts Improves Implantation Rate. *Reprod. Sci.* **2022**, *29*, 1730–1737. [CrossRef] [PubMed]
25. Luo, Z.; Sun, Y.; Qi, B.; Lin, J.; Chen, Y.; Xu, Y.; Chen, J. Human bone marrow mesenchymal stem cell-derived extracellular vesicles inhibit shoulder stiffness via let-7a/Tgfbr1 axis. *Bioact. Mater.* **2022**, *17*, 344–359. [CrossRef] [PubMed]
26. Zhang, J.; Han, L.; Chen, F. Let-7a-5p regulates the inflammatory response in chronic rhinosinusitis with nasal polyps. *Diagn. Pathol.* **2021**, *16*, 27. [CrossRef]
27. Liu, L.; Zhao, J.; Peng, Y.; Yang, M.; Zhang, L.; Jin, X. miR-let-7a-5p Inhibits Invasion and Migration of Hepatoma Cells by Regulating BZW2 Expression. *Onco Targets Ther.* **2020**, *13*, 12269–12279. [CrossRef]
28. Li, J.; Tang, Q.; Dong, W.; Wang, Y. CircBACH1/let-7a-5p axis enhances the proliferation and metastasis of colorectal cancer by upregulating CREB5 expression. *J. Gastrointest. Oncol.* **2020**, *11*, 1186–1199. [CrossRef]
29. Chen, Z.; Qiu, J.; Gao, Y.; Lu, Q.; Lin, Y.; Shi, H. Study on the mechanism of let-7a-5p in regulating the proliferation in cervical cancer cells. *Clin. Transl. Oncol.* **2022**, *24*, 1631–1642. [CrossRef]
30. Eberlé, D.; Hegarty, B.; Bossard, P.; Ferré, P.; Foufelle, F.; Eberlé, D.; Hegarty, B.; Bossard, P.; Ferré, P.; Foufelle, F. SREBP transcription factors: Master regulators of lipid homeostasis. *Biochimie* **2004**, *86*, 839–848. [CrossRef]
31. Bengoechea-Alonso, M.; Ericsson, J. The phosphorylation-dependent regulation of nuclear SREBP1 during mitosis links lipid metabolism and cell growth. *Cell Cycle* **2016**, *15*, 2753–2765. [CrossRef]
32. Bauer, S.; Wanninger, J.; Schmidhofer, S.; Weigert, J.; Neumeier, M.; Dorn, C.; Hellerbrand, C.; Zimara, N.; Schäffler, A.; Aslanidis, C.; et al. Sterol regulatory element-binding protein 2 (SREBP2) activation after excess triglyceride storage induces chemerin in hypertrophic adipocytes. *Endocrinology* **2011**, *152*, 26–35. [CrossRef] [PubMed]
33. Inoue, M.; Jiang, Y.; Barnes, R.n.; Tokunaga, M.; Martinez-Santibañez, G.; Geletka, L.; Lumeng, C.; Buchner, D.; Chun, T. Thrombospondin 1 mediates high-fat diet-induced muscle fibrosis and insulin resistance in male mice. *Endocrinology* **2013**, *154*, 4548–4559. [CrossRef] [PubMed]
34. Varma, V.; Yao-Borengasser, A.; Bodles, A.; Rasouli, N.; Phanavanh, B.; Nolen, G.; Kern, E.; Nagarajan, R.; Spencer, H.R.; Lee, M.; et al. Thrombospondin-1 is an adipokine associated with obesity, adipose inflammation, and insulin resistance. *Diabetes* **2008**, *57*, 432–439. [CrossRef] [PubMed]
35. Li, Y.; Tong, X.; Rumala, C.; Clemons, K.; Wang, S. Thrombospondin1 deficiency reduces obesity-associated inflammation and improves insulin sensitivity in a diet-induced obese mouse model. *PLoS ONE* **2011**, *6*, e26656. [CrossRef]
36. Peterson, T.; Sengupta, S.; Harris, T.; Carmack, A.; Kang, S.; Balderas, E.; Guertin, D.; Madden, K.; Carpenter, A.; Finck, B.; et al. mTOR complex 1 regulates lipin 1 localization to control the SREBP pathway. *Cell* **2011**, *146*, 408–420. [CrossRef]
37. Zhang, E.; Gao, J.; Wei, Z.; Zeng, J.; Li, J.; Li, G.; Liu, J. MicroRNA-mediated regulation of lipid metabolism in virus-infected *Emiliania huxleyi*. *ISME J.* **2022**, *16*, 2457–2466. [CrossRef]
38. Shao, J.; Wang, J.; Li, Y.; Elzo, M.; Tang, T.; Lai, T.; Ma, Y.; Gan, M.; Wang, L.; Jia, X.; et al. Growth, behavioural, serum biochemical and morphological changes in female rabbits fed high-fat diet. *J. Anim. Physiol. Anim. Nutr.* **2020**, *105*, 345–353. [CrossRef]

International Journal of
Molecular Sciences

Article

miR-889-3p Facilitates the Browning Process of White Adipocyte Precursors by Targeting the *SON* Gene

Wenqiang Sun [1,2,3,†], Xiaoxiao Zhang [1,2,3,†], Xue Bai [1,2,3], Kun Du [1,2,3], Li Chen [1,2,3], Haoding Wang [1,2,3], Xianbo Jia [1,2,3] and Songjia Lai [1,2,3,*]

1 State Key Laboratory of Swine and Poultry Breeding Industry, College of Animal Science and Technology, Sichuan Agricultural University, Chengdu 611134, China; wqsun2021@163.com (W.S.); zxx18137928846@163.com (X.Z.); baixue333work@163.com (X.B.); dukun1672@163.com (K.D.); chenl2020302120@163.com (L.C.); 19848671707@163.com (H.W.); jaxb369@sicau.edu.cn (X.J.)

2 Key Laboratory of Livestock and Poultry Multi-Omics, Ministry of Agriculture and Rural Affairs, College of Animal Science and Technology, Sichuan Agricultural University, Chengdu 611134, China

3 Farm Animal Genetic Resources Exploration and Innovation Key Laboratory of Sichuan Province, Sichuan Agricultural University, Chengdu 611134, China

* Correspondence: laisj5794@sicau.edu.cn

† These authors contributed equally to this work.

Abstract: It is well-established that beige/brown adipose tissue can dissipate stored energy through thermogenesis; hence, the browning of white adipocytes (WAT) has garnered significant interest in contemporary research. Our preceding investigations have identified a marked downregulation of miR-889-3p concurrent with the natural maturation of brown adipose tissue. However, the specific role and underlying molecular mechanisms of miR-889-3p in the browning process of white adipose tissue warrant further elucidation. In this research, we initially delved into the potential role of miR-889-3p in preadipocyte growth via flow cytometry and CCK-8 assay, revealing that miR-889-3p can stimulate preadipocyte growth. To validate the potential contribution of miR-889-3p in the browning process of white adipose tissue, we established an in vitro rabbit white adipocyte browning induction, which exhibited a significant upregulation of miR-889-3p during the browning process. RT-qPCR and Western blot analysis indicated that miR-889-3p overexpression significantly amplified the mRNA levels of UCP1, PRDM16, and CIDEA, as well as UCP1 protein levels. Furthermore, miR-889-3p overexpression fostered intracellular triglyceride accumulation. Conversely, the downregulation of miR-889-3p hindered the browning of rabbit preadipocytes. Subsequently, based on target gene prediction and luciferase reporter gene determination, we demonstrated that miR-889-3p directly targets the 3′-UTR region of *SON*. Lastly, we observed that inhibiting *SON* could facilitate the browning of rabbit preadipocytes. In conclusion, our findings suggest that miR-889-3p facilitates the browning process of white adipocyte precursors by specifically targeting the *SON* gene.

Keywords: miR-889-3p; white adipocyte; browned tissue; *SON*; UCP1

1. Introduction

Traditionally, it was believed that the body contained two types of adipose tissue. The first type, white adipose tissue (WAT), originates from Myogenic regulatory factor 5 negative (Myf5-) progenitor cells and stores energy as triglycerides [1]. The second type, brown adipose tissue, differentiates from Myogenic regulatory factor 5 positive (Myf5+) progenitor cells and highly expresses uncoupling protein 1 (UCP1), which is intimately linked to thermogenesis [2]. Beyond these conventional white and brown adipose tissues, research in 1992 discovered the existence of beige adipocytes within the white fat pool [3]. These beige adipocytes, like brown adipocytes, differentiate from Myf5- progenitor cells [3]. They contain numerous small oil droplets, have significantly more mitochondria compared to white adipocytes, and express the UCP1 gene [3]. Therefore, beige adipocytes are also called brown adipocytes. Non-shivering thermogenesis in mammals is conducted not only by classical brown adipocytes but also by these beige adipocytes [4,5]. Furthermore, the stimulation of beige adipocyte activity is linked to weight loss, insulin resistance reduction, and obesity prevention [6,7]. Hence, amplifying the number or activity of thermogenic adipocytes could potentially manipulate fat plasticity and enhance metabolic health. Understanding the molecular mechanism behind WAT browning regulation is therefore of paramount importance.

Small non-coding RNA (miRNA) can regulate gene expression at the post-transcriptional level. Recent studies have highlighted miRNA as a crucial factor in regulating white adipocyte browning [8]. However, research on miR-889-3p has primarily focused on tumor and cancer-related fields. For instance, Zhu et al. found that miR-889-3p inhibits lung cancer cell proliferation and epithelial–mesenchymal transformation by downregulating homologous domain interacting protein kinase 1 (HIPK1) [9]. Zhao et al. discovered that the overexpression of miR-889-3p can target and inhibit HOXB7, thereby impeding cervical cancer cell proliferation, migration, and invasion; enhancing cell apoptosis and radiosensitivity; and weakening tumor growth in vitro and in vivo [10]. However, studies on miR-889-3p in fat thermogenesis are limited, indicating the need for future research to elucidate its potential positive role in this process.

The *SON* DNA-binding protein (*SON*) is a widely expressed 2426-amino acid protein found in nuclear speckles featuring an RNA-binding G-patch domain, a double-stranded RNA binding motif, a forkhead N-terminal region, an arginine/serine-rich region, and amino acid repeats [11]. It facilitates the efficient splicing of challenging transcripts, playing a key role in cell cycle, DNA repair, apoptosis, and epigenetic modification [11]. Essential for cellular functions and genomic stability, *SON* also maintains the nuclear organization and function of pre-mRNA processing factors; its disruption can lead to cell cycle issues [12]. Previous studies on the *SON* protein have focused mainly on cancer [13], with not much attention given to its role in fat metabolism. Therefore, it would be valuable to explore what role *SON* might play in this area.

In this research, we initially examined the expression of miR-889-3p in brown adipose tissues. Using synthetic mimics and inhibitors of miR-889-3p, we characterized its role in browning induction by observing changes in the mRNA and protein expression of heat-producing marker genes and Oil Red O staining. The potential interaction mechanism between miR-889-3p and *SON* was explored through bioinformatics analysis and luciferase reporter gene determination. Finally, we knocked down *SON* using siRNA to further verify its role in the browning of white adipocytes.

2. Results

2.1. miR-889-3p Promotes Rabbit Preadipocyte Proliferation

In a prior study, we collected brown fat from the interscapular of Tianfu black rabbits at 0 days, 15 days, 85 days, and 2 years. We observed that the brown fat tissue at 0 days exhibited a classical brown fat structure with multi-chambered and small fat droplets, while the interscapular fat tissue at 2 years demonstrated a typical white fat structure with a single chamber and large fat droplets [14]. Intriguingly, according to our previously performed RNA-seq data, miR-889-3p was significantly downregulated during the natural growth of brown fat [14]. This finding was corroborated by PCR results (Figure 1A). These observations suggest that miR-889-3p may play a role in the browning of white adipose tissue. However, existing research on miR-889-3p primarily centers on tumor- and cancer-related fields. The effect of miR-889-3p on adipose tissue is still unknown. Our initial step was to investigate the evolutionary conservation of miR-889-3p across different species. Interestingly, we found that the sequence of miR-889-3p was 100% conserved (Figure S1). To delve into the role of miR-889-3p within adipose tissue, we first transfected miR-889-3p into white precursor adipocytes and assessed its transfection efficiency. RT-qPCR results revealed that the expression of miR-889-3p in the mimic group was significantly elevated compared to the control group (Figure 1B). CCK-8 assay results demonstrated that cell viability significantly improved following the overexpression of miR-889-3p (Figure 1C). Furthermore, cell cycle analysis via flow cytometry indicated that the overexpression of miR-889-3p resulted in fewer cells being stalled in the G0/G1 phase and more cells progressing to the S phase (Figure 1D,E). These findings suggest that miR-889-3p can stimulate rabbit preadipocyte proliferation.

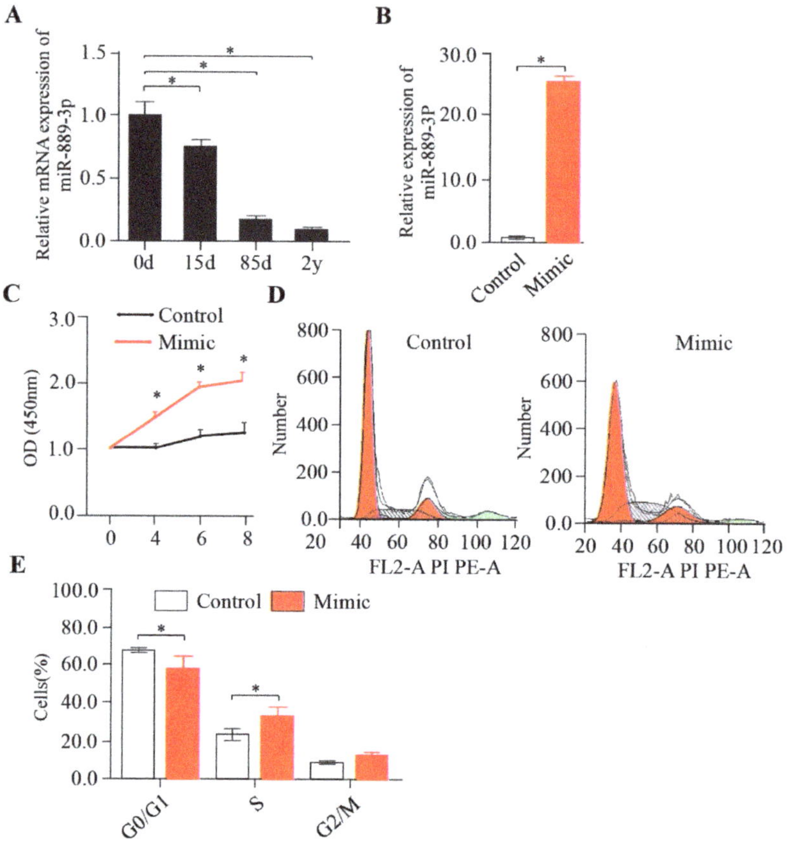

Figure 1. Effect of miR-889-3p on proliferation of rabbit white precursor adipocytes. (**A**) Relative expression of miR-889-3p in white and brown tissues. (**B**) Cell viability after overexpression of miR-889-3p. (n = 5). (**C**) Transfection efficiency of miR-889-3p. (**D**,**E**) Cell cycle changes by flow cytometry and digital conversion histogram. The red color represents the miR-889-3p mimic, while white represents the control group. The difference was significant (* p < 0.05).

2.2. Establishment of the Browning Induction of Rabbit White Preadipocytes In Vitro

To investigate whether miR-889-3p plays a role in the browning of white adipose tissue, we initially established an in vitro browning induction of rabbit white preadipocytes. As depicted in Figure 2A, with the progression of browning induction days, the cell morphology gradually transitioned from spindle-shaped to slender. The lipid droplets within the cells increased over time, and the garland-like lipid droplets around the nucleus clustered together, distributing evenly across the cell culture plate. The quantitative results of Oil Red O staining are presented in Figure 2B. During browning induction, the absorbance (at OD value = 510 nm) displayed a significant upward trend. Moreover, the expression of thermogenic marker genes UCP1, PPARα, PGC1α, PRDM16, C/EBPα, CIDEA, PPARα, and TMEM26 were significantly upregulated, as confirmed by RT-qPCR (Figure 2C–J).

These findings suggest that the browning induction of rabbit white preadipocytes was successfully established. Notably, miR-889-3p was also significantly upregulated during browning induction (Figure 2K).

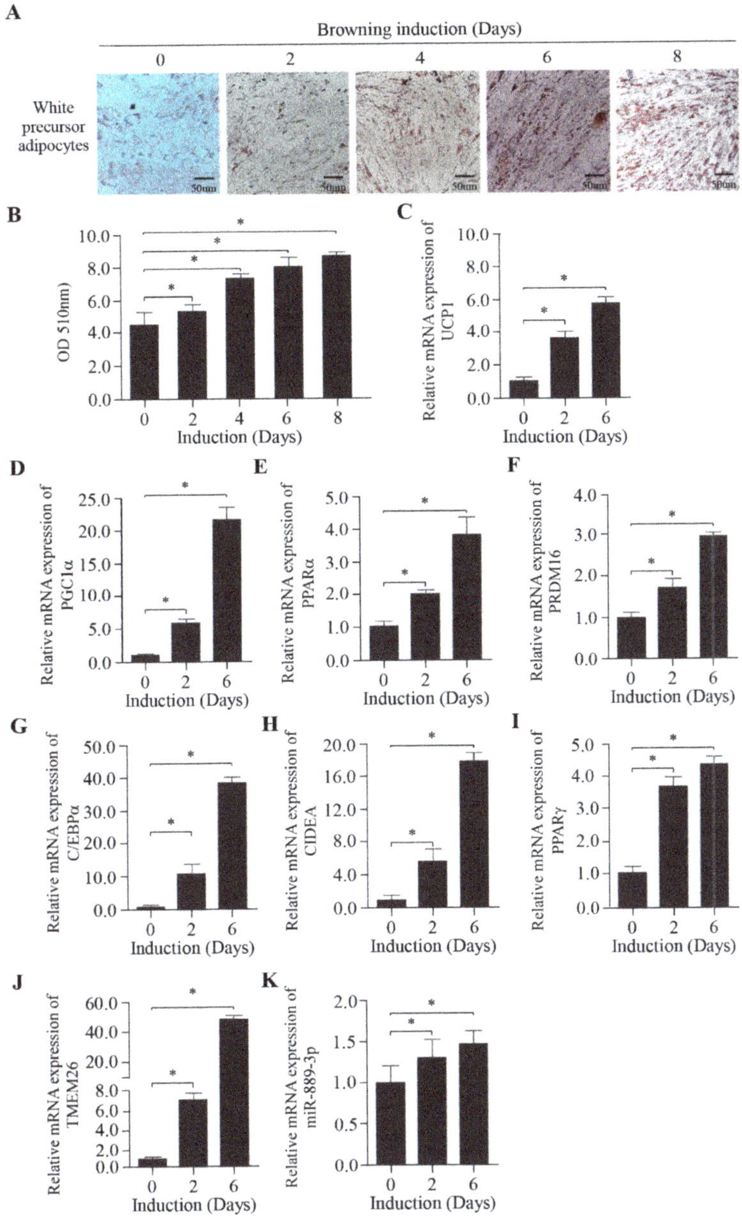

Figure 2. Establishment of browning induction of rabbit white preadipocytes invitro. (**A**) Oil Red O staining pictures of cells on days 0, 2, 4, 6, and 8 of browning induction. (**B**) Quantitative results of Oil Red O staining. (**C–J**) Expression of thermogenic marker genes (UCP1, PPARα, PGC1α, PRDM16, C/EBPα, CIDEA, PPARα, and TMEM26) in browning induction. (**K**) Expression of miR-889-3p in browning induction. The difference was significant (* $p < 0.05$).

2.3. The Overexpression of miR-889-3p Promotes the Browning of Rabbit Precursor White Adipocytes

To examine the impact of miR-889-3p on the browning of rabbit white precursor adipocytes, we transfected miR-889-3p into the cells. The Oil Red O staining results, displayed in Figure 3A, revealed that the number of fat droplets in the mimic group was significantly greater than that in the control group, with a notable increase in the number of large fat droplets. The quantification of fat droplets aligned with these observations (Figure 3B). As shown in Figure 3C,E,F, thermogenic marker genes such as UCP1, PPARα, and PRDM16 were significantly upregulated in the overexpression group compared to the control group during browning induction. Further protein analysis by Western blotting showed that UCP1 protein expression in the mimic group surpassed that in the control group (Figure 3D). These findings suggest that the overexpression of miR-889-3p can facilitate the browning of rabbit precursor white adipocytes.

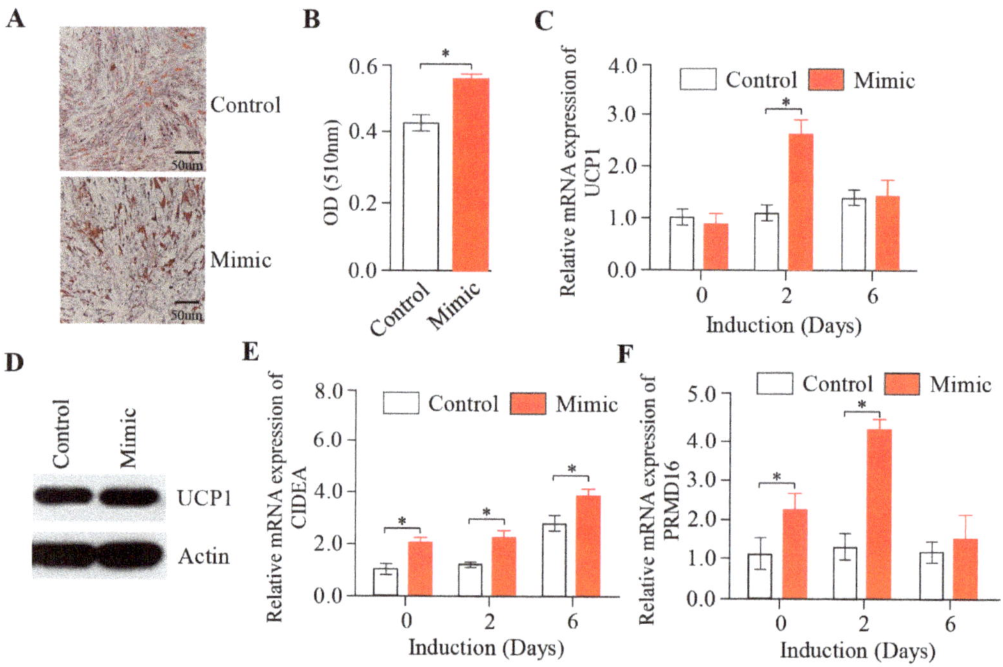

Figure 3. Effect of overexpression of miR-889-3p on browning of rabbit precursor white adipocytes. (**A**) Oil Red O staining results of miR-889-3p mimic and control. (**B**) Oil Red O quantification results. (**C**) Expression of UCP1 after transfection with mimic and control. (**D**) Protein content results of UCP1 in cells on the third day of induction after transfection with mimic and control by Western blotting. (**E**,**F**) Expression of CIDEA and PRDM16 after transfection with mimic and control. The red color represents the miR-889-3p mimic, while white represents the control group. The difference was significant (* $p < 0.05$).

2.4. The Inhibition of miR-889-3p Expression Inhibits the Browning of Rabbit Precursor White Adipocytes

To further validate the influence of miR-889-3p on the browning of rabbit white precursor adipocytes in vitro, we performed transfection with a synthetic miR-889-3p inhibitor to suppress the expression of endogenous miR-889-3p. Following transfection, the number of fat droplets in the inhibitor group was significantly fewer than that in the inhibitor control group (Figure 4A), a finding consistent with the results of fat droplet quantification (Figure 4B). On the sixth day of browning induction, the expression of UCP1, PPARα, and PRDM16 mRNA in the inhibitor group was significantly lower than that in the inhibitor control group (Figure 4C–E). These findings suggest that inhibiting miR-889-3p expression impedes the browning of rabbit precursor white adipocytes.

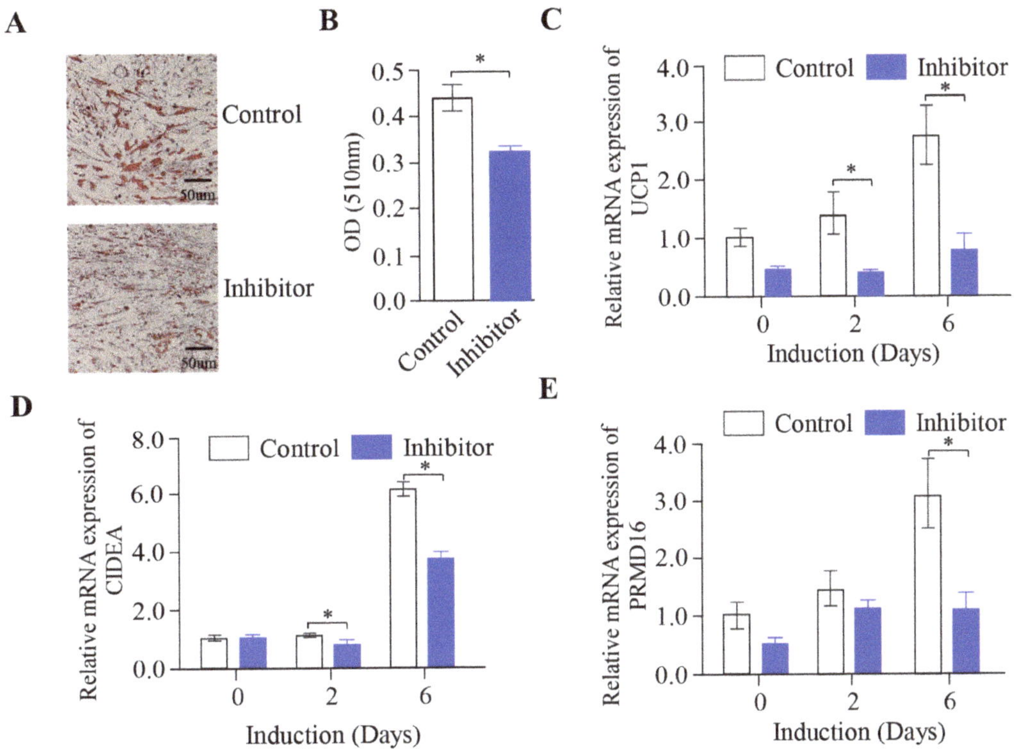

Figure 4. Effect of inhibition of miR-889-3p on browning of rabbit precursor white adipocytes. (**A**) Oil Red O staining results of miR-889-3p inhibitor and control. (**B**) Oil Red O quantification results. (**C–E**) Expression of UCP1, CIDEA, and PRDM16 after transfection with inhibitor and control. The blue color represents the miR-889-3p inhibitor, while white represents the control group. The difference was significant (* $p < 0.05$).

2.5. The Effect of Inhibiting the SON Gene on the Browning of Rabbit White Precursor Adipocytes In Vitro

To uncover the potential mechanism of miR-889-3p in the browning of rabbit precursor white adipocytes, we predicted the target gene of miR-889-3p using the miRanda. Our findings suggested that the *SON* gene could be a potential target. The luciferase reporter assay indicated that miR-889-3p could bind to the *SON* 3′UTR-WT region, but not to the *SON* 3′UTR-MUT region (Figure 5A). Furthermore, the overexpression of miR-889-3p could suppress *SON* gene expression, while the inhibition of miR-889-3p could enhance *SON* gene expression (Figure 5B,C). Further protein analysis by Western blotting showed that *SON* protein expression in the mimic group surpassed that in the control group (Figure 5D). These findings imply that *SON* is a target of miR-889-3p.

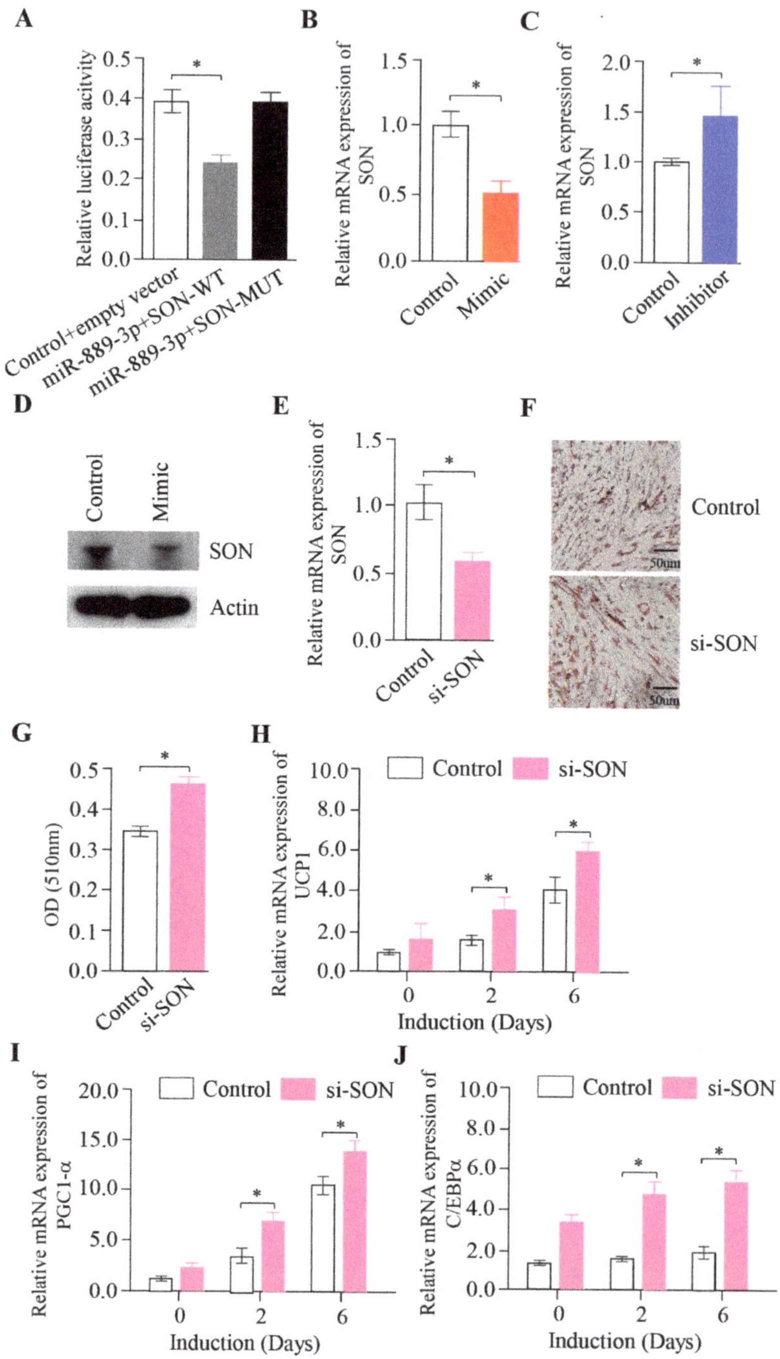

Figure 5. Effect of inhibiting *SON* gene on browning of rabbit white precursor adipocytes in vitro. (**A**) Luciferase assays were performed by co-transfection of *SON* WT and mutant plasmids with miR-8889-3p mimic and control (**B**,**C**) Expression of *SON* transfection with miR-889-3p mimic and

inhibitor, respectively. (**D**) Protein content results of *SON* in cells after transfection with mimic and control by Western blotting. (**E**) Transfection efficiency of si-*SON*. (**F**) Oil Red O staining results of si-*SON* and NC. (**G**) Oil Red O quantification results. (**H–J**) Expression of UCP1, PGC1α, and C/EBPα after transfection with si-*SON* and control. The red color represents the miR-889-3p mimic, blue represents the miR-889-3p inhibitor, pink represents si-SON, and white represents the control group. The difference was significant (* $p < 0.05$).

To examine the influence of *SON* on the browning of rabbit white precursor adipocytes, we utilized synthetic siRNA to inhibit endogenous *SON*. PCR results demonstrated that the *SON* gene content in the si-*SON* group significantly decreased when the *SON* interference analog was transfected, and browning was induced on the third day (Figure 5E). Oil Red O staining indicated that the quantity of lipid droplets in the si-*SON* group was significantly greater than that in the control group (Figure 5F). The quantification outcomes were in line with the staining results, with the si-*SON* group significantly surpassing the control group (Figure 5G). Thermogenic marker gene mRNA expression was assessed by RT-qPCR. The results revealed that UCP1, PGC1α, and CEBPA in the si-*SON* group were significantly higher than those in the control group on the second and sixth days of browning (Figure 5H–J). Taken together, these findings suggest that miR-889-3p can enhance white adipose browning by targeting *SON*.

3. Discussion

In this study, we gathered primary white fat cells from around the kidneys of rabbits and utilized DEX, Ins, IBMX, Rosi, T3, and ISO to induce the browning of white fat precursor cells. The induction method followed in this experiment was based on the browning induction method of Asano et al. [15] The browning of white precursor adipocytes essentially results from a highly sequential and precise regulation of gene expression, indicating that changes in the expression of specific genes can determine the adipocyte browning process. Hence, quantitatively detecting changes in the expression of thermogenic marker genes is a critical means of monitoring the browning process of white precursor adipocytes. In this experiment, we utilized RT-qPCR to measure the expression levels of thermogenic marker genes UCP1, PPARα, C/EBPα, PGC1α, and CIDEA at 0, 2, and 6 days, respectively. We found a significant upregulation in the expression of these thermogenic marker genes, consistent with previous research findings. The experimental results showed that the mRNA levels of heat-producing marker genes UCP1, CIDEA, PGC1α, and C/EBPα detected by PCR were significantly upregulated, aligning with the browning results of 3T3-L1 cells induced by Asano et al. [15]. Brown adipocytes are characterized by a high number of mitochondria within them, a trait shared by beige fat induced from white fat [16]. The mitochondrial membrane contains UCP1, which is almost exclusive to brown and beige fat cells. UCP1 is a proton transport protein that enables protons to leak through the mitochondrial inner membrane and dissipate and is typically used for the ATP synthesis of electrochemical gradients [17]. Over 50% of cell respiration in brown and beige adipocytes is related to the ATP synthesis electrochemical gradient [16,18]. It is noteworthy that compared to white fat, PGC1α is highly expressed in brown fat, which is primarily associated with cAMP-dependent thermogenesis in brown fat [19,20]. cAMP is the key second messenger of brown fat function and stimulates the level of key thermogenic factors. Hence, PGC1α can regulate mitochondrial biogenesis, adaptive thermogenesis, and oxidative metabolism in brown and beige fat. It associates with the PPARα complex and activates UCP1 expression by binding to PPARα response elements in the UCP1 promoter [21]. Moreover, PGC1α can activate the adaptive thermogenic gene program when expressed in white adipocytes [22,23].

Through transcriptome analysis, Seale et al. demonstrated that the expression level of C/EBPα in a brown fat pool was lower than that in a white fat pool. However, during the construction of this experimental induction, C/EBPα expression was significantly upregulated. This may suggest that browning induction did not impede the differentiation process of white adipocytes but occurred concurrently with browning [24]. Compared to white adipocytes, PRDM16 is selectively expressed in brown adipocytes and can enhance mitochondrial gene expression and mitochondrial density. The expression of UCP1 and PGC1α is further enhanced by cAMP, which strengthens the typical characteristic of brown adipocytes: uncoupled respiration. Importantly, PRDM16 activates the brown fat differentiation process in cultured white preadipocytes [24].CIDEA is a lipid droplet-related protein. Previous studies have shown that CIDEA is highly expressed in brown and beige adipocytes, but it was unexpectedly found that CIDEA acts as a heat-generating inhibitor at the molecular level [25]. Therefore, the results demonstrated that white preadipocytes treated with Dex, Ins, IBMX, Rosi, T3, and ISO were closer to mature beige adipocytes and had undergone a certain degree of browning.

miR-889-3p acts as a promoter of preadipocyte browning. Increasing research evidence indicates that miRNA, a type of short-chain non-coding RNA, can regulate the browning process of preadipocytes by targeting key genes that control white precursor fat [26–30]. In this study, we initially established an in vitro browning induction of rabbit white precursor adipocyte and measured the expression levels of miR-889-3p in the browning induction. The findings revealed a significant upregulation trend in miR-889-3p expression within the browning induction. Consequently, we further investigated the specific mechanism of miR-889-3p in the in vitro browning process of rabbit white precursor adipocytes. In previous research, Tan et al. discovered that following the overexpression of miR-669a-5p, there was a significant increase in UCP1, PGC1α, and other genes at both mRNA and protein levels. Therefore, they proposed that miR-669a-5p could promote preadipocyte browning [31]. Sun et al. suggested that miR-27 could inhibit preadipocyte browning by targeting the expression of PRDM16, PPARα, and PGC1α [32]. In our study, the overexpression of miR-889-3p demonstrated an increase in lipid droplets through Oil Red O staining. This was accompanied by an increase in the expression of thermogenic marker genes UCP1, CIDEA, and PRDM16 at the mRNA level. Western blot results showed a corresponding increase in the expression of UCP1 at the protein level. Conversely, the inhibition of miR-889-3p expression yielded opposite results. Thus, we posit that miR-889-3p acts as a promoter of preadipocyte browning.

In this study, the *SON* gene was confirmed to be the target gene of miR-889-3p through the dual-luciferase reporter gene method. The findings demonstrated that miR-889-3p could negatively regulate the expression of the *SON* gene. *SON* is an RNA-binding protein that acts as an mRNA splicing cofactor by promoting the efficient splicing of transcripts that possess weak splice sites [33]. Prior research on *SON* primarily focused on its role in cancer [34–37]. However, recent studies have revealed that when *SON* expression was knocked down, there was a significant decrease in viral particles colocalized in the late endosome labeled by CD63 (cluster differentiation 36) [38]. CD36 is a scavenger receptor involved in the high-affinity tissue uptake of long-chain fatty acids (FAs), and it promotes lipid accumulation and metabolic dysfunction under conditions of excessive fat supply [39]. Febbraio et al., in their study of CD36-deficient mice, found that these mice had normal lifespans, but fasting levels of cholesterol, non-esterified free fatty acids, and triacylglycerol were significantly elevated [40]. Thus, *SON* may play an indirect role in regulating the browning process of precursor adipocytes.

4. Materials and Methods

4.1. Animal Tissues and Sequencing

Male rabbit interscapular fat was taken for 0, 15, and 85 days and 2 years, respectively. The fat was sequenced by microRNA to detect the content of miR-889-3p in the tissue. The surplus tissue in liquid nitrogen was ground for the subsequent extraction of tissue cDNA and the content of miR-889-3p was detected by qPCR.

4.2. Cell Culture and Browning Induction

The primary preadipocytes were collected in the perirenal area of male Tianfu black rabbits using the method described earlier [41]. Briefly, preadipocytes were harvested from the perirenal adipose tissue of Tianfu black rabbits under sterile conditions. The tissues were immersed in PBS with 4% penicillin-streptomycin (Gibco) in a 6-well plate, with connective tissues and blood vessels meticulously excised. After triple washing with PBS, the adipose samples were enzymatically dissociated in a 15 mL centrifuge tube with 0.1% type I collagenase (Gibco, Carlsbad, CA, USA) at 37 °C for 1 h, with agitation at 15 min intervals. Digestion was neutralized with an equal volume of complete growth medium, followed by sequential filtration through 40 nm and 70 nm cell sieves. The isolated preadipocytes were then cultured in a T25 flask at 37 °C and 5% CO_2, with media refreshment every 1–2 days. Cells were passaged upon reaching 70–80% confluence. To stimulate the browning of rabbit white precursor adipocytes, we supplemented a high glucose medium with dexamethasone (5 mM; Sigma, Shanghai, China), insulin (0.5 mg/mL; Sigma, Shanghai, China), isobutyl-methyl-xanthine (0.5 mM; Sigma, Shanghai, China), rosiglitazone (1 mM; Sigma, Shanghai, China), triiodothyronine (T3) (1 nM; Sigma, Shanghai, China), and isoproterenol (ISO) (10 μM; Sigma, Shanghai, China). The specific culture protocol was as follows: inoculate cells into 12-well plates, 6-well plates, or culture dishes, and incubate at 37 °C in a 5% CO_2 cell incubator. Replace the growth medium every two days. Once the cell density reaches 80–90%, discard the growth medium, rinse the cells twice with preheated 37 °C PBS, and introduce a new growth medium containing 10 μg/mL insulin, 0.5 mM IBMX, 1 μM dexamethasone, and 50 nM T3 for a two-day culture period. Subsequently, rinse the cells twice with preheated 37 °C PBS and add a growth medium containing 5 μg/mL insulin, 0.5 mM IBMX, 1 μM rosiglitazone, and 50 nM T3 for a four-day culture period. Then, cleanse the cells twice with preheated 37 °C PBS and introduce a growth medium containing 0.5 mM IBMX, 1 μM rosiglitazone, and 50 nM T3 for an additional two-day culture period. Prior to each cell sample collection, ISO should be added to the culture for an advanced incubation period of 4 h. During this time, monitor the morphological changes in white precursor adipocytes and the formation and accumulation of lipid droplets to assess the adipogenic differentiation potential of white precursor adipocytes and lipid droplet formation.

4.3. RNA Extraction and RT-qPCR

The Foregene (Fuji Biotechnology Co., Ltd., Chengdu, China) RNA extraction kit was used to extract cell RNA. Then, the reverse transcription kit was used to obtain the cell cDNA. The sample was diluted 2–3 times for qRT PCR detection. All primer information is shown in Table 1.

4.4. Oil Red O Staining

After removal of the culture medium, samples were then fixed with 10% formaldehyde at room temperature and rinsed three times with PBS. To stain the lipids in the adipocytes, the cells were treated with filtered Oil Red O solution for 20 min and rinsed twice with PBS. The resulting red-stained lipid droplets were observed microscopically and extracted with isopropanol. The absorbance was measured at 540nm to quantify the residual fat content within the adipocytes.

4.5. Cell Transfection

The preadipocytes were inoculated into a 12-well or 6-well plate and transfected with Lipofectamine 3000 (Invitrogen, Carlsbad, CA, USA) according to the manufacturer's instructions after the cell density reached 90%. The final concentrations of the negative control miRNA mimic (NC miR Mimic), negative control inhibitor (NC inhibitor), and negative control *SON* siRNA (siRNA NC) were 50 nM, 100 nM, and 100 nM, respectively. Each treatment group was repeated three times independently. After transfection, cells were harvested at different time intervals to study the browning of fat.

Table 1. All primer information.

Gene Name	Primer Sequence (5′-3′)
ACTB	GTGCTTCTAGGCGGACTGTT CGGCCACATTGCAGAACTTT
UCP1	CCGGGACAATATGCGAGTGT CACAGTCCACAGTCTGCCTT
PRDM16	CCCCGGAAGAACCACGTCTA TCTCCACCCTGCTGTAGATGG
C/EBPα	CAAGAACAGCAACGAGTACCG GTCACTGGTCAACTCCAGCAC
CIDEA	TAGGGGACAACACGCACTTC CTCTGGCGATTCCCGATCTC
PPARα	ACAGGATTGATGCTGCCGAA AAACGGATGTGCTGCCCTAA
PGC1α	AAAAGCTTGACTGGCGTCAC ACTGCACCACTTGAGTCCAC
TMEM26	TCTATGCCATTCTAATTTTATGGACTTGG GATGAAGACGCTGATTCCAATGTTCC
PPARg	GAGGACATCCAGGACAACC GTCCGTCTCCGTCTTCTTT
SON	ACAGCAGTAGTGCCGAAGTC TGGCTCTAGGGCTTTTGCTC
miR-889-3p	TTAATATCGGACAACCATTGT GGAACGATACAGAGAAGATTAGC
U6	GGAACGATACAGAGAAGATTAGC TGGAACGCTTCACGAATTTGCG

4.6. Western Blotting

CIDEA and UCP1 expression was assessed by Western blot analysis and samples were normalized to β-actin. Protein extraction was blocked with 5% PBS fat-free dried milk at room temperature for 2 h and incubated at 4 °C overnight with anti-CIDEA (1:1000, Abcam), anti-UCP1 (1:1000, Abcam), anti-IgG (1:2000, Abcam), and anti-β-actin (1:3000, Santa Cruz) antibodies, respectively.

4.7. Cell Cycle Analysis

After 72 h of transfection, the cells were harvested first and then the cell suspension was digested. Then, the cells were fixed with ethanol (75%) at 4 °C for 4 h; then, the supernatant was discarded and the cells were incubated with an RNA enzyme containing iodide (PI, 40%, Sigma Aldrich). After washing cells with PBS three times, 5 μL of Annexin V-FITC and 10 μL of PI were added and mixed with cells in the dark for 5 min. Finally, the cell cycle was detected by using FACS Calibur (BD Biosciences, USA), and data analysis was conducted through FACS Diva (BD Biosciences, USA). The experiment was repeated three times.

4.8. Cell Counting Kit-8 Assay

Cell count kit 8 was used for CCK8 determination (Beyotime, Shanghai, China). In short, cells were seeded on 96-well plates. To test cell viability, a CCK8 reagent was added to each well and incubated at 37 °C for 2 h. Then, the absorption was evaluated with a microplate reader at 450 nm.

4.9. Validation of the Target Gene by Double Luciferase Report

miRanda was used to predict the target gene of miR-889-3p. *SON* was found to be the target gene of miR-889-3p. 3′UTR of *SON* containing the miR-889-3p target site was cloned into the Sac I-Xba I site of the pmirGLO vector (Promega, Madison, WI, USA) to construct luciferase report plasmid. Several cloning sites were located downstream of the luciferase gene of firefly. HeLa cells were inoculated into 24-well plates in triplicate. Then, wild-type or mutant pmirGLO *SON* 3′ UTR and synthetic miR-889-3p analogs were co-transfected into HeLa cells to achieve a cell density of 70–80%. The activity of firefly luciferase (luc2) was measured 48 h after transfection and, according to TransDetect® Description of the dual luciferase report kit (Transgen, Beijing, China), standardized to Renilla lucifer.

4.10. Statistical Analyses

The relative expression amount of all genes and miRNAs in this test was compared by the 2-ΔΔCT method. The experimental results were analyzed by single-factor ANOVA using SAS 9.0 software, and multiple comparisons were performed using the Duncan method. The experimental data were expressed as the mean ± SEM. The difference was significant at * $p < 0.05$.

5. Conclusions

In conclusion, our research indicates that miR-889-3p not only stimulates proliferation in rabbit preadipocytes but also promotes the browning of white adipose tissue by targeting *SON*. Further investigations are warranted to elucidate the role and mechanisms of *SON* in the browning process of precursor adipocytes.

Supplementary Materials: The following supporting information can be downloaded at https://www.mdpi.com/article/10.3390/ijms242417580/s1.

Author Contributions: Conceptualization, W.S., X.J., S.L. and X.Z.; software, methodology, validation, X.Z. and X.B.; writing—original draft preparation, W.S. and X.Z.; writing—review and editing, K.D., L.C. and H.W.; funding acquisition, project administration, supervision, S.L. All authors have read and agreed to the published version of the manuscript.

Funding: The current study was funded by National Natural Science Foundation of China [32102530]; national rabbit industry technology system meat rabbit variety improvement [CARS-43-A-2]; high-quality and characteristic rabbit breeding materials and method innovation and new variety breeding (breeding research project), the key R&D project of Sichuan Province [2021YFYZ0033]; Natural Science Foundation of Sichuan Province [2023NSFSC1235]; and National Natural Science Foundation of China Youth Science Foundation Project [32202670].

Institutional Review Board Statement: The animal study protocol was approved by the Institutional Review Board of the Institutional Animal Care and Use Committee from the Sichuan Agricultural University, China. (Certification No. 20200231; Approval time: 24 September 2020).

Informed Consent Statement: Not applicable.

Data Availability Statement: The data underlying this article will be shared on reasonable request to the corresponding author.

Conflicts of Interest: The authors declare that this research was conducted in the absence of any commercial or financial relationships that could be construed as a potential conflict of interest.

References

1. Seale, P.; Bjork, B.; Yang, W.; Kajimura, S.; Chin, S.; Kuang, S.; Scime, A.; Devarakonda, S.; Conroe, H.M.; Erdjument-Bromage, H.; et al. PRDM16 controls a brown fat/skeletal muscle switch. *Nature* **2008**, *454*, 961–967. [CrossRef]
2. Cancello, R.; Zingaretti, M.C.; Sarzani, R.; Ricquier, D.; Cinti, S. Leptin and UCP1 genes are reciprocally regulated in brown adipose tissue. *Endocrinology* **1998**, *139*, 4747–4750. [CrossRef]
3. Cousin, B.; Cinti, S.; Morroni, M.; Raimbault, S.; Ricquier, D.; Penicaud, L.; Casteilla, L. Occurrence of brown adipocytes in rat white adipose tissue: Molecular and morphological characterization. *J. Cell Sci.* **1992**, *103 Pt 4*, 931–942. [CrossRef]
4. Ikeda, K.; Maretich, P.; Kajimura, S. The Common and Distinct Features of Brown and Beige Adipocytes. *Trends Endocrinol. Metab.* **2018**, *29*, 191–200. [CrossRef]
5. Wu, J.; Cohen, P.; Spiegelman, B.M. Adaptive thermogenesis in adipocytes: Is beige the new brown? *Genes Dev.* **2013**, *27*, 234–250. [CrossRef]
6. Kajimura, S.; Spiegelman, B.M.; Seale, P. Brown and Beige Fat: Physiological Roles beyond Heat Generation. *Cell Metab.* **2015**, *22*, 546–559. [CrossRef]
7. Wang, Q.A.; Tao, C.; Gupta, R.K.; Scherer, P.E. Tracking adipogenesis during white adipose tissue development, expansion and regeneration. *Nat. Med.* **2013**, *19*, 1338–1344. [CrossRef]

8. Gharanei, S.; Shabir, K.; Brown, J.E.; Weickert, M.O.; Barber, T.M.; Kyrou, I.; Randeva, H.S. Regulatory microRNAs in Brown, Brite and White Adipose Tissue. *Cells* **2020**, *9*, 2489. [CrossRef]
9. Zhu, Q.; Li, Y.; Li, L.; Guo, M.; Zou, C.; Xu, Y.; Yang, Z. MicroRNA-889-3p restrains the proliferation and epithelial-mesenchymal transformation of lung cancer cells via down-regulation of Homeodomain-interacting protein kinase 1. *Bioengineered* **2021**, *12*, 10945–10958. [CrossRef]
10. Zhao, X.; Dong, W.; Luo, G.; Xie, J.; Liu, J.; Yu, F. Silencing of hsa_circ_0009035 Suppresses Cervical Cancer Progression and Enhances Radiosensitivity through MicroRNA 889-3p-Dependent Regulation of HOXB7. *Mol. Cell. Biol.* **2021**, *41*, e0063120. [CrossRef]
11. Hickey, C.J.; Kim, J.H.; Ahn, E.Y. New discoveries of old SON: A link between RNA splicing and cancer. *J. Cell. Biochem.* **2014**, *115*, 224–231. [CrossRef]
12. Sharma, A.; Takata, H.; Shibahara, K.; Bubulya, A.; Bubulya, P.A. Son is essential for nuclear speckle organization and cell cycle progression. *Mol. Biol. Cell* **2010**, *21*, 650–663. [CrossRef]
13. Furukawa, T.; Tanji, E.; Kuboki, Y.; Hatori, T.; Yamamoto, M.; Shimizu, K.; Shibata, N.; Shiratori, K. Targeting of MAPK-associated molecules identifies SON as a prime target to attenuate the proliferation and tumorigenicity of pancreatic cancer cells. *Mol. Cancer* **2012**, *11*, 88. [CrossRef]
14. Du, K.; Bai, X.; Yang, L.; Shi, Y.; Chen, L.; Wang, H.; Cai, M.; Wang, J.; Chen, S.; Jia, X.; et al. De Novo Reconstruction of Transcriptome Identified Long Non-Coding RNA Regulator of Aging-Related Brown Adipose Tissue Whitening in Rabbits. *Biology* **2021**, *10*, 1176. [CrossRef]
15. Asano, H.; Kanamori, Y.; Higurashi, S.; Nara, T.; Kato, K.; Matsui, T.; Funaba, M. Induction of beige-like adipocytes in 3T3-L1 cells. *J. Vet. Med. Sci.* **2014**, *76*, 57–64. [CrossRef]
16. Prusiner, S.B.; Cannon, B.; Ching, T.M.; Lindberg, O. Oxidative metabolism in cells isolated from brown adipose tissue. 2. Catecholamine regulated respiratory control. *Eur. J. Biochem.* **1968**, *7*, 51–57. [CrossRef]
17. Klingenberg, M. Uncoupling protein—A useful energy dissipator. *J. Bioenerg. Biomembr.* **1999**, *31*, 419–430. [CrossRef]
18. Nedergaard, J.; Cannon, B.; Lindberg, O. Microcalorimetry of isolated mammalian cells. *Nature* **1977**, *267*, 518–520. [CrossRef]
19. Lin, J.; Wu, P.H.; Tarr, P.T.; Lindenberg, K.S.; St-Pierre, J.; Zhang, C.Y.; Mootha, V.K.; Jager, S.; Vianna, C.R.; Reznick, R.M.; et al. Defects in adaptive energy metabolism with CNS-linked hyperactivity in PGC-1alpha null mice. *Cell* **2004**, *119*, 121–135. [CrossRef]
20. Uldry, M.; Yang, W.; St-Pierre, J.; Lin, J.; Seale, P.; Spiegelman, B.M. Complementary action of the PGC-1 coactivators in mitochondrial biogenesis and brown fat differentiation. *Cell Metab.* **2006**, *3*, 333–341. [CrossRef]
21. Liu, C.; Lin, J.D. PGC-1 coactivators in the control of energy metabolism. *Acta Biochim. Biophys. Sin.* **2011**, *43*, 248–257. [CrossRef]
22. Puigserver, P.; Wu, Z.; Park, C.W.; Graves, R.; Wright, M.; Spiegelman, B.M. A cold-inducible coactivator of nuclear receptors linked to adaptive thermogenesis. *Cell* **1998**, *92*, 829–839. [CrossRef] [PubMed]
23. Tiraby, C.; Tavernier, G.; Lefort, C.; Larrouy, D.; Bouillaud, F.; Ricquier, D.; Langin, D. Acquirement of brown fat cell features by human white adipocytes. *J. Biol. Chem.* **2003**, *278*, 33370–33376. [CrossRef] [PubMed]
24. Seale, P.; Kajimura, S.; Yang, W.; Chin, S.; Rohas, L.M.; Uldry, M.; Tavernier, G.; Langin, D.; Spiegelman, B.M. Transcriptional control of brown fat determination by PRDM16. *Cell Metab.* **2007**, *6*, 38–54. [CrossRef] [PubMed]
25. Son, Y.; Choi, C.; Song, C.; Im, H.; Cho, Y.K.; Son, J.S.; Joo, S.; Joh, Y.; Lee, Y.J.; Seong, J.K.; et al. Development of CIDEA reporter mouse model and its application for screening thermogenic drugs. *Sci. Rep.* **2021**, *11*, 18429. [CrossRef] [PubMed]
26. Pan, J.A.; Lin, H.; Yu, J.Y.; Zhang, H.L.; Zhang, J.F.; Wang, C.Q.; Gu, J. MiR-21-3p Inhibits Adipose Browning by Targeting FGFR1 and Aggravates Atrial Fibrosis in Diabetes. *Oxid. Med. Cell. Longev.* **2021**, *2021*, 9987219. [CrossRef] [PubMed]
27. Vonhogen, I.G.C.; El Azzouzi, H.; Olieslagers, S.; Vasilevich, A.; de Boer, J.; Tinahones, F.J.; da Costa Martins, P.A.; de Windt, L.J.; Murri, M. MiR-337-3p Promotes Adipocyte Browning by Inhibiting TWIST1. *Cells* **2020**, *9*, 1056. [CrossRef] [PubMed]
28. Zhang, H.; Zhu, L.; Bai, M.; Liu, Y.; Zhan, Y.; Deng, T.; Yang, H.; Sun, W.; Wang, X.; Zhu, K.; et al. Exosomal circRNA derived from gastric tumor promotes white adipose browning by targeting the miR-133/PRDM16 pathway. *Int. J. Cancer* **2019**, *144*, 2501–2515. [CrossRef]
29. Saha, P.K.; Hamilton, M.P.; Rajapakshe, K.; Putluri, V.; Felix, J.B.; Masschelin, P.; Cox, A.R.; Bajaj, M.; Putluri, N.; Coarfa, C.; et al. miR-30a targets gene networks that promote browning of human and mouse adipocytes. *Am. J. Physiol. Endocrinol. Metab.* **2020**, *319*, E667–E677. [CrossRef]
30. Ye, C.; Duan, J.; Zhang, X.; Yao, L.; Song, Y.; Wang, G.; Li, Q.; Wang, B.; Ai, D.; Wang, C.; et al. Cold-induced Yes-associated-protein expression through miR-429 mediates the browning of white adipose tissue. *Sci. China Life Sci.* **2021**, *64*, 404–418. [CrossRef]
31. Tan, X.; Zhu, T.; Zhang, L.; Fu, L.; Hu, Y.; Li, H.; Li, C.; Zhang, J.; Liang, B.; Liu, J. miR-669a-5p promotes adipogenic differentiation and induces browning in preadipocytes. *Adipocyte* **2022**, *11*, 120–132. [CrossRef] [PubMed]
32. Sun, L.; Trajkovski, M. MiR-27 orchestrates the transcriptional regulation of brown adipogenesis. *Metabolism* **2014**, *63*, 272–282. [CrossRef] [PubMed]
33. Lu, X.; Ng, H.H.; Bubulya, P.A. The role of SON in splicing, development, and disease. *Wiley Interdiscip. Rev. RNA* **2014**, *5*, 637–646. [CrossRef] [PubMed]
34. Chiu, C.; Loth, S.; Kuhlen, M.; Ginzel, S.; Schaper, J.; Rosenbaum, T.; Pietsch, T.; Borkhardt, A.; Hoell, J.I. Mutated SON putatively causes a cancer syndrome comprising high-risk medulloblastoma combined with cafe-au-lait spots. *Fam. Cancer* **2019**, *18*, 353–358. [CrossRef] [PubMed]

35. Kim, J.H.; Baddoo, M.C.; Park, E.Y.; Stone, J.K.; Park, H.; Butler, T.W.; Huang, G.; Yan, X.; Pauli-Behn, F.; Myers, R.M.; et al. SON and Its Alternatively Spliced Isoforms Control MLL Complex-Mediated H3K4me3 and Transcription of Leukemia-Associated Genes. *Mol. Cell* **2016**, *61*, 859–873. [CrossRef] [PubMed]
36. Tokita, M.J.; Braxton, A.A.; Shao, Y.; Lewis, A.M.; Vincent, M.; Kury, S.; Besnard, T.; Isidor, B.; Latypova, X.; Bezieau, S.; et al. De Novo Truncating Variants in SON Cause Intellectual Disability, Congenital Malformations, and Failure to Thrive. *Am. J. Hum. Genet.* **2016**, *99*, 720–727. [CrossRef] [PubMed]
37. Kim, J.H.; Shinde, D.N.; Reijnders, M.R.F.; Hauser, N.S.; Belmonte, R.L.; Wilson, G.R.; Bosch, D.G.M.; Bubulya, P.A.; Shashi, V.; Petrovski, S.; et al. De Novo Mutations in SON Disrupt RNA Splicing of Genes Essential for Brain Development and Metabolism, Causing an Intellectual-Disability Syndrome. *Am. J. Hum. Genet.* **2016**, *99*, 711–719. [CrossRef]
38. Sun, C.T.; Lo, W.Y.; Wang, I.H.; Lo, Y.H.; Shiou, S.R.; Lai, C.K.; Ting, L.P. Transcription repression of human hepatitis B virus genes by negative regulatory element-binding protein/SON. *J. Biol. Chem.* **2001**, *276*, 24059–24067. [CrossRef]
39. Pepino, M.Y.; Kuda, O.; Samovski, D.; Abumrad, N.A. Structure-function of CD36 and importance of fatty acid signal transduction in fat metabolism. *Annu. Rev. Nutr.* **2014**, *34*, 281–303. [CrossRef]
40. Febbraio, M.; Abumrad, N.A.; Hajjar, D.P.; Sharma, K.; Cheng, W.; Pearce, S.F.; Silverstein, R.L. A null mutation in murine CD36 reveals an important role in fatty acid and lipoprotein metabolism. *J. Biol. Chem.* **1999**, *274*, 19055–19062. [CrossRef]
41. Luo, G.; Hu, S.; Lai, T.; Wang, J.; Wang, L.; Lai, S. MiR-9-5p promotes rabbit preadipocyte differentiation by suppressing leptin gene expression. *Lipids Health Dis.* **2020**, *19*, 126. [CrossRef] [PubMed]

International Journal of
Molecular Sciences

MDPI

Article

Domestication Gene *Mlx* and Its Partner *Mondo* Are Involved in Controlling the Larval Body Size and Cocoon Shell Weight of *Bombyx mori*

Xiaoxuan Qin [1,2,3], Liang Jiang [1,2,3], Ping Zhao [1,2,3], Ying Lin [1,2,3], Yi Zhang [4,*] and Qingyou Xia [1,2,3,*]

1 Integrative Science Center of Germplasm Creation in Western China (Chongqing) Science City, Biological Science Research Center, Southwest University, Chongqing 400715, China; qinxiaox@foxmail.com (X.Q.); jiangliang@swu.edu.cn (L.J.); zhaop@swu.edu.cn (P.Z.); ly908@swu.edu.cn (Y.L.)

2 Key Laboratory for Germplasm Creation in Upper Reaches of the Yangtze River, Ministry of Agriculture and Rural Affairs, Chongqing 400715, China

3 Chongqing Key Laboratory of Sericultural Science, Chongqing Engineering and Technology Research Center for Novel Silk Materials, Southwest University, Chongqing 400715, China

4 State Key Laboratory of Resource Insects, Institute of Sericulture and Systems Biology, Southwest University, Chongqing 400715, China

* Correspondence: zhangyi2@swu.edu.cn (Y.Z.); xiaqy@swu.edu.cn (Q.X.)

Abstract: *Bombyx mori* was domesticated from *Bombyx mandarina*. The long-term domestication of the silkworm has brought about many remarkable changes to its body size and cocoon shell weight. However, the molecular mechanism underlying the improvement in the economic characteristics of this species during domestication remains unclear. In this study, we found that a transposable element (TE)—Bm1—was present in the upstream regulatory region of the *Mlx* (Max-like protein X) gene in wild silkworms but not in all domesticated silkworms. The absence of Bm1 caused an increase in the promoter activity and mRNA content of *Mlx*. Mlx and its partner Mondo belong to the bHLHZ transcription factors family and regulate nutrient metabolism. RNAi of *Mlx* and *Mondo* decreased the expression and promoter activity of glucose metabolism-related genes (*trehalose transport* (*Tret*), *phosphofructokinase* (*PFK*), and *pyruvate kinase* (*PK*)), lipogenic genes (*Acetyl-CoA carboxylase* (*ACC*) and *fatty acid synthase* (*FAS*)), and glutamine synthesis gene (*Glutamine synthase 2,* (*GS2*)). Furthermore, the transgenic overexpression of *Mlx* and *Mondo* in the fat body of silkworms increased the larval body size, cocoon shell weight, and egg number, but the silencing of the two genes resulted in the opposite phenotypes. Our results reveal the molecular mechanism of *Mlx* selection during domestication and its successful use in the molecular breeding of *Bombyx mori*.

Keywords: silkworm; domestication; transcription factor; *Mlx*; *Mondo*

1. Introduction

The domesticated silkworm, *Bombyx mori*, is the only truly domesticated lepidopteran insect [1,2], and has important economic value. The gene recombination rate and polymorphism of domesticated silkworms have decreased, and the homozygosity of domesticated silkworms has increased during more than 5000 years of purposeful selection [3,4]. Long-term domestication leads to remarkable changes in body color and size [5], however, the molecular mechanism responsible for these improvements remains unknown. Many works have revealed that transposable elements (TEs) play important roles in building eukaryotic genome evolution and function, after McClintock B. proposed for the first time that transposable genes control the color of maize grains [6–8]. TEs constitute an important part of the genome of most multicellular eukaryotes and can profoundly affect their evolution [9]. For example, in plants, the insertion of hopscotch transposon causes corn straw to change from multiple branches to single branches [10]. Blood orange takes advantage of a Copia retrotransposon adjacent to a gene-encoding Sicilian Ruby-related anthocyanin [11].

Many studies on TE movement and function in metazoans have been documented in *Drosophila* [12,13]. In animals, TE insertions are associated with body size and alteration in the hair color of dogs and cats [14–17]. In *Bombyx mori*, the Taguchi transposon retained during domestication is located in the regulatory region of the ecdysone oxidase gene (*EO*) and enhances its expression, inhibiting the premature maturation of larva [18]. However, more cases of TE in *Bombyx mori* domestication are still needed.

Sugar provides energy for the normal physiological activities of multicellular organisms and is the basis for the survival and reproduction of organisms. There are other regulatory factors along with the insulin pathway involved in the effect of sugar on organisms and cells [19]. Some studies show that Mondo-Mlx can sense phosphorylated hexoses such as glucose-6-phosphate in cells [20–22]. The transcription factors Mlx and Mondo belong to the bHLHZ family [23–26]. MondoA and ChREBP (carbohydrate response element-binding protein, MLXIPL, or MondoB) are paralogs in mammals in which MondoA is mainly expressed in the skeletal muscle [24], while ChREBP plays a role in the liver, adipose tissue, and pancreatic β cells [26–28]. In invertebrates, there is generally a single ortholog called Mondo.

Mondo-Mlx binds to the ChoRE (carbohydrate response element) motif as heterodimers to regulate most of the global glucose-induced transcriptional responses and many of their target genes involved in carbohydrate, lipid, and amino acid metabolism [29–36]. These target genes include *Phosphofructokinase 2* (*PFK2*), *Aldehyde dehydrogenase type III* (*Adh III*), *Glucose-6-phosphate dehydrogenase* (*G6PD*), *FAS*, *ACC*, *GS2*, and so on in flies [29–34,36,37]. Meanwhile, Mondo-Mlx regulates sugar transport GLUT and the carbohydrate digestion gene *Amy-p* [33]. In addition, Mondo-Mlx can also indirectly regulate transcription factors to modulate metabolic homeostasis. The Krüppel-like transcription factor Cabut, a Mondo-Mlx target, inhibits *pepck* and the circadian cycling of metabolic genes to coordinate energy metabolism and the circadian clock in response to sugar sensing [31,38]. Moreover, Mondo-Mlx regulates the Activin ligand Dawdle and Gli-Similar transcription factor Sugarbabe to mediate organismal sugar sensing [33].

In order to clarify the biological roles of Mondo-Mlx, many studies have been carried out using loss-of-function organizations or individuals. The lifespan of C. elegan is shortened by a lack of mml-1(Mondo) or mxl-2(Mlx) [39]. Mlx null mutant flies display dramatic sugar intolerance, including delayed pupation and reduced pupal survival on a high-sugar diet [31]. ChREBP−/− mice have difficulty surviving on a high-sugar diet [40]. ChREBP/MondoA-Mlx is associated with metabolic disease disorders in humans. A novel short ChREBP isoform (ChREBPβ) is related to circulating triglyceride levels, and decontrolled ChREBP is involved in severe obesity in adipose and liver tissue [41–43]. In other reports, ChREBP plays a key role in maintaining fructose tolerance in liver and intestine tissues [44,45]. Moreover, ChREBP/MondoA-Mlx is involved in regulating metabolic processes related to the proliferation of cancer cells and tumorigenesis by deregulating the Myc oncogene [46–49].

The absorption and utilization of carbohydrates by organisms require the coordination of a variety of biological processes, including gene regulation and protein–protein interactions in metabolic networks. The effective utilization and transformation of nutrients in mulberry leaves by silkworms not only form the material basis for growth and development but also influence the yield of cocoons and eggs in silkworms. Silkworms absorb carbohydrates from mulberry leaves to consume them as energy and synthesize storage carbohydrates, such as glycogen, lipids, and chitin. The absorption and utilization of nutrients in mulberry leaves are important traits in sericulture production, and their regulation network exists under the pressure of artificial selection. Previous studies have shown that the key transcription factor Mlx regulating carbohydrate metabolism, is located in the genomic regions of domesticated selection signals, indicating that *Mlx* may have been selected [5,50,51].

In this study, we explored the molecular adaption of the transcription factor Mlx in the process of silkworm domestication, and investigated the physiological functions of the

Mondo-Mlx complex in the fat body of silkworms. We observed the presence of a TE (Bm1) in the upstream regulatory region of *Mlx* in wild silkworms, contrasting its total absence in the domesticated silkworm strains. The absence of Bm1 in the upstream regulatory region may be involved in inhibiting *Mlx*, probably due to removing the repression of Cabut. The upstream region of *Mlx* is selected during domestication and TE is removed in domesticated silkworms. Mlx and Mondo also regulate metabolic genes and transcription factors in silkworm. We generated transgenic knocking down and overexpression strains in silkworm to better study the physiological roles of Mondo-Mlx. Mondo-Mlx is related to the individual size of silkworms, the quality of cocoons, and the number of silkworm eggs. Overexpressing *Mlx* or *Mondo* can increase the content of triglyceride, increase the mass of the fat body, and thus improve the weight of silkworms and economic characteristics. Interfering with *Mlx* or *Mondo* reduces the weight of silkworms and their quality of economic characteristics. Our works suggest that the domestication gene *Mlx* and its partner *Mondo* can be used as molecular targets to promote growth and production in silkworm breeding.

2. Results

2.1. Domestication Selection Analysis of a TE Deletion in the Upstream Regulatory Region of Mlx Gene in Domesticated Silkworm

According to the analysis of the resequencing results of domesticated and Chinese wild silkworm strains, it was predicted that *Mlx* is one of the domestication candidate genes [5,50,51]. To study the differences in and the role of the *Mlx* gene between domesticated and wild silkworms, further research was carried out. We amplified the upstream region of the *Mlx* translation start codon (~1000 bp) from four domesticated and two wild strains. There was a single fragment (~1000 bp) in all domesticated silkworms, however, another longer fragment (~1500 bp) appeared in wild silkworms (Figure 1a). We found that there was no short band in lane 2 of WS1 and no long band in lane 1 of WS2. We speculate that if the genome of wild silkworm is homozygous, there is only a long band, and when the genome is heterozygous, short and long bands will appear. We sequenced the short and long fragments from these six strains. The major difference between the sequences from domesticated and wild silkworms is the existence of an approximately 450 bp (in W1 and W2) fragment. This 450 bp fragment was predicted to be a short interspersed nuclear element (SINE), Bm1 (Figure S1), located in ~380 bp upstream of the transcription start site (TSS) of the *Mlx* gene (Figure 1b). The amplified sequences are shown in Figure S2. The results shown that a retrotransposon was absent in the upstream regulating region of the *Mlx* gene in domesticated silkworms. Thus, the retrotransposon deletion might be adaptive during domestication.

We searched for the domestication selection signatures in the upstream region of *Mlx* translation start codon to determine the deletion of TE in domesticated silkworms (eighteen domesticated and four wild silkworm strains, Table S1). In this region, the polymorphism was lower in the domesticated strains (0.00842 vs. 0.01751), and the polymorphism was also lower in the non-TE sequences (0.02301 vs. 0.06996) (Table 1). This shows that the genetic diversity of domesticated silkworms was simpler than that of wild silkworms in this region. Meanwhile, Tajima's D value was negative (−2.62506) in domesticated silkworms, but it was positive in wild silkworms (0.34000) (Table 1). This indicates that this region was under pressure from directional selection and selective sweep. This evidence indicates that the region that changes from wild silkworms to domesticated silkworms is under pressure from directional selection, and the deletion of Bm1 is the result of artificial domestication selection.

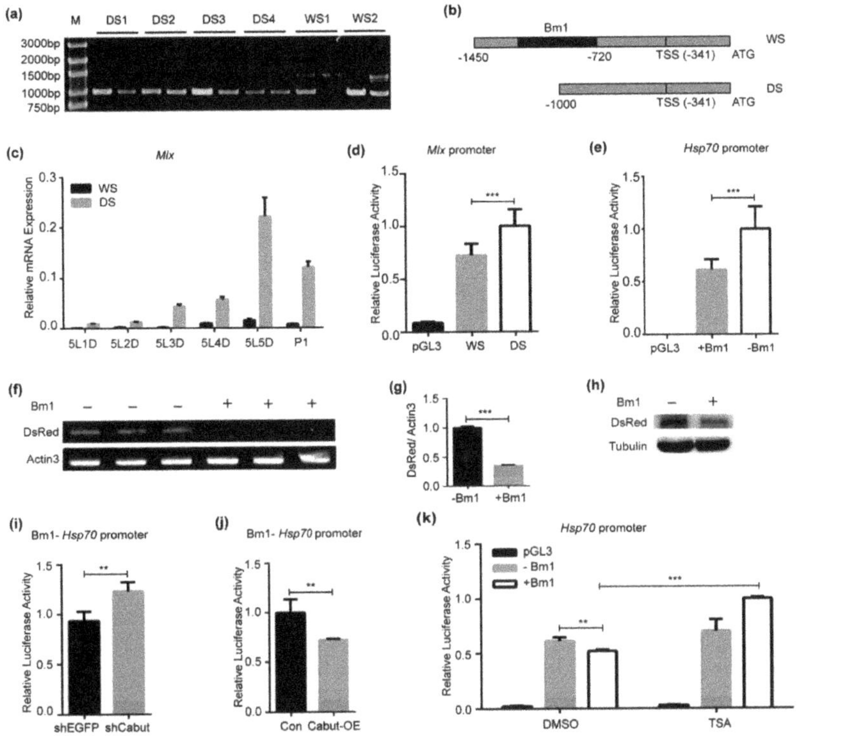

Figure 1. Diversity and activity analyses of *Mlx* upstream regulatory sequence. (**a**) Cloning of *Mlx* upstream regulatory region in domesticated silkworms (DS1-4) and wild silkworms (WS1-2). (**b**) Diagram of *Mlx* upstream regulatory region in domesticated and wild silkworms. (**c**) Comparing the expression of *Mlx* gene in the 5th instar and pupal stage of the fat body in domesticated and wild silkworms. 5L1D: the 1st day of the 5th instar, 5L2D: the 2nd day of the 5th instar, 5L3D: the 3rd day of the 5th instar, 5L4D: the 4th day of the 5th instar, 5L5D: the 5th day of the 5th instar, P1: the 1st day of the pupal stage. (**d**) Activity of *Mlx* upstream regulatory region was examined via a dual-luciferase reporter assay in domesticated and wild silkworms. (**e**) Presence of Bm1 repressed the transcriptional activity of the *Hsp70* promoter. Bm1 was inserted before the luciferase gene driven by the *Hsp 70* promoter. (**f–h**) Bm1 decreased the expression of *DsRed* detected using semiquantitative PCR and Western blot. Bm1 was inserted before the *DsRed* gene, driven by the *A4* promoter. (**i,j**) Cabut affected the transcriptional repression associated with Bm1. Decreasing the expression of *Cabut* increased the transcriptional activity of the *Hsp 70* promoter when Bm1 was present, increasing the expression of *Cabut* has the opposite effect. (**k**) Treatment with 1 μM TSA can release the transcriptional repression associated with Bm1. Error bars show the SD. Significant differences: ** $p < 0.01$, *** $p < 0.001$.

Table 1. Nucleotide diversity and neutral test.

Test Set	5′-Flanking Region (~3 kb)					
	S	π	θw	Tajima's D	Fu and Li's D	Fu and Li's F
Domestic (n = 18)	232	0.00842	0.02202	−2.62506	−4.00135	−4.1804
Wild (n = 8)	123	0.01751	0.01622	0.34000	0.97307	0.92106
With TE (n = 4)	377	0.06996	0.06066	1.5464	1.56333	1.68282
Without TE (n = 4)	126	0.02301	0.02326	−0.35185	−0.27742	−0.31234

2.2. Retrotransposon Bm1 Is Involved in Modulating the Expression of Mlx

To confirm the fact that Bm1 deletion affected the expression of nearby genes, we compared the expression of *Mlx* in the fat body of domesticated and wild silkworms. We found that the expression of *Mlx* was much higher in domesticated silkworms than wild silkworms (Figure 1c). Moreover, the expression of some metabolism-related genes and factors, such as *ACC*, *FAS* and *Cabut*, was higher in domesticated silkworms, which could be regulated by Mlx (Figure S3). Then, we constructed luciferase reporter systems including *Mlx* gene upstream sequences without Bm1 (-Bm1, in domesticated silkworm) or with Bm1 (+Bm1, in wild silkworm). The luciferase activity is higher in the sequences without Bm1 than in those with Bm1 (Figure 1d). To verify the universally and broad transcriptional repression of Bm1, we utilized a luciferase reporter system containing Bm1 before the *Hsp70* promoter. The absence of Bm1 was associated with higher transcriptional activity of the *Hsp70* promoter, while its presence was associated with lower *Hsp70* promoter activity (Figure 1e). Moreover, this association was also supported by elevated transcription and translation of DsRed (Figure 1f–h). The results shows that the presence of Bm1 was associated with transcriptional repression.

We predicted the transcription factors binding to the Bm1 sequence to further clarify how Bm1 is associated with transcriptional repression. We found six GC-rich motifs between the 25th and 35th bp of Bm1, representing a potential KLF10 (or SP1) transcription factor binding site (Table S2). The KLF10/SP1 transcription factor is a kind of transcription factor containing zinc finger structures in mammals [52–54]. KLF10/SP1 can recruit the HDAC1-Sin3A complex and bind to GC-rich DNA region, leading the chromatin concentration to inhibit the transcription of downstream genes [55–59]. The homologous protein of the KLF10/SP1 transcription factor is Cabut in flies [38,60,61]. The luciferase activity was increased after interfering with *Cabut* via RNAi when Bm1 was present, meaning that the transcriptional repression associated with Bm1 was weakened (Figure 1i). In contrast, the overexpression of *Cabut* produced the opposite result (Figure 1j). Acetylation and deacetylation modifications of histones can affect gene expression. HAT (histone acetyltransferase) acetylates the histones, forming an "open" chromatin structure to facilitate transcription. On the contrary, HDAC (histone deacetylase) deacetylates histones and forms a closed structure in chromatin, leading to gene silencing [62–70]. Trichostatin A (TSA) is a commonly used broad-spectrum inhibitor of HDAC family. TSA can strongly and selectively inhibit HDAC. The inhibition of HDAC by TSA can increase histone acetylation levels and lead to increased transcription levels of some genes [71,72]. After treatment with TSA, the lesser repression in the presence of Bm1 in the upstream regulatory region supports our hypothesis that additional deacetylation in Bm1 affects the gene expression level (Figure 1k).

The above results show that *Mlx* and some metabolism-related genes were more highly expressed in domesticated than in wild silkworms. The presence of Bm1 in the regulatory region of *Mlx* is associated with repressing *Mlx* expression. This repression may be mediated by Cabut binding and histone deacetylation.

2.3. Mondo-Mlx Regulates the Expression of Genes Related to Trehalose Transport and Glycolysis

Trehalose is the blood glucose of the silkworm, and its direct energy source [73]. The trehalose transporter Tret can transfer trehalose to maintain the homeostasis of blood glucose. After interfering with *Mlx* or *Mondo* (Figure 2a,b), we found that the expression of *Tret* decreased (Figure 2c), and the transcriptional activity of the *Tret* promoter decreased (Figure 2d). ChIP-PCR results showed that Mlx or Mondo could bind to the *Tret* promoter to directly activate it (Figure 2e,f). This indicates that Mondo-Mlx is involved in the metabolic homeostasis of blood glucose by regulating *Tret* and improving sugar tolerance.

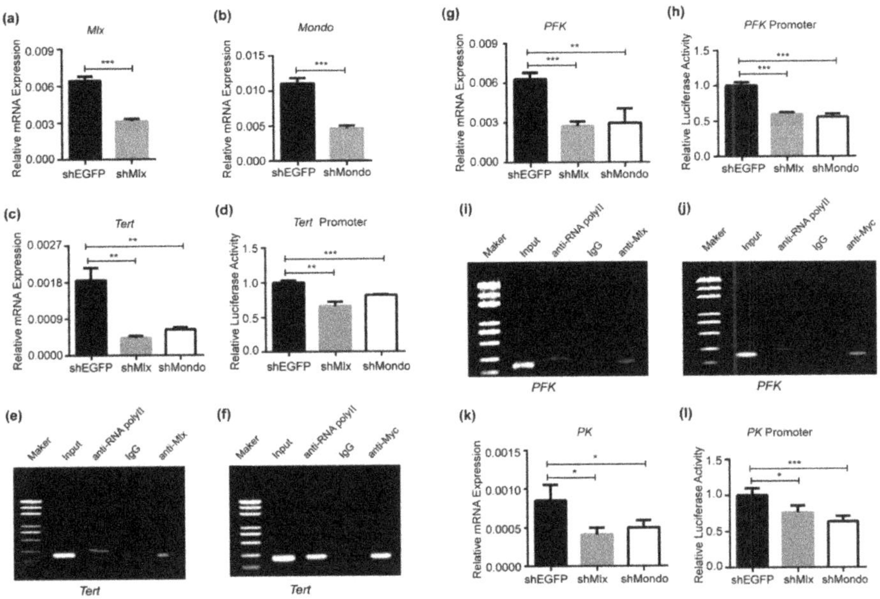

Figure 2. Mlx and Mondo regulate trehalose transport and glycolysis genes. (**a**,**b**) Verifying the interference effects of *Mlx* or *Mondo* in BmNs cells by means of qRT-PCR. (**c**,**d**) Expression and promoter transcriptional activity of *Tret* were reduced via RNAi of *Mlx* or *Mondo* in BmNs cells. (**e**,**f**) Mlx or Mondo being bound to the *Tret* promoter was verified by ChIP-PCR. Specific primers covering the ChoRE element of the *Tret* promoter for ChIP-PCR were designed. Genome fragments were precipitated by Mlx or Myc antibodies after *Mlx* or *Mondo-Myc* overexpression. The IgG antibody was used as negative control and RNA polyII antibody was used as the positive control. (**g**,**h**) Expression and promoter transcriptional activity of *PFK* were decreased via RNAi of *Mlx* or *Mondo* in BmNs cells. (**i**,**j**) ChIP-PCR confirmed Mlx or Mondo being bound to *PFK* promoter. See above for experimental details. (**k**,**l**) RNAi of *Mlx* or *Mondo* decreased the expression and the promoter activity of *PK*. Error bars show SD. Significant differences: * $p < 0.05$, ** $p < 0.01$, *** $p < 0.001$.

Previous research with mammals and flies has demonstrated that the downstream target genes of ChREBP/MondoA-Mlx involve many intermediate enzymes of the glycolysis process [29,32–34]. In silkworm, the glycolytic pathway is like that of flies and mammals. After knocking down *Mlx* and *Mondo*, the expression of *PFK* (Figure 2g) and *PK* (Figure 2k) decreased, which were the key rate limiting enzymes in the glycolysis process. The transcriptional activities of the *PFK* (Figure 2h) and *PK* (Figure 2l) promoter were inhibited with a decrease in the expression of *Mlx* and *Mondo*. ChIP-PCR was used to verify that Mlx and Mondo directly regulate *PFK*. The results shows that Mlx and Mondo can directly bind to the promoter of *PFK* (Figure 2i,j), indicating that Mondo-Mlx is involved in regulating glycolysis by activating *PFK* and *PK*.

2.4. Direct Regulation of Lipogenic Genes by Mondo-Mlx

As one of the three major nutrient types, fatty acids can not only provide energy for organisms but are also the raw materials for synthesizing other substances. ChREBP-Mlx in mammals and flies regulate transcription of the lipogenic genes *ACC* and *FAS* [29–31,33,36,37]. When RNAi was performed for *Mlx* or *Mondo* in BmNs cells, the expression of *ACC* was decreased, which is the rate-limiting enzyme in the fatty acid synthesis pathway (Figure 3a). The result of dual-luciferase activity detection showed that the transcriptional activity of the *ACC* promoter was repressed when the expression of *Mlx* and *Mondo* was knocked down (Figure 3b). Next, we checked that Mlx and Mondo regulated *ACC* directly via ChIP-PCR. The results reveal that Mlx and Mondo can specifically immunoprecipitated the ChoRE motif of the *ACC* promoter, demonstrating that Mlx and Mondo can directly bind to the *ACC* promoter (Figure 3c,d). Furthermore, EMSA was carried out to prove the direct binding for Mlx in vitro. The result indicates that Mlx can bind to the biotin probe in a dose-dependent manner (Figure 3e). This binding can be competitively inhibited by unlabeled cold probes, and a mutated cold probe cannot compete with this binding (Figure 3e). This indicates that Mondo-Mlx regulates *ACC* gene by specifically binding to the *ACC* promoter.

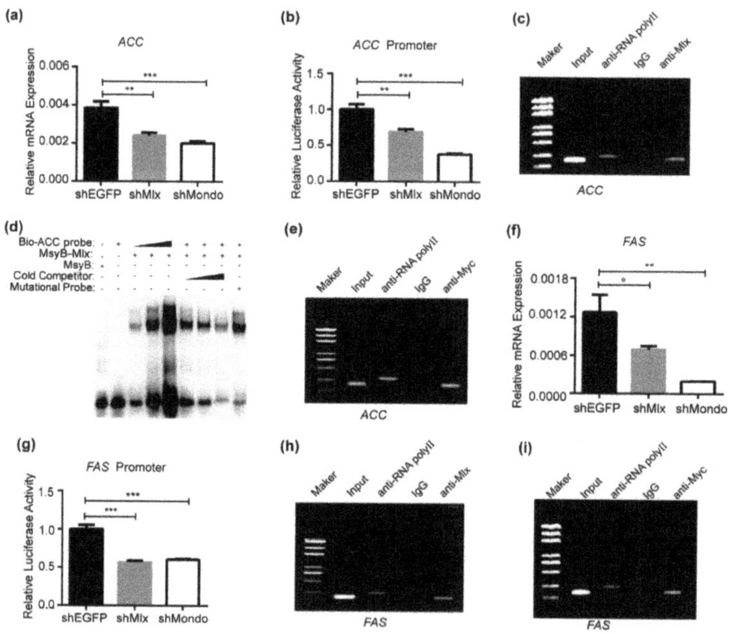

Figure 3. Mondo-Mlx regulates the expression of lipogenic genes. (**a**,**b**) RNAi of *Mlx* or *Mondo* reduced the expression and promoter transcriptional activity of *ACC* in BmNs cells. (**c**,**d**) ChIP-PCR and ENSA with *Mlx* overexpression in BmNs cells. Specific primers covering the ChoRE element of the *ACC* promoter were used for ChIP-PCR. Prokaryotically expressed Mlx-MsyB fusion protein was purified, and specific biotinylated probe based on the ChoRE element of the *ACC* promoter for EMSA was designed. (**e**) ChIP-PCR assay in BmNs cells with *Mondo-Myc* overexpression. Specific primers covering ChoRE element of the *FAS* promoter were used. (**f**,**g**) Expression and promoter transcriptional activity of *FAS* was decreased when *Mlx* or *Mondo* interfered with the shRNA in BmNs cells. (**h**,**i**) ChIP-PCR assay of the direct binding of *Mlx* or *Mondo* to the ChoRE element of the *FAS* promoter in BmNs cells. Error bars show SD. Significant difference: * $p < 0.05$, ** $p < 0.01$, *** $p < 0.001$.

Meanwhile, the expression of *FAS*, another rate-limiting enzyme gene in fatty acid synthesis, was reduced after the RNAi of *Mlx* or *Mondo* (Figure 3f). Similarly, the dual-luciferase reporter assay confirmed that the transcription activity of the *FAS* promoter was decreased (Figure 3g). ChIP-PCR revealed that Mlx or Mondo could bind to the *FAS* promoter (Figure 3h,i). These data together demonstrate that Mondo-Mlx activates the expression of the lipogenic genes *ACC* and *FAS* in *Bombyx mori*.

2.5. Mondo-Mlx Regulates Directly the Expression of Glutamine Synthase 2

The non-essential amino acids glutamate and glutamine play an essential role in amino acid and energy metabolism [74]. Mattila J. found that glutamate/glutamine biosynthetic genes were induced by Mlx in high sugar feeding, and the promoter of the *GS2* was bound by Mlx in flies [33]. When interfering with Mlx or Mondo in BmNs cells, the expression of GS2 was decreased (Figure 4a). The dual-luciferase reporter assay showed that *GS2* promoter activity is regulated by Mlx and Mondo (Figure 4b). After the overexpression of *Mlx* or *Mondo-Myc*, the ChIP-PCR results showed that genome fragments could be immunoprecipitated by Mlx or Myc antibodies, respectively, and amplified a single specific band (Figure 4c,d), indicating that Mondo-Mlx can specifically bind to the *GS2* promoter. The above results show that Mondo-Mlx can directly regulate the transcription of *GS2*, thereby regulating the synthesis of glutamine. Taken together, the biosynthesis of glutamine needs to be coordinated with sugar metabolism in order to maintain the optimal growth state of silkworm. Mondo-Mlx is a central regulator in this process.

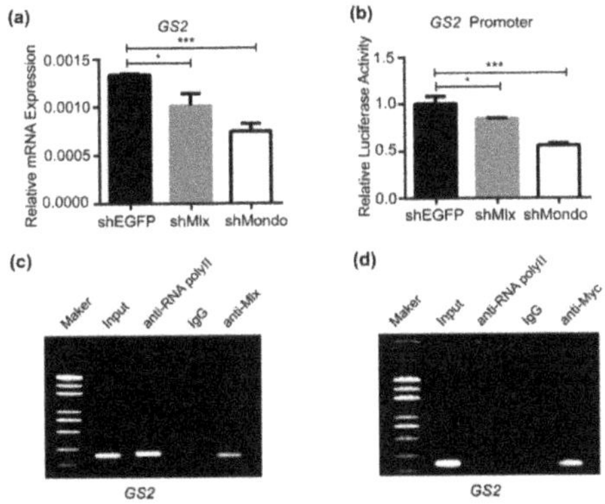

Figure 4. Mlx and Mondo are involved in regulation of *GS2* expression. (**a**,**b**) RNAi of *Mlx* or *Mondo* reduced the expression and promoter transcriptional activity of *GS2* in BmNs cells. (**c**,**d**) Binding of Mlx or Mondo to the ChoRE element of the *GS2* promoter was examined by means of ChIP-PCR. Error bars show SD. Significant differences: * $p < 0.05$, *** $p < 0.001$.

2.6. Mondo-Mlx Controls Cabut, Cycle and FAMeT5 Expression

In *Drosophila*, Mondo-Mlx activates the expression of Krüppel-like transcription factor Cabut [31,38], which can inhibit the expression of lipolysis, glycogenesis and glycerol metabolism-related genes such as *Brummer* (*Bmm*), *Phosphoenolpyruvate carboxykinase* (*PEPCK*), *Fructose-1,6-bisphophase* (*FBP*), and *Glycerol kinase* (*Gyk*) [38]. In addition, *Cabut* is regulated by the core circadian clock protein Cycle [38] which can link the metabolism regulation with the circadian rhythm. In BmNs cells, Mlx and Mondo were knocked down. The qRT-PCR result showed that the expression of *Cabut* decreased with knocking down *Mlx* or *Mondo* (Figure 5a), and the transcriptional activity of *Cabut* promoter was reduced when Mlx or Mondo was decreased (Figure 5b). Moreover, ChIP-PCR analysis revealed that Mondo-Mlx can bind the *Cabut* promoter to directly regulate Cabut (Figure 5c,d) in silkworms.

In order to find new downstream target genes in silkworms, *Mlx* or *Mondo* was knocked down in BmNs cells, and samples were sent for transcriptome sequencing. The results show that the knock down of *Mlx* has 89 upregulated genes and 217 downregulated genes (Table S3), while the RNAi of *Mondo* has 85 upregulated genes and 228 downregulated genes (Table S4). There are 49 genes co-activated and 116 genes co-inhibited by Mlx and Mondo (Figure S4, Table S5). We found that the core circadian clock gene *Cycle* (Table S4, BGIBMGA003869) is one of the genes regulated by Mondo. When knocking down Mondo via shRNA, the expression and promoter transcriptional activity of *Cycle* was reduced (Figure 5e,f). Farnesoic acid O-methyltransferase (FAMeT) is considered to play a catalytic role in the biochemical process of farnesoic acid (FA) being methylated to methyl farnesoate (MF) in arthropods [75–77]. In our results, we found that *FAMeT5* (Tables S3 and S4, BGIBMGA006318) was also a target gene of Mlx and Mondo, and the transcription of *FAMeT5* was regulated by Mlx and Mondo (Figure 5g–i). In summary, we demonstrated that Mondo-Mlx regulated two transcription factor genes (*Cabut* and *Cycle*) and a potential enzyme gene *FAMeT5*.

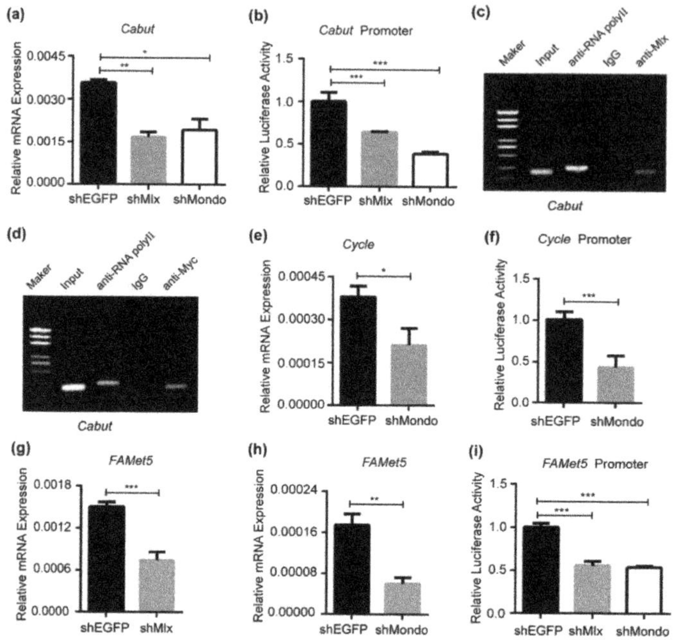

Figure 5. Effects of Mlx and Mondo on *Cabut*, *Cycle* and *FAMeT5*. (**a**) Knock down of *Mlx* or *Mondo* repressed the expression of *Cabut* in BmNs cells. (**b**) When knocking down *Mlx* or *Mondo* in BmNs cells, the transcriptional activity of *Cabut* promoter was inhibited. (**c**,**d**) Binding of Mlx or Mondo to the *Cabut* promoter was examined by means of ChIP-PCR. (**e**) Decreased expression of *Mondo* led to reduced expression of *Cycle* in BmNs cells. (**f**) Transcriptional activity of *Cycle* promoter was inhibited when *Mondo* was reduced in BmNs cells. (**g**,**h**) Expression of *FAMeT5* decreased after the knock down of *Mlx* or *Mondo* in BmNs cells. (**i**) Knock down of *Mlx* or *Mondo* repressed *FAMeT5* promoter activity in BmNs cells. Error bars show SDs. Significant differences: * $p < 0.05$, ** $p < 0.01$, *** $p < 0.001$.

2.7. Manipulating the Expression of Mondo-Mlx Alters Body Size and Cocoon Weight

In order to investigate the biological function of Mondo-Mlx in individuals, we employed transgenic interfered and overexpressed vectors with a piggyBac (pBac) transposon [78], and we obtained overexpressed and knocked down transgenic silkworms (Figure S5). With the overexpression of *Mlx* or *Mondo* in the fat body, their target genes were upregulated (Figures S6 and S7). According to the observations and phenotype statistics, the larva body, pupal and cocoon size of the overexpression strain became larger than those of WT (Figures 6a–d and 7a–d). The cocoon shell weight was increased in the overexpression strain (Figures 6e and 7e). However, the cocoon shell ratio displayed no significant differences (Figures 6f and 7f). After spawning, the number of eggs in the overexpression strain was raised (Figures 6g,h and 7g,h). In addition, the cell size of the fat body was larger in the overexpression strain (Figures 6i,j and 7i,j), and we found that the triglyceride content increased significantly the overexpression strain (Figures 6k and 7k). In sum, Mondo-Mlx can promote fat accumulation, increase fat mass, and improve the economic characteristics of silkworms.

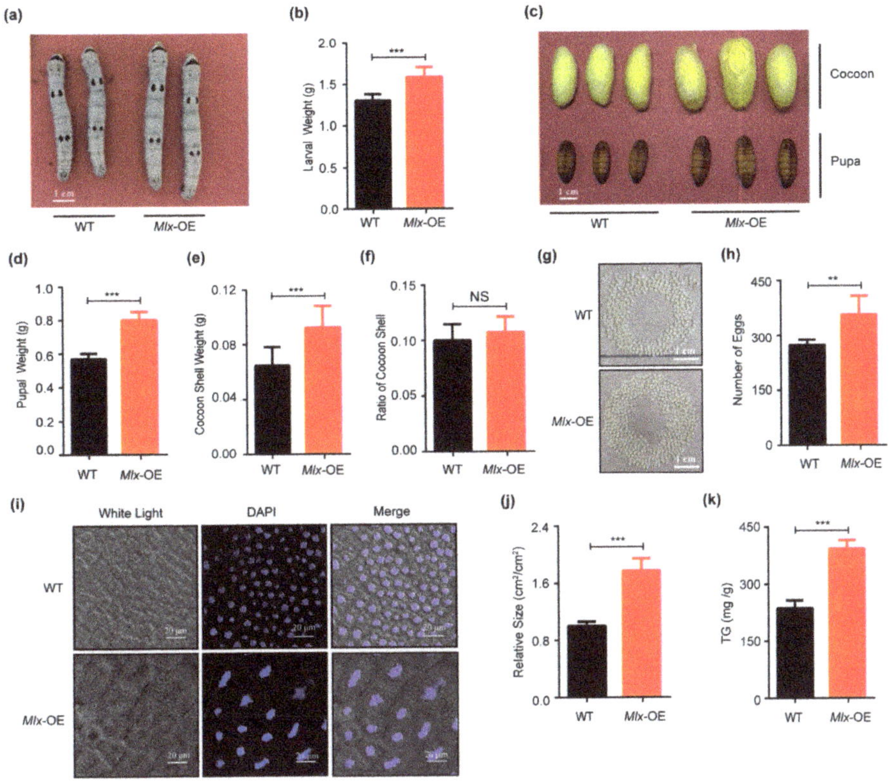

Figure 6. Effects of overexpression of *Mlx* (*Mlx*-OE) in transgenic silkworms. (**a**,**b**) Comparison of body size and larval weight of wild-type silkworm (WT) and *Mlx*-OE on the fifth day of fifth instar. (**c**–**h**): Investigation of pupal and cocoon size (**c**), pupal weight (**d**), cocoon shell weight (**e**), cocoon shell ratio (**f**), and egg number between (**g**,**h**) *Mlx*-OE and WT. (**i**–**k**) Analysis of size (**i**,**j**) and triglyceride content (**k**) of fat body cells in *Mlx*-OE and WT on the fifth day of the fifth instar. Error bars show SD. Significant differences: ** $p < 0.01$; *** $p < 0.001$; NS, no significant difference.

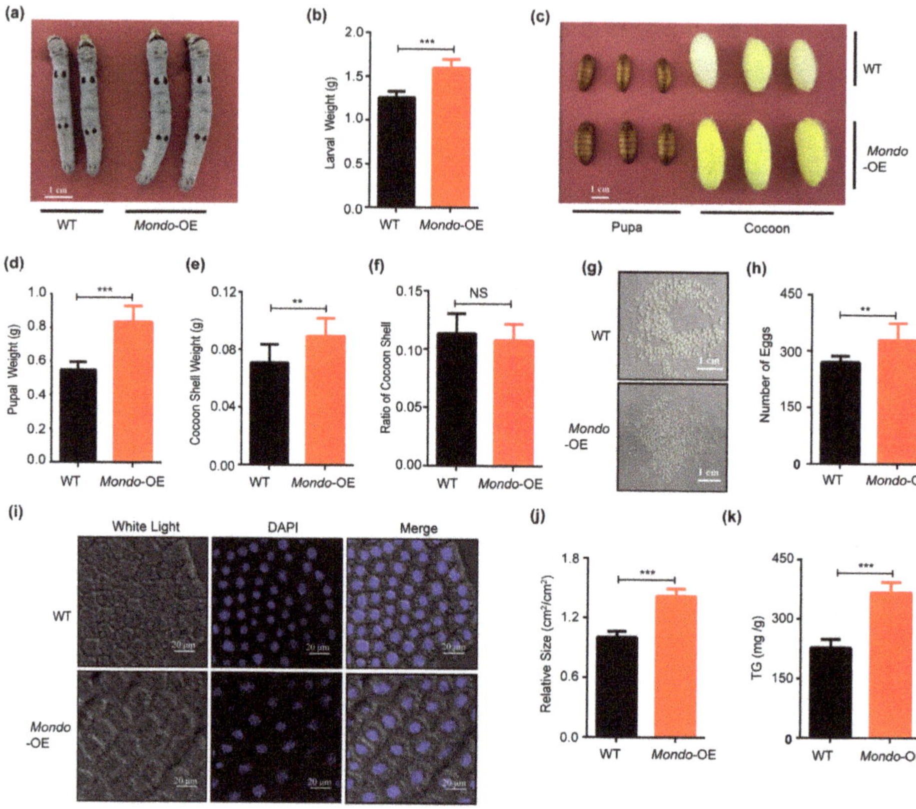

Figure 7. Phenotypic analysis of *Mondo* overexpression (*Mondo*-OE) in transgenic silkworms. (**a–h**) Investigation of larval growth and body weight (**a**,**b**), pupal and cocoon size (**c**), pupal weight, cocoon shell weight and ratio (**e**,**f**), and egg number (**g**,**h**) of *Mondo*-OE and WT. (**i–k**) Detection of size (**i**,**j**) and triglyceride content (**k**) of fat body cells in *Mondo*-OE and WT on the fifth day of the fifth instar. Error bars show SDs. Significant differences: ** $p < 0.01$; *** $p < 0.001$; NS, no significant difference.

Meanwhile, when *Mlx* or *Mondo* was knocked down in the fat body, target genes of Mondo-Mlx were repressed (Figures S8 and S9), and the larvae became smaller in the knocked down strain (Figures 8a,b and 9a,b). The pupae and cocoons also became significantly smaller (Figures 8c and 9c). The weight of the pupa and cocoon shell decreased (Figures 8d,e and 9d,e), but the cocoon shell ratio displayed no significant difference (Figures 8f and 9f). The number of eggs was reduced in knocked down strain (Figures 8g,h and 9g,h). Moreover, the fat body cell size of the knocked down strain was smaller than that of WT (Figures 8i,j and 9i,j), and the triglyceride content of fat body cells decreased in the knocked down strain (Figures 8k and 9k). In sum, the knock down of Mondo-Mlx can decrease fat mass accumulation and negatively impact the economic characteristics of silkworms.

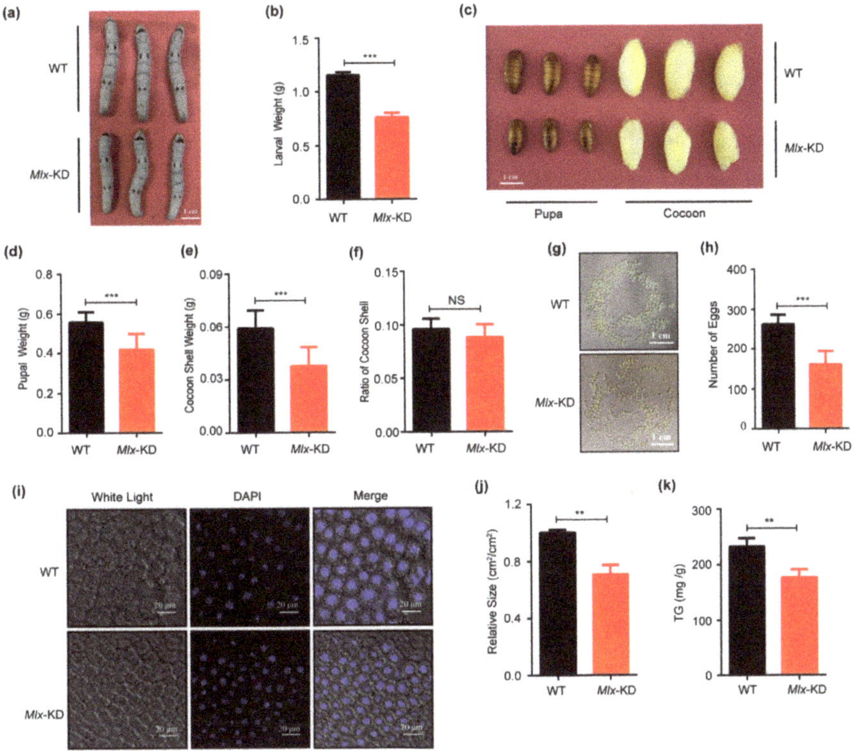

Figure 8. Effects of *Mlx* knock-down (*Mlx*-KD) in transgenic silkworms. (**a**–**h**): Analysis of larval size and weight (**a**,**b**), pupal and cocoon size (**c**), pupal weight (**d**), cocoon shell weight and ratio (**e**,**f**), and egg number (**g**,**h**) between *Mlx*-KD and WT. (**i**–**k**) Test of size (**i**,**j**) and triglyceride content (**k**) of fat body cells in *Mlx*-KD and WT on the fifth day of the fifth instar. Error bars show SD. Significant differences: ** $p < 0.01$; *** $p < 0.001$; NS, no significant difference.

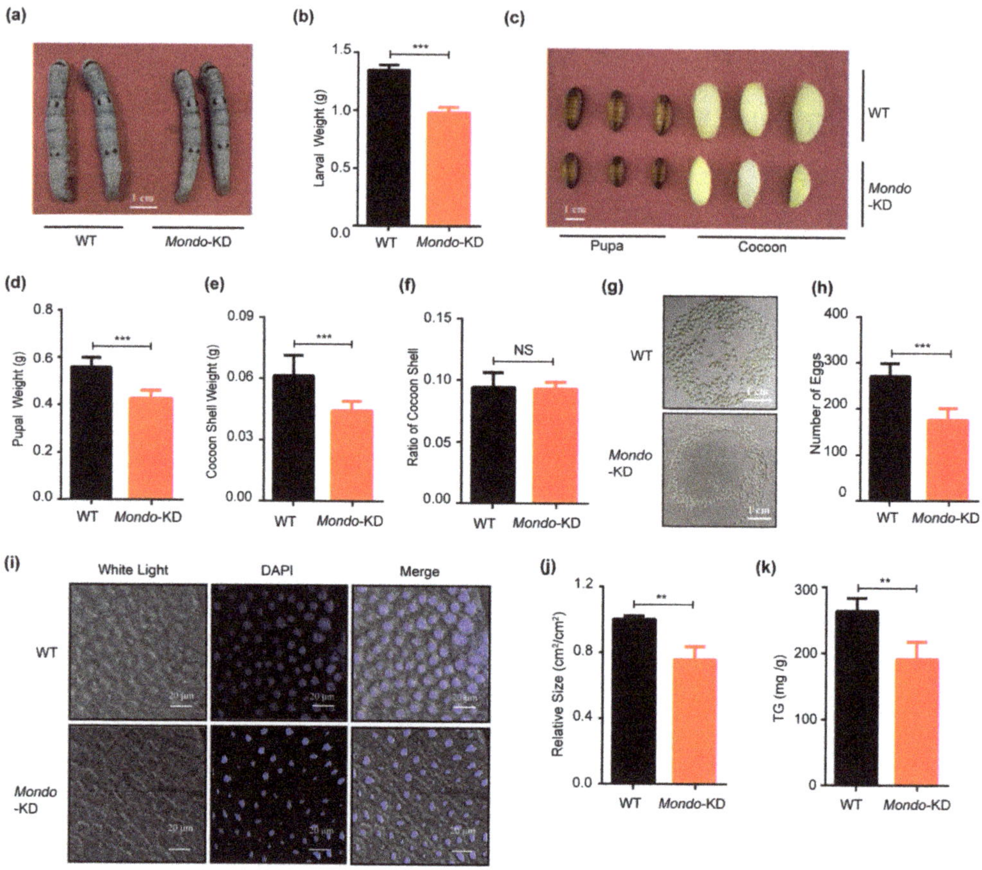

Figure 9. Interference effect of *Mondo* (*Mondo*-KD) in transgenic silkworms. (**a–h**) Larval size and weight (**a**,**b**), pupal and cocoon size (**c**), pupal weight (**d**), cocoon shell weight and ratio (**e**,**f**), and egg number (**g**,**h**) between *Mondo*-KD and WT. (**i–k**) Measurement of size (**i**,**j**) and triglyceride content (**k**) of fat body cells in *Mondo*-KD and WT on the fifth day of the fifth instar. Error bars show SDs. Significant differences: ** $p < 0.01$; *** $p < 0.001$; NS, no significant difference.

3. Discussion

In the long-term process of domestication and selection, major changes have taken place in the biological processes of domesticated silkworms, including energy metabolism. Energy metabolism is the basic physiological process of silkworms and is an important characteristic that affects the production of sericulture. Mlx, a factor of the regulatory network for energy metabolism, is one of 354 candidate genes located in genomic regions of selective signals [50], and is under pressure from artificial domestication and selection in *Bombyx mori*. In this study, we found the domesticated molecular mechanism of Mlx. An adaptive TE (Bm1) deletion in the upstream regulating region of the *Mlx* gene can increase the expression of *Mlx* and the subsequent metabolism regulating genes in domesticated silkworms. The transcriptional repression associated with Bm1 could be partly mediated by Cabut and histone deacetylation. DNA methylation of Bm1 may also inhibit *Mlx* in wild silkworms because a CpG island exists in Bm1 (Figure S10). Based on the above results, we hypothesize that Mlx is involved in improving the individual growth and economic characteristics of silkworms during domestication. Moreover, Bm1 has multiple copies in the whole genome, and some are also inserted adjacent to the regulatory region of coding proteins [79,80], which may also alter the expression of nearby genes. Although this study can provide an example of Bm1 affecting nearby genes, the exact effects of Bm1 relative to other nearby genes must be examined further.

Previous studies have determined that Mondo-Mlx has preserved and physiological significance in model animal mutants (*Drosophila* and *Rattus*) [31,40]. Much evidence has also proven that the physiological function of Mondo-Mlx contributes to the regulation of lipid, carbohydrate and amino acid metabolic genes in mice or flies [20,29–31,33,37,41,81]. Organisms can control the intake and excretion of nutrients to maintain nutrient homeostasis. Our results reveal that Mondo-Mlx can control *Tret*, *PFK*, and *PK* involved in blood glucose transport and glycolysis, potentially to balance sugar levels in circulation and to avoid excessive sugar. It is speculated that FAMeT may not be the key enzyme in the JH biosynthesis pathway in silkworms. FAMeT could be related to the absorption and digestion of nutrients in the larval stage, and could be related to the transportation and excretion of metabolic substances in the pupal stage [82]. Mondo-Mlx regulates *FAMet5*, potentially being involved in nutrient absorption and digestion. We also confirmed that lipogenic genes (*FAS* and *ACC*) are regulated by Mondo-Mlx. Glutamine provides biosynthetic materials for cell growth and division. Glutamine is also an essential nitrogen source for the synthesis of non-essential amino acids and nucleotides [74]. Mondo-Mlx is the intermediate linker of sugar and glutamine metabolism by regulating the *GS2* gene. Moreover, we found that not only does *Cabut* establish coordinating nutrition metabolism and circadian clock, but also core the circadian clock gene *Cycle* is regulated by Mondo-Mlx. Mondo-Mlx could play a central role in coordinating the biological processes of nutrient metabolism and circadian rhythm through controlling *Cabut* and *Cycle*. This further confirms the importance of Mondo-Mlx in energy metabolism and the circadian rhythm pathway. To clarify the physiological function of Mondo-Mlx, we established transgenic overexpressed and knocked down strains using a promoter mainly expressed in the fat body. We found that Mondo-Mlx can promote fat accumulation, increase larval, pupal and cocoon shell weight, and the number of eggs in order to improve economic characteristics and reproduction.

In conclusion, through the comprehensive analysis of Mondo-Mlx, we investigated the adaptive molecular mechanism during the domestication of *Mlx*. In domesticated strains, the expression of *Mlx* increased and Mondo-Mlx upregulated genes related to nutrient metabolism pathways. Thus, Mondo-Mlx may promote the metabolic utilization of nutrients for growth and reproduction. Mondo-Mlx can coordinate the nutrition pathway with other pathways such as circadian rhythm by regulating downstream genes in order to maintain the link between a variety of complex physiological processes. There are currently no reports on the functions of Mondo-Mlx in lepidoptera insects. Our research reveals the biological functions of Mondo-Mlx for the first time in silkworms, although bioinformatic analysis has been reported on Mlx and Mondo in *Spodoptera litura*, such as phylogenetic analysis, chromosomal location, and expression profiles [83]. Our work reveals the adaptive mechanism of gene and the exceptional role of gene regulation during the domestication of *Bombyx mori*, as well as providing new targets for improved strains through genetic breeding.

4. Materials and Methods

4.1. Statistics of Sequence Polymorphism and Neutrality Tests

All of the sequences were combined. The sequences were aligned using DANMAN 6.0 (https://www.lynnon.com/index.html, accessed on 7 February 2024). Polymorphism parameters including S (the number of segregating sites), π (the mean number of nucleotide differences per site), and θ_w (Watterson's estimator of 4Nem) were analyzed using by DnaSP 5.10 (http://www.ub.edu/dnasp/index_v5.html, accessed on 7 February 2024) [84]. To test deviation from the neutral evolution model and to reveal the evolutionary history between the TE deleted or inserted in domesticated and wild silkworm strains, DnaSP 5.10 was utilized to analyze Tajima's D test and Fu and Li's D test.

4.2. Cells Culture and Transfection

The BmNs cell line was maintained at 28 °C and cultured in TC-100 insect medium (Gibco, Grand Island, NY, USA) containing 2 billion units per liter of penicillin and streptomycin (Gibco, Grand Island, NY, USA), and 10% (*v*/*v*) fetal bovine serum (Gibco, Grand Island, NY, USA). Cell transfection was conducted with at last 75% cell density. The cells in each well were incubated for 6–8 h in a mixture with 1 μg of plasmid, 2 μL of X-tremeGENE HP DNA Transfection Reagent (Roche, Mannheim, Germany), and TC-100 insect medium serum- and antibiotic-free, then we replaced the medium with serum and antibiotics for 48 h at 28 °C before mRNA extraction, the luciferase assay or the chromatin immunoprecipitation PCR (ChIP-PCR) assay.

4.3. Total RNA Extraction and Quantitative Real-Time PCR (qPCR)

Cells were collected after transfecting 48 h, and the total RNA was extracted using TRIzol™ reagent (Invitrogen, Carlsbad, CA, USA). Total RNA was used as templates for reverse transcription reaction. cDNA synthesis was performed to using the reverse transcription kit (Promega, Madison, WI, USA). The primers used for qPCR were listed in the Table S6. qPCR reactions were conducted in a 20 μL reaction volume with Takara TB Green® Premix Ex Taq™ II (Takara, Dalian, Liaoning, China) by Applied Biosystems7500 (ABI, Carlsbad, CA, USA). The qPCR conditions were 95 °C for 1 min, and 40 cycles of 95 °C for 30 s, 95 °C for 3 s, and 65 °C for 30 s. The last step was a melting curve analysis (from 65 to 95 °C with a slope of 0.1 °C/s). Each reaction was replicated at least three times.

4.4. Dual Luciferase Reporter Assay

Promoters of downstream genes were cloned into the pMD-19 vector (Takara, Dalian, Liaoning, China) and then inserted into the pGL3-basic vector (Promega, Madison, WI, USA). The primers used for cloning were listed in the Table S7. We constructed a shRNA interference vector containing a specific 21 bp fragment initiated by U6 using the method designed by Tanaka [85] (GFP: GAAGGTTATGTACAGGAAAGT, Mlx: GGCTGACGGACCAAGACAAAT, Mondo: GGATCGACTAGCACAAGTAGC). The RNAi vector was a 21 bp shRNA fragment that was ligated into the U6-pEYFP vector (Solarbio, Beijing, China). The sequence and orientation were confirmed via sequencing. The renilla luciferase gene driven by the IE1 promoter was the reference plasmid. The luciferase activity was standardized according to the renilla luciferase activity. At 12 h post-transfection, the cells were treated with DMSO or TSA (Absin, Shanghai, China) at 1 μM for 24 h, then the cells were collected for the luciferase assay. The luciferase activity was conducted to using the Dual-Luciferase® Reporter Assay System (Promega, Madison, WI, USA) after the transfected cells were washed with phosphate-buffered saline (PBS, Beyotime, Shanghai, China). The data were normalized. The ratio of luciferase activity of the control group (shGFP)was defined as 100%, and the experimental group was adjusted by corresponding multiples. All assays were repeated at least three times.

4.5. Electrophoretic Mobility Shift Assay (EMSA)

All probes (biotin, cold, and mutation probe) for EMSA were synthesized by Sangon Biotech (Sangon, Shanghai, China). The probes used for EMSA were listed in the Table S8. The Mlx purified protein fused with the Msyb tag was expressed at 16 °C, with 0.3% (*w*/*v*) IPTG (Beyotime, Shanghai, China). EMSA was performed using the Chemiluminescent EMSA Kit (Beyotime, Shanghai, China). All the binding reactions, including the negative, sample, cold competition, and mutation cold competition reactions, were carried out according to the EMSA kit instructions. The corresponding amount of 5 × EMSA/gel shift binding buffer and nuclease free water, 1 μg of purified protein, 1 μL of the biotin-unlabeled cold or mutation probe (only competition assay), and 1 μL of the biotin-labeled probe were added into a 10 μL reaction volume. To eliminate the non-specific binding of the probe and protein, we mixed the remaining components before adding the biotin-labeled probe and left the mixture to sit at room temperature for 10 min, or allowed the cold probe to react first. Then, we added the biotin-labeled probe, mixed the mixture well and left it to sit at room temperature for 20 min. After adding the loading buffer into the reaction, electrophoresis was performed with 6% polyacrylamide gel on ice at 100 V, then we stopped electrophoresis until the blue dye bromophenol blue in the loading buffer reached 1/4 of the lower edge of the gel. The samples on the gel were transferred to the positively charged nylon membrane. Then, the membrane was cross-linked at 254 nm UV wavelength and 120 MJ/cm^2 for 45–60 s using the HL-2000-HybriLinker (UVP, Upland, CA, USA). Finally, the biotin-labeled probe was detected via chemiluminescence according to the Super Signal West Pico Chemiluminescent Substrate (Thermo Fisher, Carlsbad, CA, USA).

4.6. ChIP-PCR

The CDS of *Mlx* and *Mondo* were inserted into the A4-p1180 and IE1-pEYFP over-expression plasmid, respectively. BmNs cells were transfected with Mlx and Mondo overexpression plasmids, respectively. After 48 h, cells were collected for ChIP. ChIP was performed according to the manufacturer's instructions for the ChIP Assay Kit (Beyotime, Shanghai, China). The primers used for ChIP-PCR were listed in the Table S8. Briefly, cells were incubated with 1% formaldehyde at room temperature for 10 min and the reaction was terminated with glycine solution. PBS containing 1% PMSF (Beyotime, Shanghai, China) was used to clean the cells three times, and then the cells were lysed with ChIP Dilution Buffer. The genome was broken into 250–500 bp fragments via ultrasound. The supernatant was collected via centrifugation and incubated with antibodies at 4 °C overnight. The supernatant was precipitated with protein G magnetic beads (Pierce, Rockford, IL, USA) and purified for PCR detection.

4.7. RNA-Seq

When interfering with *Mlx* or *Mondo* in BmNs cells, their expression level was significantly reduced by qRT-PCR. We collected control and RNAi cells with Trizol, and then sent them off for transcriptome sequencing and analysis. RNA-seq was performed with Illumina Hiseq technology with paired-end sequencing. Reads were mapped with Ensemble to the standard *B. mori* reference genome (*Bombyx mori*. ASM15162v1. dna. toplevel. fa). See Supplementary Materials for data analysis.

4.8. Transgenic Strains Generated

The 947 bp enhancer Hr3, the 1111 bp promoter of the LP3 gene, CDS with Myc tag of Mlx (or Mondo) and SV40 were amplified. Then, the Hr3, the LP3 promoter, Mlx (or Mondo) and SV40 were orderly conjugated with the pSL1180 vector to form pSL1180-Hr3- [LP3-Mlx (or Mondo)-SV40]. Subsequently, Hr3- [LP3-Mlx (or Mondo)-SV40] was inserted into pBac- [3× P3-Red (or GFP)-SV40] to generate pBac- [3× P3-Red-SV40-Hr3-LP3-Mlx-SV40] and pBac- [3× P3-GFP-SV40-Hr3-LP3-Mondo-SV40]. The primers were listed in the Table S9. The construction of transgenic interference vectors was similar to the above method, except that the CDS was replaced with a specific inverted repeat fragment. We then microinjected the plasmid into the non-diapause eggs of D9L with germline transformation within 2 h of egg-laying. Next, the microinjected eggs were incubated at room temperature and approximately 75% humidity. The hatched larvae (G0) were fed with fresh mulberry leaves and sibmated after mothing. Lastly, the positive G1 eggs were selected using fluorescence microscope (Zeiss, Oberkochen, Baden-Württemberg, Germany).

4.9. Insect Rearing and Statistics of Economic Characteristics

The wild-type and G2 generations were fed with fresh mulberry leaves. Female silkworms were fed separately, and the weight of larvae was investigated every day, with more than 30 silkworms in each group. The pupal weight, whole cocoon weight, cocoon weight, and cocoon shell ratio were measured, and the number of eggs was counted after laying. The female cocoons from the fifth day after pupation were randomly selected. The larval body weight, cocoon shell weight, pupal weight and cocoon shell ratio of WT and transgenic silkworms were measured.

4.10. Measure of Triglyceride Content

The fat body on the fifth day of the fifth instar was taken and freeze-dried. The samples of the same mass were weighed and measured for triglyceride content according to the Micro Triglyceride Content Assay Kit (Solarbio, Beijing, China).

4.11. Statistical Analysis

Statistical significance was determined with the F test or Student's *t*-test, * $p < 0.05$; ** $p < 0.01$; *** $p < 0.001$. Statistical data are presented as the means ± SDs of at least three independent biological replicates.

Supplementary Materials: The following supporting information can be downloaded at: https://www.mdpi.com/article/10.3390/ijms25063427/s1.

Author Contributions: Conceptualization, X.Q., P.Z., Y.Z. and Q.X.; methodology, X.Q. and Y.Z.; formal analysis, X.Q.; investigation, X.Q.; software, X.Q.; writing—original draft preparation, X.Q.; writing—review and editing, Y.Z. and L.J.; visualization, X.Q.; supervision, P.Z. and Y.L.; project administration, P.Z. and Y.L.; funding acquisition, P.Z. and Q.X. All authors have read and agreed to the published version of the manuscript.

Funding: This work was supported by the National Natural Science Foundation of China (No.32030103) and the Natural Science Foundation of Chongqing, China (No. cstc2020jcyj-cxttX0001).

Institutional Review Board Statement: Not applicable.

Informed Consent Statement: Not applicable.

Data Availability Statement: Data are contained within the article and Supplementary Materials.

Conflicts of Interest: The authors declare no conflicts of interest.

References

1. Hubbell, S. *Shrinking the Cat: Genetic Engineering Before We Knew About Genes*; Houghton Mifflin: Boston, MA, USA, 2001; p. 175.
2. Goldsmith, M.R.; Shimada, T.; Abe, H. The genetics and genomics of the silkworm, *Bombyx mori*. *Annu. Rev. Entomol.* **2005**, *50*, 71–100. [CrossRef]
3. Underhill, A.P. Current issues in Chinese neolithic archaeology. *J. World Prehistory* **1997**, *11*, 103–160. [CrossRef]
4. Xiang, Z.H.; Huang, J.T.; Xia, J.G.; Lu, C. *Biology of Sericulture*; China Forestry Publishing House: Beijing, China, 2005.
5. Xiang, H.; Liu, X.; Li, M.; Zhu, Y.; Wang, L.; Cui, Y.; Liu, L.; Fang, G.; Qian, H.; Xu, A.; et al. The evolutionary road from wild moth to domestic silkworm. *Nat. Ecol. Evol.* **2018**, *2*, 1268–1279. [CrossRef] [PubMed]
6. Mc, C.B. Maize genetics. In *Year Book*; Carnegie Institution of Washington: Washington, DC, USA, 1946; Volume 45, pp. 176–186.
7. Mc, C.B. The origin and behavior of mutable loci in maize. *Proc. Natl. Acad. Sci. USA* **1950**, *36*, 344–355.
8. McClintock, B. Intranuclear systems controlling gene action and mutation. *Brookhaven Symp. Biol.* **1956**, *8*, 58–74.
9. Burns, K.H.; Boeke, J.D. Human transposon tectonics. *Cell* **2012**, *149*, 740–752. [CrossRef] [PubMed]
10. Studer, A.; Zhao, Q.; Ross-Ibarra, J.; Doebley, J. Identification of a functional transposon insertion in the maize domestication gene tb1. *Nat. Genet.* **2011**, *43*, 1160–1163. [CrossRef]
11. Butelli, E.; Licciardello, C.; Zhang, Y.; Liu, J.; Mackay, S.; Bailey, P.; Reforgiato-Recupero, G.; Martin, C. Retrotransposons control fruit-specific, cold-dependent accumulation of anthocyanins in blood oranges. *Plant Cell* **2012**, *24*, 1242–1255. [CrossRef]
12. Engels, W.R. The P family of transposable elements in *Drosophila*. *Annu. Rev. Genet.* **1983**, *17*, 315–344. [CrossRef]
13. Lewis, A.P.; Brookfield, J.F.Y. Movement of *Drosophila* melanogaster transposable elements other than P elements in a P-M hybrid dysgenic cross. *Mol. Gen. Genet.* **1987**, *208*, 506–510. [CrossRef]
14. Klutsch, C.F.; de Caprona, M.D. The IGF1 small dog haplotype is derived from Middle Eastern grey wolves: A closer look at statistics, sampling, and the alleged Middle Eastern origin of small dogs. *BMC Biol.* **2010**, *8*, 119. [CrossRef] [PubMed]
15. Clark, L.A.; Wahl, J.M.; Rees, C.A.; Murphy, K.E. Retrotransposon insertion in SILV is responsible for merle patterning of the domestic dog. *Proc. Natl. Acad. Sci. USA* **2006**, *103*, 1376–1381. [CrossRef] [PubMed]
16. Dreger, D.L.; Schmutz, S.M. A SINE insertion causes the black-and-tan and saddle tan phenotypes in domestic dogs. *J. Hered.* **2011**, *102* (Suppl. S1), S11–S18. [CrossRef] [PubMed]
17. David, V.A.; Menotti-Raymond, M.; Wallace, A.C.; Roelke, M.; Kehler, J.; Leighty, R.; Eizirik, E.; Hannah, S.S.; Nelson, G.; Schaffer, A.A.; et al. Endogenous retrovirus insertion in the KIT oncogene determines white and white spotting in domestic cats. *G3* **2014**, *4*, 1881–1891. [CrossRef] [PubMed]

18. Sun, W.; Shen, Y.H.; Han, M.J.; Cao, Y.F.; Zhang, Z. An adaptive transposable element insertion in the regulatory region of the EO gene in the domesticated silkworm, *Bombyx mori*. *Mol. Biol. Evol.* **2014**, *31*, 3302–3313. [CrossRef] [PubMed]
19. Foufelle, F.; Gouhot, B.; Pegorier, J.P.; Perdereau, D.; Girard, J.; Ferre, P. Glucose Stimulation of Lipogenic Enzyme Gene-Expression in Cultured White Adipose-Tissue-a Role for Glucose-6-Phosphate. *J. Biol. Chem.* **1992**, *267*, 20543–20546. [CrossRef] [PubMed]
20. Dentin, R.; Tomas-Cobos, L.; Foufelle, F.; Leopold, J.; Girard, J.; Postic, C.; Ferre, P. Glucose 6-phosphate, rather than xylulose 5-phosphate, is required for the activation of ChREBP in response to glucose in the liver. *J. Hepatol.* **2012**, *56*, 199–209. [CrossRef] [PubMed]
21. Li, M.V.; Chen, W.; Harmancey, R.N.; Nuotio-Antar, A.M.; Imamura, M.; Saha, P.; Taegtmeyer, H.; Chan, L. Glucose-6-phosphate mediates activation of the carbohydrate responsive binding protein (ChREBP). *Biochem. Biophys. Res. Commun.* **2010**, *395*, 395–400. [CrossRef]
22. Stoltzman, C.A.; Kaadige, M.R.; Peterson, C.W.; Ayer, D.E. MondoA senses non-glucose sugars: Regulation of thioredoxin-interacting protein (TXNIP) and the hexose transport curb. *J. Biol. Chem.* **2011**, *286*, 38027–38034. [CrossRef]
23. Billin, A.N.; Eilers, A.L.; Queva, C.; Ayer, D.E. Mlx, a novel Max-like BHLHZip protein that interacts with the Max network of transcription factors. *J. Biol. Chem.* **1999**, *274*, 36344–36350. [CrossRef]
24. Billin, A.N.; Eilers, A.L.; Coulter, K.L.; Logan, J.S.; Ayer, D.E. MondoA, a novel basic helix-loop-helix-leucine zipper transcriptional activator that constitutes a positive branch of a max-like network. *Mol. Cell. Biol.* **2000**, *20*, 8845–8854. [CrossRef] [PubMed]
25. Cairo, S.; Merla, G.; Urbinati, F.; Ballabio, A.; Reymond, A. WBSCR14, a gene mapping to the Williams--Beuren syndrome deleted region, is a new member of the Mlx transcription factor network. *Hum. Mol. Genet.* **2001**, *10*, 617–627. [CrossRef] [PubMed]
26. Yamashita, H.; Takenoshita, M.; Sakurai, M.; Bruick, R.K.; Henzel, W.J.; Shillinglaw, W.; Arnot, D.; Uyeda, K. A glucose-responsive transcription factor that regulates carbohydrate metabolism in the liver. *Proc. Natl. Acad. Sci. USA* **2001**, *98*, 9116–9121. [CrossRef] [PubMed]
27. Wang, H.; Wollheim, C.B. ChREBP rather than USF2 regulates glucose stimulation of endogenous L-pyruvate kinase expression in insulin-secreting cells. *J. Biol. Chem.* **2002**, *277*, 32746–32752. [CrossRef] [PubMed]
28. He, Z.; Jiang, T.; Wang, Z.; Levi, M.; Li, J. Modulation of carbohydrate response element-binding protein gene expression in 3T3-L1 adipocytes and rat adipose tissue. American journal of physiology. *Endocrinol. Metab.* **2004**, *287*, E424–E430.
29. Ma, L.; Tsatsos, N.G.; Towle, H.C. Direct role of ChREBP.Mlx in regulating hepatic glucose-responsive genes. *J. Biol. Chem.* **2005**, *280*, 12019–12027. [CrossRef] [PubMed]
30. Ma, L.; Robinson, L.N.; Towle, H.C. ChREBP*Mlx is the principal mediator of glucose-induced gene expression in the liver. *J. Biol. Chem.* **2006**, *281*, 28721–28730. [CrossRef]
31. Havula, E.; Teesalu, M.; Hyotylainen, T.; Seppala, H.; Hasygar, K.; Auvinen, P.; Oresic, M.; Sandmann, T.; Hietakangas, V. Mondo/ChREBP-Mlx-regulated transcriptional network is essential for dietary sugar tolerance in *Drosophila*. *PLoS Genet.* **2013**, *9*, e1003438. [CrossRef]
32. Ishii, S.; Iizuka, K.; Miller, B.C.; Uyeda, K. Carbohydrate response element binding protein directly promotes lipogenic enzyme gene transcription. *Proc. Natl. Acad. Sci. USA* **2004**, *101*, 15597–15602. [CrossRef]
33. Mattila, J.; Havula, E.; Suominen, E.; Teesalu, M.; Surakka, I.; Hynynen, R.; Kilpinen, H.; Vaananen, J.; Hovatta, I.; Kakela, R.; et al. Mondo-Mlx Mediates Organismal Sugar Sensing through the Gli-Similar Transcription Factor Sugarbabe. *Cell Rep.* **2015**, *13*, 350–364. [CrossRef]
34. Sans, C.L.; Satterwhite, D.J.; Stoltzman, C.A.; Breen, K.T.; Ayer, D.E. MondoA-Mlx heterodimers are candidate sensors of cellular energy status: Mitochondrial localization and direct regulation of glycolysis. *Mol. Cell. Biol.* **2006**, *26*, 4863–4871. [CrossRef] [PubMed]
35. Stoltzman, C.A.; Peterson, C.W.; Breen, K.T.; Muoio, D.M.; Billin, A.N.; Ayer, D.E. Glucose sensing by MondoA:Mlx complexes: A role for hexokinases and direct regulation of thioredoxin-interacting protein expression. *Proc. Natl. Acad. Sci. USA* **2008**, *105*, 6912–6917. [CrossRef] [PubMed]
36. Jeong, Y.S.; Kim, D.; Lee, Y.S.; Kim, H.J.; Han, J.Y.; Im, S.S.; Chong, H.K.; Kwon, J.K.; Cho, Y.H.; Kim, W.K.; et al. Integrated expression profiling and genome-wide analysis of ChREBP targets reveals the dual role for ChREBP in glucose-regulated gene expression. *PLoS ONE* **2011**, *6*, e22544. [CrossRef] [PubMed]
37. Musselman, L.P.; Fink, J.L.; Ramachandran, P.V.; Patterson, B.W.; Okunade, A.L.; Maier, E.; Brent, M.R.; Turk, J.; Baranski, T.J. Role of fat body lipogenesis in protection against the effects of caloric overload in *Drosophila*. *J. Biol. Chem.* **2013**, *288*, 8028–8042. [CrossRef] [PubMed]
38. Bartok, O.; Teesalu, M.; Ashwall-Fluss, R.; Pandey, V.; Hanan, M.; Rovenko, B.M.; Poukkula, M.; Havula, E.; Moussaieff, A.; Vodala, S.; et al. The transcription factor Cabut coordinates energy metabolism and the circadian clock in response to sugar sensing. *EMBO J.* **2015**, *34*, 1538–1553. [CrossRef] [PubMed]
39. Johnson, D.W.; Llop, J.R.; Farrell, S.F.; Yuan, J.; Stolzenburg, L.R.; Samuelson, A.V. The Caenorhabditis elegans Myc-Mondo/Mad complexes integrate diverse longevity signals. *PLoS Genet.* **2014**, *10*, e1004278. [CrossRef] [PubMed]
40. Iizuka, K.; Bruick, R.K.; Liang, G.; Horton, J.D.; Uyeda, K. Deficiency of carbohydrate response element-binding protein (ChREBP) reduces lipogenesis as well as glycolysis. *Proc. Natl. Acad. Sci. USA* **2004**, *101*, 7281–7286. [CrossRef]
41. Herman, M.A.; Peroni, O.D.; Villoria, J.; Schon, M.R.; Abumrad, N.A.; Bluher, M.; Klein, S.; Kahn, B.B. A novel ChREBP isoform in adipose tissue regulates systemic glucose metabolism. *Nature* **2012**, *484*, 333–338. [CrossRef]

42. Kathiresan, S.; Melander, O.; Guiducci, C.; Surti, A.; Burtt, N.P.; Rieder, M.J.; Cooper, G.M.; Roos, C.; Voight, B.F.; Havulinna, A.S.; et al. Six new loci associated with blood low-density lipoprotein cholesterol, high-density lipoprotein cholesterol or triglycerides in humans. *Nat. Genet.* **2008**, *40*, 189–197. [CrossRef]
43. Kooner, J.S.; Chambers, J.C.; Aguilar-Salinas, C.A.; Hinds, D.A.; Hyde, C.L.; Warnes, G.R.; Gomez Perez, F.J.; Frazer, K.A.; Elliott, P.; Scott, J.; et al. Genome-wide scan identifies variation in MLXIPL associated with plasma triglycerides. *Nat. Genet.* **2008**, *40*, 149–151. [CrossRef]
44. Shi, J.H.; Lu, J.Y.; Chen, H.Y.; Wei, C.C.; Xu, X.; Li, H.; Bai, Q.; Xia, F.Z.; Lam, S.M.; Zhang, H.; et al. Liver ChREBP Protects Against Fructose-Induced Glycogenic Hepatotoxicity by Regulating L-Type Pyruvate Kinase. *Diabetes* **2020**, *69*, 591–602. [CrossRef]
45. Zhang, D.; Wang, S.; Ospina, E.; Shabandri, O.; Lank, D.; Akakpo, J.Y.; Zhao, Z.; Yang, M.; Wu, J.; Jaeschke, H.; et al. Fructose Protects Against Acetaminophen-Induced Hepatotoxicity Mainly by Activating the Carbohydrate-Response Element-Binding Protein alpha-Fibroblast Growth Factor 21 Axis in Mice. *Hepatol. Commun.* **2021**, *5*, 992–1008. [CrossRef] [PubMed]
46. Carroll, P.A.; Diolaiti, D.; McFerrin, L.; Gu, H.; Djukovic, D.; Du, J.; Cheng, P.F.; Anderson, S.; Ulrich, M.; Hurley, J.B.; et al. Deregulated Myc requires MondoA/Mlx for metabolic reprogramming and tumorigenesis. *Cancer Cell* **2015**, *27*, 271–285. [CrossRef] [PubMed]
47. Tong, X.; Zhao, F.; Mancuso, A.; Gruber, J.J.; Thompson, C.B. The glucose-responsive transcription factor ChREBP contributes to glucose-dependent anabolic synthesis and cell proliferation. *Proc. Natl. Acad. Sci. USA* **2009**, *106*, 21660–21665. [CrossRef] [PubMed]
48. Vander Heiden, M.G.; Cantley, L.C.; Thompson, C.B. Understanding the Warburg effect: The metabolic requirements of cell proliferation. *Science* **2009**, *324*, 1029–1033. [CrossRef]
49. Zhang, J.; Lu, J.Y.; Yan, X.D.; Su, K.; Cao, D.M.; Zhang, W.P. Fructose intolerance induced by intestinal ChREBP deficiency in mice. *Chin. J. Pathophysiol.* **2021**, *37*, 586–592.
50. Xia, Q.; Guo, Y.; Zhang, Z.; Li, D.; Xuan, Z.; Li, Z.; Dai, F.; Li, Y.; Cheng, D.; Li, R.; et al. Complete resequencing of 40 genomes reveals domestication events and genes in silkworm (*Bombyx*). *Science* **2009**, *326*, 433–436. [CrossRef] [PubMed]
51. Tong, X.; Han, M.J.; Lu, K.; Tai, S.; Liang, S.; Liu, Y.; Hu, H.; Shen, J.; Long, A.; Zhan, C.; et al. High-resolution silkworm pan-genome provides genetic insights into artificial selection and ecological adaptation. *Nat. Commun.* **2022**, *13*, 5619. [CrossRef]
52. McConnell, B.B.; Yang, V.W. Mammalian Kruppel-like factors in health and diseases. *Physiol. Rev.* **2010**, *90*, 1337–1381. [CrossRef]
53. Pearson, R.; Fleetwood, J.; Eaton, S.; Crossley, M.; Bao, S. Kruppel-like transcription factors: A functional family. *Int. J. Biochem. Cell Biol.* **2008**, *40*, 1996–2001. [CrossRef]
54. Suske, G.; Bruford, E.; Philipsen, S. Mammalian SP/KLF transcription factors: Bring in the family. *Genomics* **2005**, *85*, 551–556. [CrossRef] [PubMed]
55. Tetreault, M.P.; Yang, Y.; Katz, J.P. Kruppel-like factors in cancer. Nature reviews. *Cancer* **2013**, *13*, 701–713. [PubMed]
56. Hirota, T.; Kon, N.; Itagaki, T.; Hoshina, N.; Okano, T.; Fukada, Y. Transcriptional repressor TIEG1 regulates Bmal1 gene through GC box and controls circadian clockwork. *Genes Cells Devoted Mol. Cell. Mech.* **2010**, *15*, 111–121. [CrossRef]
57. Lomberk, G.; Urrutia, R. The family feud: Turning off Sp1 by Sp1-like KLF proteins. *Biochem. J.* **2005**, *392 Pt 1*, 1–11. [CrossRef] [PubMed]
58. Shields, J.M.; Yang, V.W. Identification of the DNA sequence that interacts with the gut-enriched Kruppel-like factor. *Nucleic Acids Res.* **1998**, *26*, 796–802. [CrossRef] [PubMed]
59. Spittau, B.; Wang, Z.; Boinska, D.; Krieglstein, K. Functional domains of the TGF-beta-inducible transcription factor Tieg3 and detection of two putative nuclear localization signals within the zinc finger DNA-binding domain. *J. Cell. Biochem.* **2007**, *101*, 712–722. [CrossRef] [PubMed]
60. Brown, J.L.; Grau, D.J.; DeVido, S.K.; Kassis, J.A. An Sp1/KLF binding site is important for the activity of a Polycomb group response element from the *Drosophila* engrailed gene. *Nucleic Acids Res.* **2005**, *33*, 5181–5189. [CrossRef] [PubMed]
61. Belacortu, Y.; Weiss, R.; Kadener, S.; Paricio, N. Transcriptional activity and nuclear localization of Cabut, the *Drosophila* ortholog of vertebrate TGF-beta-inducible early-response gene (TIEG) proteins. *PLoS ONE* **2012**, *7*, e32004. [CrossRef]
62. Grunstein, M. Histone acetylation in chromatin structure and transcription. *Nature* **1997**, *389*, 349–352. [CrossRef]
63. Eberharter, A.; Becker, P.B. Histone acetylation: A switch between repressive and permissive chromatin. Second in review series on chromatin dynamics. *EMBO Rep.* **2002**, *3*, 224–229. [CrossRef]
64. Vidal, M.; Gaber, R.F. RPD3 encodes a second factor required to achieve maximum positive and negative transcriptional states in Saccharomyces cerevisiae. *Mol. Cell. Biol.* **1991**, *11*, 6317–6327. [PubMed]
65. Yang, W.M.; Inouye, C.; Zeng, Y.; Bearss, D.; Seto, E. Transcriptional repression by YY1 is mediated by interaction with a mammalian homolog of the yeast global regulator RPD3. *Proc. Natl. Acad. Sci. USA* **1996**, *93*, 12845–12850. [CrossRef] [PubMed]
66. Yang, W.M.; Yao, Y.L.; Sun, J.M.; Davie, J.R.; Seto, E. Isolation and characterization of cDNAs corresponding to an additional member of the human histone deacetylase gene family. *J. Biol. Chem.* **1997**, *272*, 28001–28007. [CrossRef] [PubMed]
67. Kao, H.Y.; Downes, M.; Ordentlich, P.; Evans, R.M. Isolation of a novel histone deacetylase reveals that class I and class II deacetylases promote SMRT-mediated repression. *Genes Dev.* **2000**, *14*, 55–66. [CrossRef] [PubMed]
68. Kao, H.Y.; Lee, C.H.; Komarov, A.; Han, C.C.; Evans, R.M. Isolation and characterization of mammalian HDAC10, a novel histone deacetylase. *J. Biol. Chem.* **2002**, *277*, 187–193. [CrossRef] [PubMed]
69. Zhou, X.; Richon, V.M.; Rifkind, R.A.; Marks, P.A. Identification of a transcriptional repressor related to the noncatalytic domain of histone deacetylases 4 and 5. *Proc. Natl. Acad. Sci. USA* **2000**, *97*, 1056–1061. [CrossRef] [PubMed]

70. Seto, E.; Yoshida, M. Erasers of histone acetylation: The histone deacetylase enzymes. *Cold Spring Harb Perspect Biol* **2014**, *6*, a018713. [CrossRef] [PubMed]
71. Yoshida, M.; Kijima, M.; Akita, M.; Beppu, T. Potent and specific inhibition of mammalian histone deacetylase both in vivo and in vitro by trichostatin A. *J. Biol. Chem.* **1990**, *265*, 17174–17179. [CrossRef]
72. Graff, J.; Tsai, L.H. The potential of HDAC inhibitors as cognitive enhancers. *Annu. Rev. Pharmacol. Toxicol.* **2013**, *53*, 311–330. [CrossRef]
73. Zhang, G.Z.; Li, M.W.; Lu, C.; Ai, J.W.; Bi, L.H.; Chen, J.; Chen, P.; Chen, T.; Cheng, D.J. *Chinese Sericulture*; Shanghai Scientific & Technical Publishers: Shanghai, China, 2020.
74. DeBerardinis, R.J.; Cheng, T. Q's next: The diverse functions of glutamine in metabolism, cell biology and cancer. *Oncogene* **2010**, *29*, 313–324. [CrossRef]
75. Silva Gunawardene, Y.I.; Chow, B.K.; He, J.G.; Chan, S.M. The shrimp FAMeT cDNA is encoded for a putative enzyme involved in the methylfarnesoate (MF) biosynthetic pathway and is temporally expressed in the eyestalk of different sexes. *Insect Biochem. Mol. Biol.* **2001**, *31*, 1115–1124. [CrossRef] [PubMed]
76. Wang, Z.; Ding, Q.; Yagi, K.J.; Tobe, S.S. Terminal stages in juvenile hormone biosynthesis in corpora allata of Diploptera punctata: Developmental changes in enzyme activity and regulation by allatostatins. *J. Insect Physiol.* **1994**, *40*, 217–233. [CrossRef]
77. Williamson, M.; Lenz, C.; Winther, A.M.; Nassel, D.R.; Grimmelikhuijzen, C.J. Molecular cloning, genomic organization, and expression of a B-type (cricket-type) allatostatin preprohormone from *Drosophila* melanogaster. *Biochem. Biophys. Res. Commun.* **2001**, *281*, 544–550. [CrossRef] [PubMed]
78. Tamura, T.; Thibert, C.; Royer, C.; Kanda, T.; Abraham, E.; Kamba, M.; Komoto, N.; Thomas, J.L.; Mauchamp, B.; Chavancy, G.; et al. Germline transformation of the silkworm *Bombyx mori* L. using a piggyBac transposon-derived vector. *Nat. Biotechnol.* **2000**, *18*, 81–84. [CrossRef] [PubMed]
79. Adams, D.S.; Eickbush, T.H.; Herrera, R.J.; Lizardi, P.M. A highly reiterated family of transcribed oligo(A)-terminated, interspersed DNA elements in the genome of *Bombyx mori*. *J. Mol. Biol.* **1986**, *187*, 465–478. [CrossRef]
80. Osanai-Futahashi, M.; Suetsugu, Y.; Mita, K.; Fujiwara, H. Genome-wide screening and characterization of transposable elements and their distribution analysis in the silkworm, *Bombyx mori*. *Insect Biochem. Mol. Biol.* **2008**, *38*, 1046–1057. [CrossRef] [PubMed]
81. Benhamed, F.; Denechaud, P.D.; Lemoine, M.; Robichon, C.; Moldes, M.; Bertrand-Michel, J.; Ratziu, V.; Serfaty, L.; Housset, C.; Capeau, J.; et al. The lipogenic transcription factor ChREBP dissociates hepatic steatosis from insulin resistance in mice and humans. *J. Clin. Investig.* **2012**, *122*, 2176–2194. [CrossRef] [PubMed]
82. Meng, M.; Cheng, D.J.; Peng, J.; Kang, L.X.; Wang, Y.H.; Qian, W.L.; Xia, Q.Y. Identification and Expression Profiling of Farnesoic Acid O-methyltransferase Gene in Silkworm, *Bombyx mori*. *Sci. Seric.* **2013**, *39*, 0680–0688.
83. Zhang, J.; Zhu, M.; Li, Q.; Tang, T.; Wen, L.; Zhong, J.; Zhang, R.; Yu, X.Q.; Lu, Y. Genome-wide identification and characterization of basic helix-loop-helix transcription factors in *Spodoptera litura* upon pathogen infection. *Insect Sci.* **2022**, *29*, 977–992. [CrossRef]
84. Librado, P.; Rozas, J. DnaSP v5: A software for comprehensive analysis of DNA polymorphism data. *Bioinformatics* **2009**, *25*, 1451–1452. [CrossRef]
85. Tanaka, H. Construction of shRNA expression plasmids for silkworm cell lines using single-stranded DNA and Bst DNA polymerase. *Methods Mol. Biol.* **2013**, *942*, 347–355.

International Journal of
Molecular Sciences

MDPI

Article

Impact of Obesity and Ageing on the Expression of Key Mediators of SARS-CoV-2 Infection in Human Adipose Tissue

Maria Salazar [1,2], Mariana Ferreira [1], Sandra Marisa Oliveira [3], Francisca Saraiva [3], Carlos Pinho [4], Mariana Jarnalo [4,5], Inês Correia-Sá [5], Inês Falcão-Pires [3], Adelino Leite-Moreira [3], Delminda Neves [1,2], Henrique Almeida [1,2], Adriana R. Rodrigues [1,2,6] and Alexandra M. Gouveia [1,2,*]

1 Unidade de Biologia Experimental, Departamento de Biomedicina, Faculdade de Medicina da Universidade do Porto, Alameda Prof. Hernâni Monteiro, 4200-319 Porto, Portugal; mariajsalazar88@gmail.com (M.S.); adrod@med.up.pt (A.R.R.)
2 i3S—Instituto de Investigação e Inovação em Saúde, Universidade do Porto/IBMC—Instituto de Biologia Molecular e Celular, Universidade do Porto, Rua Alfredo Allen 208, 4200-135 Porto, Portugal
3 Cardiovascular R&D Centre—UnIC@RISE, Department of Surgery and Physiology, Faculty of Medicine, University of Porto, Alameda Prof. Hernâni Monteiro, 4200-319 Porto, Portugal
4 Department of Plastic and Reconstructive Surgery, Instituto Português de Oncologia do Porto FG, Rua Dr. António Bernardino de Almeida, 4200-072 Porto, Portugal
5 Department of Plastic, Reconstructive and Aesthetic Surgery, and Burn Unit, Centro Hospitalar e Universitário de São João, University of Porto, Alameda Prof. Hernâni Monteiro, 4200-319 Porto, Portugal
6 Faculty of Nutrition and Food Sciences of the University of Porto, Rua do Campo Alegre 823, 4150-180 Porto, Portugal
* Correspondence: agouveia@med.up.pt

Abstract

Increased body mass index (BMI) and age are associated with COVID-19 severity. SARS-CoV-2 infection occurs through ACE2 binding, with TMPRSS2, ADAM17, and NRP1 facilitating this process. This study describes how adipose tissue (AT) location, BMI, age, and obesity affect these proteins' expression. AT was collected from subcutaneous (abdominal superficial [AS], abdominal deep [AD], thigh [T]) and visceral (epiploon [E]) areas from middle-aged women without obesity (BMI 23.9 kg/m^2, age 48.3 years). Subcutaneous AT was also obtained from middle-aged women with previous obesity (BMI 24.8 kg/m^2, previously 41.7 kg/m^2, age 46.9 years), older women with obesity (BMI 32.3 kg/m^2, age 70.8 years), and older women without obesity (BMI 23.7 kg/m^2, age 70.6 years). ACE2, TMPRSS2, ADAM17, and NRP1 expression was evaluated by qPCR and Western blotting. All proteins were more expressed in visceral AT. ACE2, TMPRSS2, and NRP1 positively correlated with BMI in AS and/or E, while NRP1 correlated with age in T. In subcutaneous AT, ACE2 and NRP1 were more influenced by obesity while TMPRSS2 was more age-dependent. In women with previous obesity, ACE2 and NRP1 levels decreased, while TMPRSS2 and ADAM17 remained unchanged. These findings highlight the differential influence of visceral AT, obesity, and age on the expression of SARS-CoV-2 cell entry mediators, potentially contributing to COVID-19 severity.

Keywords: obesity; ageing; adipose tissue; COVID-19; SARS-CoV-2 receptors; ACE2; TMPRSS2; ADAM17; NRP1

1. Introduction

The COVID-19 global pandemic, caused by SARS-CoV-2 virus infection, has emerged as a significant health concern. Several risk factors, such as age, hypertension, type 2 diabetes, and obesity are associated with COVID-19 severity [1–4]. Obesity is characterized by an excessive deposition of adipose tissue and is associated with several comorbidities, such as type 2 diabetes, hypertension, and heart diseases. According to the World Health Organization, in 2022, 2.5 billion adults were overweight and, of these, 890 million present with obesity. A body mass index (BMI) over 25 kg/m^2 is considered overweight, while over 30 kg/m^2 corresponds to obesity. Obesity induces a profound remodelling of the subcutaneous and visceral adipose tissue (AT), including an increase in inflammatory molecule secretion [5]. In fact, patients with obesity and COVID-19 present higher hospital admission rates, increased disease severity, often requiring mechanical ventilation, and higher mortality rates [6–9]. This outcome is the likely result of the compression of the diaphragm and lungs caused by abdominal obesity and/or to the chronic inflammation associated with obesity-induced adipose tissue (AT) dysfunction [10]. Another explanation relates to the SARS-CoV-2 virus's high affinity for its main receptor, the human angiotensin-converting enzyme 2 (ACE2), a member of the renin–angiotensin system [11]. ACE2 is an 805 amino acid protein composed of an extracellular N-terminal domain to which SARS-CoV-2 binds, a transmembrane region, and a short intracellular C-terminal domain. This receptor is expressed in human AT in higher levels compared to the lung tissue [11–13], highlighting the importance of AT to SARS-CoV-2 infection. Moreover, the susceptibility of AT to SARS-CoV-2 infection depends on its anatomical location. In vitro, adipocytes from the visceral compartment present higher levels of ACE2 expression and infection rates compared to those from the subcutaneous depot [14]. It was reported that SARS-CoV-2 directly infects mature adipocytes and AT-resident macrophages [15,16].

SARS-CoV-2 infection is not only mediated by ACE2 as other proteins are also implicated in virus cell entry. The SARS-CoV-2 virus enters host cells through the binding of its viral spike proteins to cellular ACE2 receptors. The S protein must be primed by cell proteases, such as transmembrane protease serine 2 (TMPRSS2) and A disintegrin and metalloproteinase 17 (ADAM17) [17,18]. TMPRSS2 and ADAM17 also cleave the ACE2 receptor itself [19]. ADAM17 contributes to the shedding of ACE2 within amino acids 652 to 659 and the release of an extracellular soluble form of ACE2 [20], which binds to extracellular viral particles, preventing their entry into cells. TMPRSS2 sheds ACE2 within amino acids 697 to 716 to promote the fusion of SARS-CoV-2 with the host cell membrane [21]. In addition to the ACE2 receptor, neuropilin-1 (NRP1) was identified as a SARS-CoV-2 co-receptor [22]. It binds to the virus spike protein and, when co-expressed with ACE2 and TMPRSS2, potentiates SARS-CoV-2 infection [22,23].

Age-related changes in AT composition and function, such as impaired adipocyte progenitor function, increased senescent cells, and pro-inflammatory adipocytokine production [24], are also contributors to the severity of COVID-19.

Given the interplay between AT function, obesity, and ageing, it is critical to further investigate how these conditions influence the mechanisms of AT SARS-CoV-2 infection. Specifically, we aim to characterize the expression patterns of the key SARS-CoV-2 cell entry mediators ACE2, TMPRSS2, ADAM17, and NRP1 across distinct subcutaneous and visceral AT domains in order to understand their modulation by weight gain, ageing, and weight loss. These findings could have significant implications for understanding COVID-19 pathophysiology and for future provision of preventive actions and the development of targeted therapeutic strategies for high-risk populations, including individuals with obesity and advanced age.

2. Results

2.1. Anthropometric and Biochemical Characterization of All Individuals

To characterize the susceptibility of the different AT depots to SARS-CoV-2 entry and understand why obesity and ageing predispose to a higher risk of severe COVID-19, we examined AT from various patient groups. These groups were categorized according to BMI and age as follows: middle-aged women without obesity (control group, BMI 19.88–29.48 kg/m^2, age 39–56 years); middle-aged women with previous obesity (BMI 22–27.2 kg/m^2, previously 30.9–52.6 kg/m^2, age 31–60 years); older women with obesity (BMI 30.43–35.52 kg/m^2, age 61–81 years); and older women without obesity (BMI 19.56–26.30 kg/m^2, age 62–81 years).

In terms of BMI, statistically significant differences ($p < 0.0001$) were observed between the group composed of women with obesity (32.26 kg/m^2) and the control group (23.9 kg/m^2), the group with previous obesity (24.8 kg/m^2), and the older group without obesity (23.7 kg/m^2) (Figure 1A). Age differs significantly ($p < 0.0001$) between the middle-aged groups (control group and group with previous obesity, with average ages of 48.3 and 46.9 years, respectively) and the older groups (the group with obesity and the group without obesity, with average ages of 70.8 and 70.6 years, respectively) (Figure 1B).

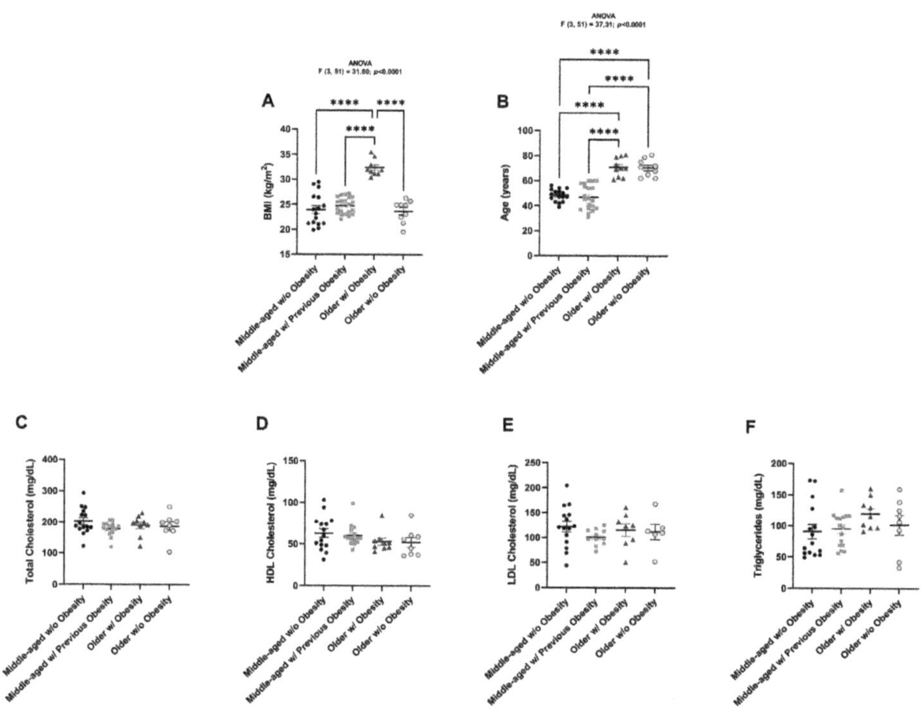

Figure 1. Graphical representation of the anthropometric and biochemical parameters from all women studied. BMI (kg/m^2) (**A**), age (years) (**B**), total cholesterol levels (mg/dL) (**C**), HDL cholesterol levels (mg/dL) (**D**), LDL cholesterol levels (mg/dL) (**E**), and triglyceride levels (mg/dL) (**F**) of middle-aged women without obesity (n = 15–16), middle-aged women with previous obesity (n = 11–21), older women with obesity (n = 8–10) and older women without obesity (n = 6–9). The results are represented as mean ± S.E.M. **** $p < 0.0001$, according to the one-way ANOVA, followed by Tukey's multiple comparisons test.

Regarding medical conditions that could potentially increase the risk of severe COVID-19, the group of middle-aged women without obesity had almost no comorbidities, except for four individuals with hypertension and one with type 2 diabetes. Similarly, in the group of women with previous obesity, only one presented with dyslipidaemia. In contrast, in the older groups, women presented with more comorbidities. The group with obesity includes ten women with dyslipidaemia, nine with hypertension, four with type 2 diabetes, and one with chronic lung disease. Among the older women without obesity, four had hypertension, five dyslipidaemia, one type 2 diabetes and three chronic lung disease. All patients were under appropriate medication to manage these conditions. As expected, women in the older groups were highly medicated; for example, almost all of them in both older groups, with or without obesity, were taking antihypertensive medication.

No differences were found in total cholesterol, HDL cholesterol, LDL cholesterol, and triglyceride levels among these groups (Figure 1C–F).

2.2. SARS-CoV-2 Cell Infection Mediators Have Increased Expression in Visceral Adipose Tissue of Middle-Aged Women Without Obesity

ACE-2, TMPRSS2, ADAM17, and NRP1 are the main cell membrane mediators that promote entrance of SARS-CoV-2 viruses into cells. To clarify if their expression pattern depends on the anatomical location of AT, the expression of these cell entry factors was evaluated in subcutaneous (AS, AD, and T) and visceral (E) compartments of AT from the middle-aged control patients.

ACE2 protein expression in epiploon AT (E) was 7-fold higher than in subcutaneous AS and AD (p = 0.0005 and 0.0003, respectively) and 2.3-fold higher (p = 0.0314) than in the T depot (Figure 2A). TMPRSS2 expression was 2-fold increased in the E depot relative to AS (p = 0.0259) (Figure 2B). ADAM17 levels were increased 2-fold in E when compared to AS (although this result is not statistically significant, it is very close to reaching significance; p = 0.0806) (Figure 2C). The NRP1 protein expression pattern parallels the one observed in ACE2: it is increased in E relative to all subcutaneous depots (AS: p = 0.002, AD: p = 0.0187 and T: p = 0.0154) (Figure 2D). Globally, the protein expression of these SARS-CoV-2 cell entry factors, evaluated by Western blot, was increased in visceral AT when compared with the subcutaneous counterparts (Figure 2). No differences were found within subcutaneous AT depots. A similar pattern was observed in the subcutaneous AT of patients with previous obesity, with no significant differences found between fat depots from the abdominal (A), thigh (T), and arm areas (Figure S1).

Regarding ACE2 mRNA expression, levels were higher in visceral tissue (E) when compared to the subcutaneous AD fraction (p = 0.0124) and no differences were obtained when comparing with AS and T AT samples (Figure 2E). ADAM17 mRNA is equally expressed in all fat tissues (Figure 2F). Surprisingly, the mRNA expression of NRP1 was decreased in E when compared to T (p = 0.0131) (Figure 2G), contrary to the results obtained at the protein expression level.

2.3. The Protein Levels of SARS-CoV-2 Cell Entry Mediators Positively Correlate with BMI

Within the control group, the protein expression levels of ACE2, TMPRSS2, ADAM17, and NRP1 in all AT compartments (AS, AD, T, and E) were correlated with the subject's BMI. Generally, there is a tendency for a positive correlation, though statistically significant results are only observed for ACE2, TMPRSS2, and NRP1 in specific fat compartments, as highlighted in Figure 3. ACE2 protein levels positively correlate with BMI in AS and E depots (p = 0.0187 and 0.0164, respectively) (Figure 3A). Additionally, there is a statistically significant positive correlation in the protein expression of TMPRSS2 in E (p = 0.0236) and of NRP1 in AS (p = 0.0296) (Figure 3B,D) with BMI. Moreover, two correlations almost reached significance, the protein levels of TMPRSS2 in AS and NRP1 in AD (p = 0.0759 and p = 0.0685, respectively) (Figure 3B,D).

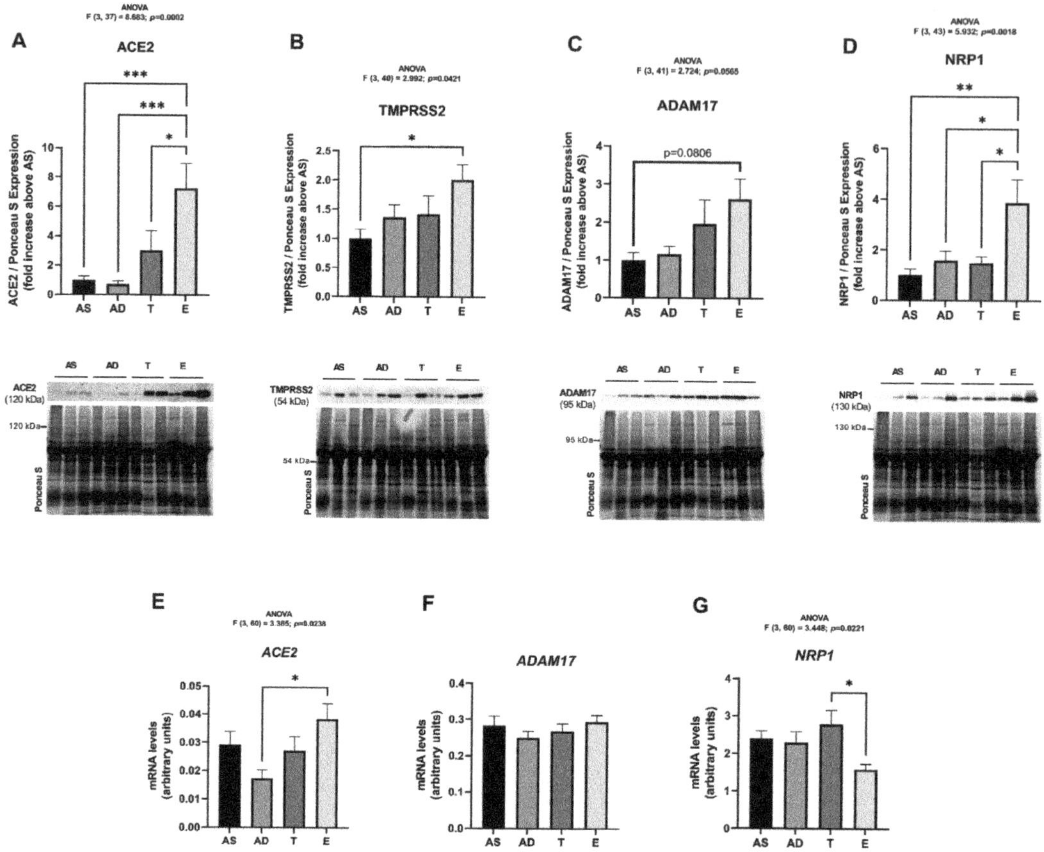

Figure 2. Expression of SARS-CoV-2 cell entry factors in human subcutaneous (AS—abdominal superficial; AD—abdominal deep; and T—thigh) and visceral (E—epiploon) adipose tissue from middle-aged individuals without obesity (BMI from 19.88 to 29.48 kg/m^2), quantified by Western blot and real-time PCR. ACE2 (**A**), TMPRSS2 (**B**), ADAM17 (**C**), and NRP1 (**D**) protein expression in AS (n = 10–12), AD (n = 11–12), T (n = 10–12) and E (n = 9–12). ACE2 (**E**), ADAM17 (**F**), and NRP1 (**G**) mRNA expression in AS (n = 16), AD (n = 16), T (n = 16), and E (n = 16). The results are represented as mean ± S.E.M. * $p < 0.05$, ** $p < 0.01$, and *** $p < 0.001$, according to the one-way ANOVA, followed by Tukey's multiple comparisons test.

Age is known to be an important factor in the modulation of SARS-CoV-2 infection [1]. To address this, correlations between the levels of all SARS-CoV-2 cell entry mediators and age were also determined in the control group (Figure 4). Although there is also a general tendency towards positive correlations, only NRP1 levels in the T depot significantly increase with ageing (p = 0.0308, Figure 4D). In conclusion, the expression of analysed receptors and coreceptors appears to positively correlate more prominently with BMI than with age.

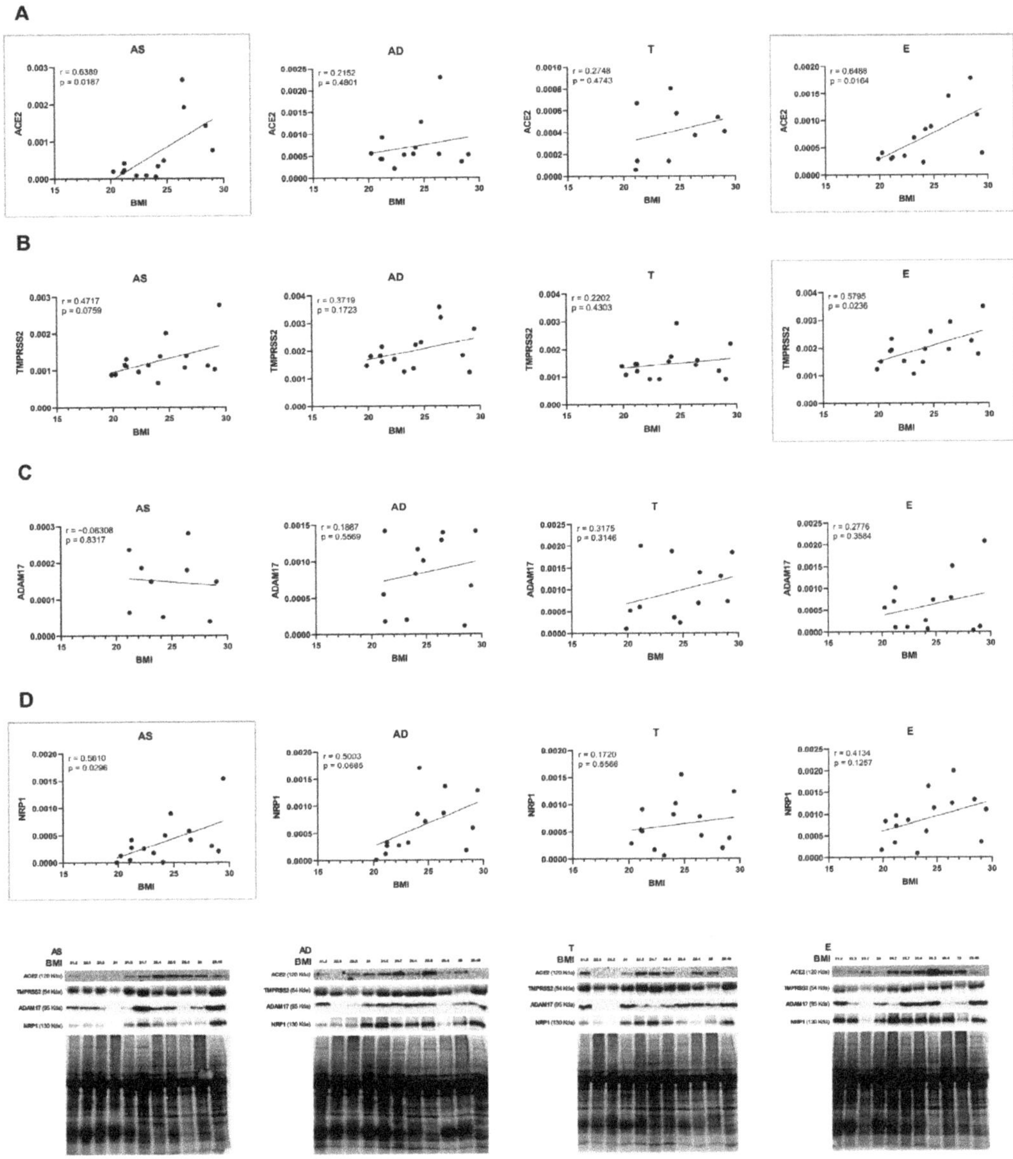

Figure 3. Correlations of SARS-CoV-2 receptor and co-receptor protein levels with BMI, quantified in human subcutaneous (AS—abdominal superficial; AD—abdominal deep; and T—thigh) and visceral (E—epiploon) adipose tissue from middle-aged individuals without obesity (BMI from 19.88 to 29.48 kg/m^2) by Western blot. Respective representative blots are presented in the figure. ACE2 (**A**), TMPRSS2 (**B**), ADAM17 (**C**), and NRP1 (**D**) protein level correlation with BMI in AS (n = 9–15), AD (n = 12–15), T (n = 9–15), and E (n = 13–15). *p*-values lower than 0.05 are considered statistically significant, according to the Pearson correlation coefficient.

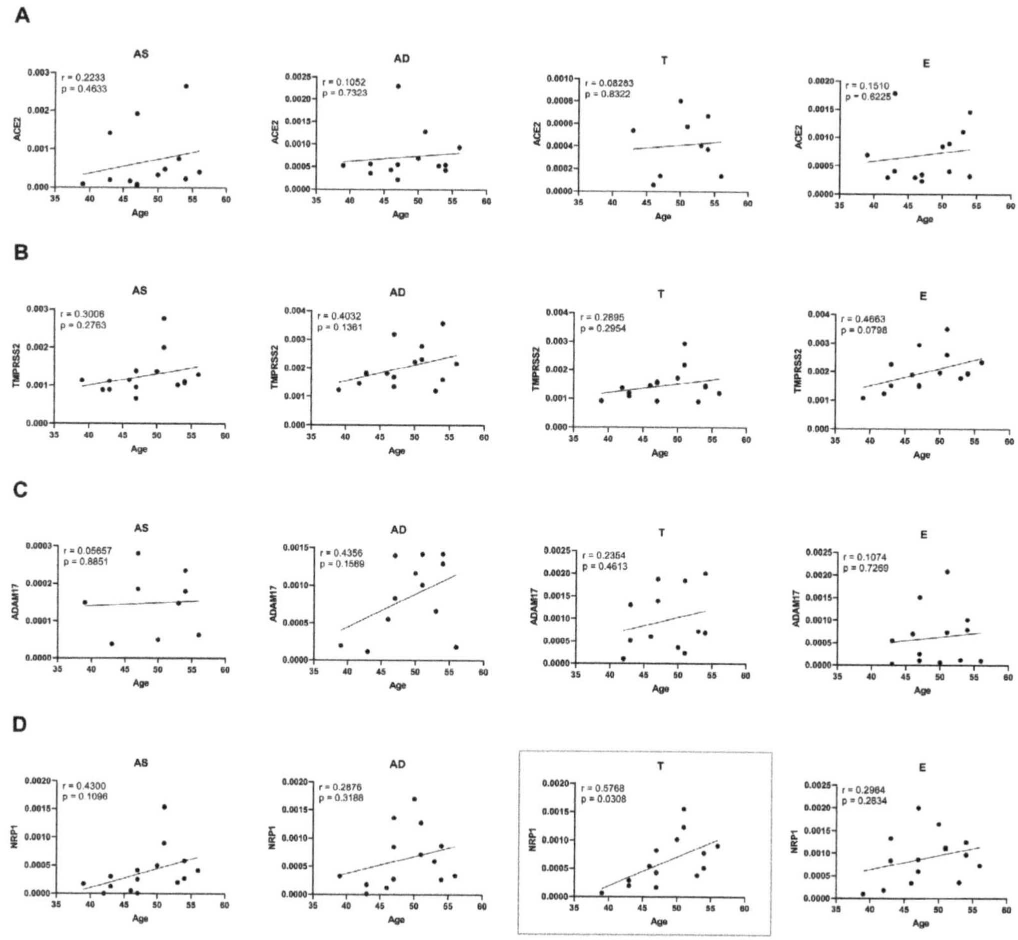

Figure 4. Correlations of SARS-CoV-2 receptor and co-receptor protein levels with age, quantified in human subcutaneous (AS—abdominal superficial; AD—abdominal deep; and T—thigh) and visceral (E—epiploon) adipose tissue from middle-aged individuals without obesity (BMI from 19.88 to 29.48 kg/m^2) by Western blot. Representative blots are the same as in Figure 3. ACE2 (**A**), TMPRSS2 (**B**), ADAM17 (**C**), and NRP1 (**D**) protein level correlation with age in AS (n = 9–15), AD (n = 12–15), T (n = 9–15), and E (n = 13–15). *p*-values lower than 0.05 are considered statistically significant, according to the Pearson correlation coefficient.

Unexpectedly, mRNA levels of ACE2 decrease with increasing BMI ($p = 0.0488$) in the E depot (Figure 5A), contrary to the protein levels, which showed a positive correlation with BMI (Figure 3A). No statistically significant differences were observed in mRNA expression of the other cell entry proteins evaluated. However, a tendency for mRNA expression to decrease with BMI in specific fat depots was noted.

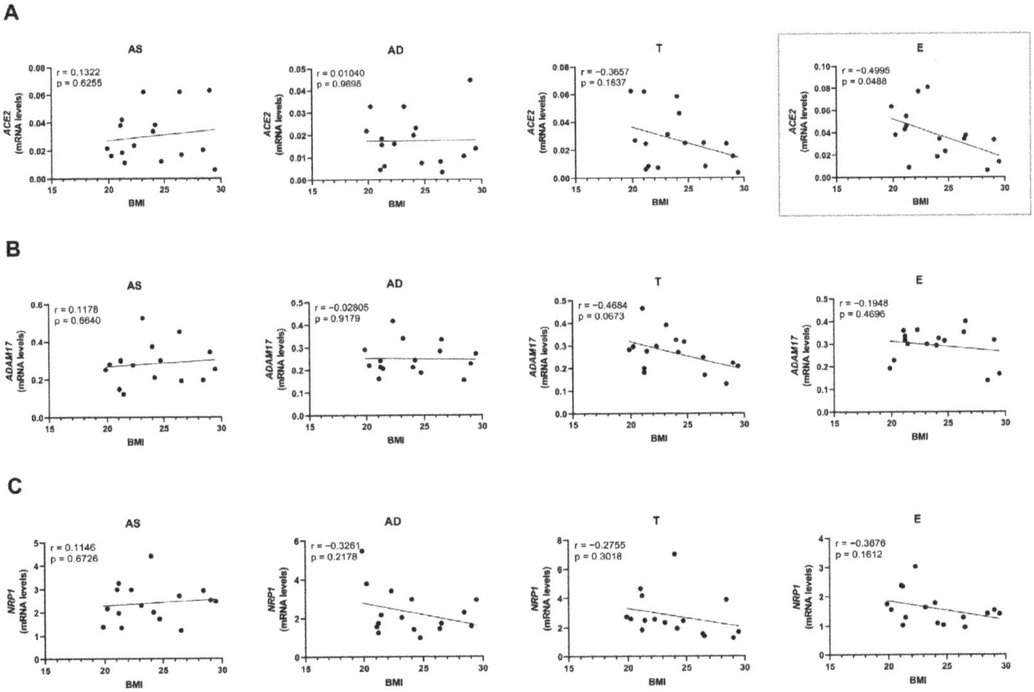

Figure 5. Correlation of SARS-CoV-2 receptor and co-receptor mRNA levels with BMI, quantified in human subcutaneous (AS—abdominal superficial; AD—abdominal deep; and T—thigh) and visceral (E—epiploon) adipose tissue from middle-aged individuals without obesity (BMI from 19.88 to 29.48 kg/m^2) by real-time PCR. ACE2 (**A**), ADAM17 (**B**), and NRP1 (**C**) mRNA level correlation with BMI in AS (n = 16), AD (n = 16), T (n = 16), and E (n = 16). p-values lower than 0.05 are considered statistically significant, according to the Pearson correlation coefficient.

2.4. Ageing and Obesity Differently Influence the Expression of the SARS-CoV-2 Cell Entry Factors in Subcutaneous Adipose Tissue

To further explore the combined effect of ageing and obesity on SARS-CoV-2 infection, the expression of ACE2, TMPRSS2, ADAM17, and NRP1 was evaluated in subcutaneous AT from older women with or without obesity. In addition, a group of samples collected from middle-aged patients with previous obesity was included to evaluate the effect of weight loss on the expression of the proteins.

In subcutaneous AT, older women with obesity have 4.2-fold higher ACE2 protein levels compared to older individuals without obesity (p = 0.0208) (Figure 6A). Thus, a clear effect of obesity on ACE2 receptor protein expression is observed in older patients. Moreover, in the group with previous obesity, the levels of ACE2 are significantly reduced compared to the group with obesity (p = 0.0028). In addition, ACE2 levels are significantly higher in older women with obesity when comparing to the middle-aged women without obesity (p = 0.001). Since no differences were detected between the middle-aged and older women without obesity, it is possible to conclude that the ageing effect on ACE2 expression is not as pronounced as obesity (Figure 6A).

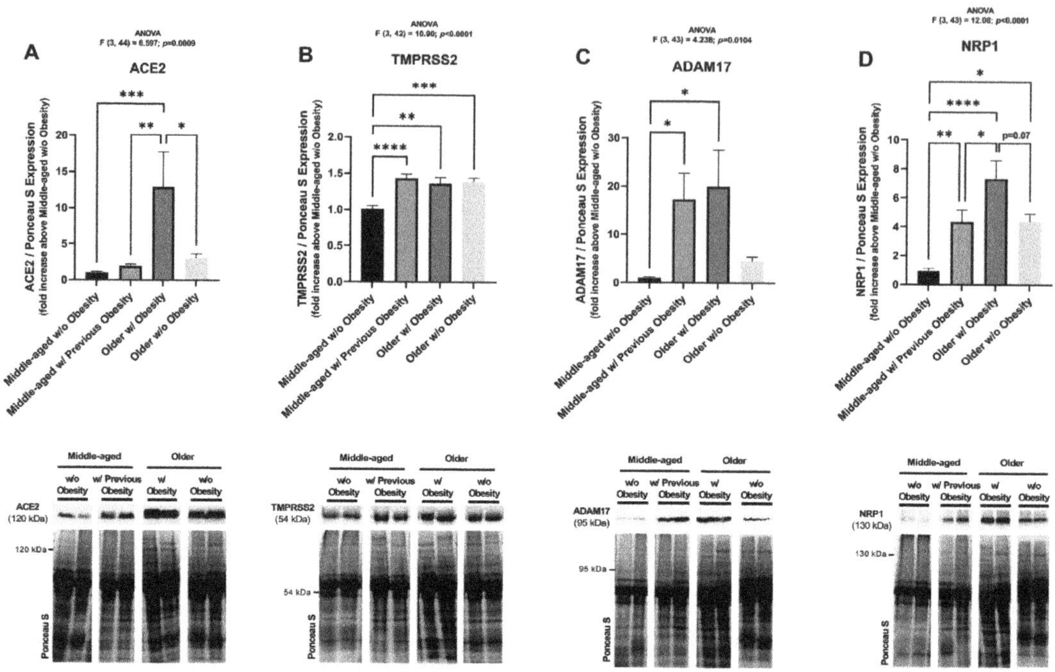

Figure 6. Expression of SARS-CoV-2 cell entry factors in human subcutaneous adipose tissue from middle-aged women without obesity (BMI from 19.88 to 29.48 kg/m^2, average age of 48.2 years), middle-aged women with previous obesity (BMI from 22.0 to 27.2 kg/m^2, previously 33.5 to 52.6 kg/m^2, average age of 46.9 years), older women with obesity (BMI from 30.43 to 35.52 kg/m^2, average age of 70.8 years), and older women without obesity (BMI from 19.56 to 26.3 kg/m^2, average age of 70.6 years), quantified by Western blot. Representative blots are indicated in the figure. ACE2 (**A**), TMPRSS2 (**B**), ADAM17 (**C**), and NRP1 (**D**) protein expression in middle-aged women without obesity (n = 13–15), middle-aged women with previous obesity (n = 14), older women with obesity (n = 10) and older women without obesity (n = 9). The results are represented as mean ± S.E.M. * $p < 0.05$, ** $p < 0.01$, *** $p < 0.001$, and **** $p < 0.0001$, according to the one-way ANOVA, followed by Tukey's multiple comparisons test.

Both TMPRSS2 and ADAM17 protein levels increase in older women with obesity, when compared to controls (p = 0.0013 and p = 0.0274, respectively), and do not differ significantly from middle-aged individuals with prior obesity (Figure 6B,C). TMPRSS2 levels increase with age (p = 0.0009, older group without obesity) to levels similar to those observed in obesity or after weight loss (Figure 6B).

Age alone induces a 4-fold increase in NRP1 protein levels (p = 0.0179 between control and older women without obesity groups). When advanced age is combined with obesity, those levels are almost twice as high. Although not statistically significant, they were close to reaching significance (p = 0.07). On the other hand, NRP1 protein levels decreased by almost half (p = 0.04) when comparing the groups of women with previous obesity and the older women with obesity, but still did not reach those observed in the control group, remaining elevated 4-fold and significantly higher (p = 0.0063) (Figure 6D).

Overall, the protein levels of all of the analysed SARS-CoV-2 cell entry factors are increased in older women with obesity (Figure 6) but these two factors (ageing and obesity) differently influence their expression.

3. Discussion

The present study highlights the importance of AT, specifically visceral AT, in the modulation of the clinical response to SARS-CoV-2 infection. In fact, it demonstrates that the expression of the SARS-CoV-2 cell entry mediators ACE2, TMRPSS2, ADAM17, and NRP1 is increased in human visceral AT. Accordingly, a computed tomography analysis of patients with symptomatic COVID-19 found an association between higher visceral AT area and severe forms of COVID-19, whereas subcutaneous fat levels were not related [25]. Moreover, Saccon et al. found increased expression of ACE2 receptor in primary cultured visceral adipocytes, which were more prone to SARS-CoV-2 infection compared with the subcutaneous [14]. Both TMPRSS2 and ADAM17 cleave ACE2 and viral S protein to facilitate SARS-CoV-2 fusion with host cell membranes [17,26], promoting a crucial environment conducive to viral entry and replication [17,27]. Conversely, ADAM17 may have a protective role against COVID-19 infection by releasing a soluble ACE2 isoform that binds to circulating viral particles, preventing their entry into host cells [20]. NRP1 is expressed in macrophages [28], more abundant in human visceral than in subcutaneous AT [29]. This supports the current findings of higher NRP1 protein expression in the visceral depot. NRP1 is also involved in immune function, which is particularly relevant given the exaggerated immune responses associated with severe COVID-19 [30]. Therefore, the increased levels of SARS-CoV-2 cell entry factors in the visceral adipose tissue suggests that this depot may act as a reservoir for viral particles, potentially contributing to systemic viral spread that ultimately leads to the disruption in immune signalling and severe inflammatory responses observed in COVID-19 patients with a higher volume of visceral fat [25].

Obesity, a condition accompanied by increased serum ACE2 levels [31], and excess weight are known risk factors for severe forms of COVID-19 [6,32,33]. This is supported by the positive correlation between the protein expression levels of ACE2, TMPRSS2, and NRP1 and BMI, both in subcutaneous and visceral AT, herein described. Similar findings have been reported: Patel et al. described increased ACE2 protein levels in epicardial AT of patients with obesity and diet-induced obese mice [34]; higher ACE2 gene expression was reported in epicardial and subcutaneous AT of individuals with obesity [35]; and Frühbeck and colleagues showed that ACE2 mRNA increases only in visceral and not in the subcutaneous depot of patients with obesity [36]. Only a single, sample-limited immunofluorescence study reported a lack of variation in ACE2 protein levels in human subcutaneous AT from lean or obese individuals [37]. TMPRSS2 gene expression was found to be increased in lung epithelial cells and in pancreatic islets of subjects with obesity [38,39]. As well, rats fed a high fat and sugar diet also present elevated protein expression of TMPRSS2 in the lungs compared with the controls [40]. NRP1 mRNA expression significantly increases in the subcutaneous AT of patients with obesity and type 2 diabetes. In patients with obesity but without type 2 diabetes, its levels also increase, although without statistical significance [36], as shown in the current work. Conversely, in immunofluorescence studies, Steenblock et al. found no differences in NRP1 protein expression in subcutaneous AT between individuals with or without obesity [37]. The different methodologies implied for protein quantification (Western blotting versus immunofluorescence) may justify the divergency in results.

Interestingly, here, a different pattern of expression of the receptors (ACE2 and NRP1) and co-receptors (ADAM17 and TMRPSS2) with obesity and ageing is reported. In older individuals without obesity, protein levels of subcutaneous ACE2 and NRP1 receptors, to which SARS-CoV-2 binds directly, did not increase as markedly as they did in individuals of the same age with obesity. Thus, an evident effect of obesity in modulation of ACE2 and NPR1 expression was demonstrated. The ageing effect on the expression of these proteins is controversial in the literature since some studies report no correlation of ACE2 and NRP1 gene expression with age in subcutaneous AT [41,42] but in a gene set database analysis, ACE2 gene expression was found to be increased in older individuals [43].

ADAM17 expression in subcutaneous AT seems to depend simultaneously on ageing and obesity, since the isolated factors did not significantly influence ADAM17 protein levels. Dou et al. found that age and obesity are associated with increased ADAM17 enzyme activity in human mediastinal AT blood vessels and adipocytes [44], which aligns with the present data, showing an increase in ADAM17 protein expression in subcutaneous AT from older individuals with obesity, a condition that parallels the enhanced expression of ADAM17 mRNA in the subcutaneous AT of obese mice [45]. In omental AT of patients with morbid obesity, ADAM17 is considered a predictor of insulin resistance [46]. Moreover, transgenic mice with ADAM17 overexpression presented increased macrophage infiltration and fibrosis when fed an HFD [47], whereas ADAM17 gene silencing in visceral macrophages improved AT inflammation and type 2 diabetes [48].

In turn, TMPRSS2 expression appears to be more dependent on ageing rather than obesity, since its expression remains virtually the same in both older groups (with or without obesity), while it increases in comparison to the middle-aged control group. Moreover, the expression of TMPRSS2 and ADAM17 remains unchanged in individuals with previous obesity. Since these are two proteases that are involved in cancer and inflammation processes [49,50], and subcutaneous AT from individuals with prior obesity retains molecular characteristics of obese AT [51], this may explain why there is no reversion in TMRPSS2 and ADAM17 receptor protein levels after weight loss. Patients with previous obesity, after undergoing bariatric surgery and stabilizing their weight for one year, underwent abdominoplasty, during which adipose tissue samples were collected. It is, however, unclear if, over a longer period, these patients would reach the levels of TMPRSS2 and ADAM17 observed in individuals without obesity. Conversely, the protein expression levels of ACE2 and NRP1, receptors to which SARS-CoV-2 directly binds, decrease in individuals with previous obesity. Other studies that access gene expression are in line with these findings; ACE2 levels have been shown to decrease in subcutaneous AT two years after bariatric surgery [52] and also after weight loss without surgical interventions [38,41]. Cauwenberghs et al. showed a decrease in the soluble form of ACE2, through proteomic measurements in human plasma, after 6 months of a restrictive diet [53], and Soll and colleagues described a decreased NRP1 mRNA expression in subcutaneous AT after weight loss [42].

The importance of the study of the diverse cell entry factors herein described is highlighted by the evolving infection behaviour of the SARS-CoV-2 variants [54–56]. It is known that when ACE2 levels are low, some co-factors, like ADAM9, promote SARS-CoV-2 infection, particularly in alpha (B.1.1.7), delta (B.1.617.2), and omicron (B.1.1.529) variants [57]. In fact, ADAM9 and other proteases, such as ADAM17 and TMPRSS2, interact with the spike protein to facilitate virus entry, especially in cells with limited ACE2 expression [57,58]. Additionally, new SARS-CoV-2 lineages such as KP.2, KP.3, and LB.1, although still dependent on ACE2, have subtle differences in receptor use in different species and tissues [59]. As shown herein, the concomitant expression of TMPRSS2, ADAM17, and NRP1 in adipose tissue—receptors that become elevated in obesity and, in the case of TMPRSS2 and ADAM17, remain elevated after weight loss—likely amplifies SARS-CoV-2 tropism and infectivity in adipose tissue, even when ACE2 levels decrease following weight loss. This observation is consistent with recent findings by Li et al., who demonstrated that in women with gestational diabetes, metabolic memory can sustain increased SARS-CoV-2 susceptibility, as metabolic alterations persist postpartum along with sustained low-grade inflammation [60]. Moreover, AT itself preserves an obesity-related epigenetic memory after weight loss [61]. This supports the notion that the persistent expression of these proteases observed in the adipose tissue samples after weight loss may reflect a broader metabolic memory effect, potentially relevant across different tissues and physiological contexts.

In this work, it was consistently observed that mRNA data did not align with the protein expression results. In accordance, Wang and colleagues found significant discrepancies between mRNA and protein levels of ACE2 within and throughout tissues when analysing proteomic and RNA-seq datasets [62]. These authors concluded that protein levels provide a more precise measurement of ACE2 abundance compared to mRNA levels. Regarding NRP1, an increase in its protein levels was observed in visceral fat compared to subcutaneous depots. Again, the opposite was observed in NRP1 mRNA levels, as already reported by Saccon et al. [14]. The regulation of NRP1 involves complex interactions with various proteins and microRNAs. Studies on breast and gastric cancer have shown that RNA binding proteins such as PUM2 and Lin28B can bind to NRP1 mRNA, increasing its stability and expression, thereby affecting protein levels [63,64]. Conversely, in glioblastoma and breast cancer studies, NRP1 is targeted by miR-124-3p and miR-376a, which suppress its expression [65,66]. These mechanisms illustrate the complex regulation of NRP1 at both mRNA and protein levels and suggest that a particular regulatory mechanism exists in AT. Apart from all of the mentioned interactions, other post-transcriptional mechanisms such as alternative splicing, RNA editing, translation efficiency, and protein degradation [67] could also contribute to the discrepancies between mRNA and protein levels observed in this study, which reinforces that protein abundance should be assessed to better reflect functional effects.

This study has some limitations on the retrospective nature of data collection and the homogeneity in the types of AT samples collected across all groups, much on account of the nature of the surgeries performed. Considering this, all adipose tissue samples were collected from women and, for this reason, the results obtained could be gender specific. The inclusion of adipose tissue samples from men would be important to the generalization of the conclusions.

Although the interference of comorbidities and medications with ACE2 expression is controversial [68–73], in the present study, consistent differences in the protein levels of the SARS-CoV-2 mediators were observed between groups with or without obesity, despite comparable comorbidity profiles and similar patterns of medication use. Therefore, the observed variations in ACE2, ADAM17, and NRP1 expression are primarily attributable to obesity itself rather than to confounding effects from underlying conditions or treatments.

4. Materials and Methods

4.1. Human Adipose Tissue Samples

Control AT samples were collected from 16 middle-aged women without obesity undergoing breast reconstruction surgeries after mastectomy, in the context of breast cancer, at the Instituto Português de Oncologia do Porto (IPO-Porto), from 4 different anatomical locations: subcutaneous abdominal superficial (AS), abdominal deep (AD), thigh (T), and visceral epiploon (E). At Centro Hospitalar Universitário São João (CHUSJ), subcutaneous AT samples were collected from 30 middle-aged women, with previous obesity, that underwent bariatric surgery and were subsequently referred to abdominoplasty, arm, and thigh lift surgeries. Samples were collected from the patient's abdominal (A), arm, and thigh (T) regions. Subcutaneous AT from 10 older women with obesity and 9 older women without obesity were collected during cardiac surgeries (coronary, valvular, and ascending aorta) at the Cardiothoracic Surgery Service of CHUSJ. Those were harvested from the pre-sternal region. All samples were stored at −80 °C for subsequent protein and RNA extraction.

The patients' age, BMI, cholesterol (total, HDL, and LDL) levels, and triglyceride levels were obtained by consulting the biochemical analysis performed nearest the time of respective surgery, when the tissues were collected. Unfortunately, it was not possible to have access to data on specific markers of inflammation.

A flow chart summarizing the inclusion criteria, established groups, sample collection sites, and analysis performed is presented in Figure 7.

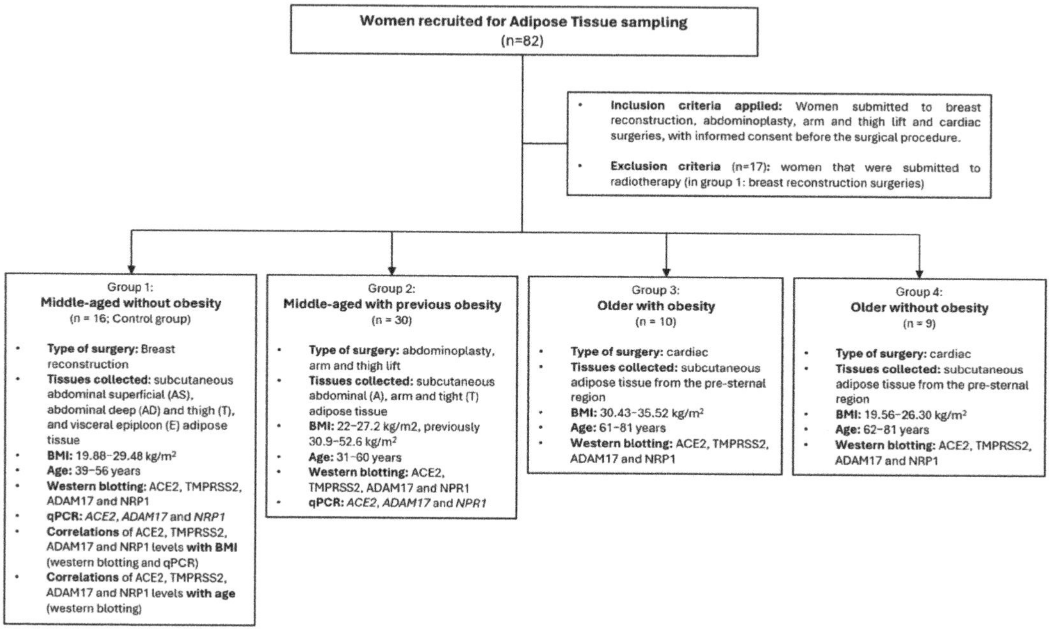

Figure 7. Flow chart summarizing all groups of patients recruited to this study divided according to age and obesity status. Inclusion/exclusion criteria, type of surgeries and respective AT anatomical location are also indicated in the figure.

4.2. Real-Time PCR

For RNA extraction, the TripleXtractor directRNA kit was used (#GK23.0100, Grisp, Porto, Portugal). A total of 100 to 200 mg of AT was weighed and placed in tubes with magnetic BulkBeads (Precellys, Montigny-le-Bretonneux, France) and homogenised in 700 µL of TripleXtractor reagent in a MagNA Lyser instrument (Roche, Mannheim, Germany) at 6500 rpm for 30 s. Then, samples were centrifuged at 15,000× g for 5 min at room temperature to remove the upper fat layer. RNA was extracted from the supernatant, following the RNA extraction kit manufacturer's instructions. RNA concentration and purity were analysed using NanoDrop One equipment (ThermoFisher Scientific, Waltham, MA, USA). Then, agarose gel electrophoresis was performed to verify the integrity of the RNAs. A total of 300 ng of each RNA sample was run on a 1% agarose gel for 15 to 20 min at 100 mA. The 28S and 18S rRNA bands were visualized on a Chemidoc UV transilluminator (Chemi-DocTM XRS, Bio-Rad, Hercules, CA, USA). cDNA was then obtained from 400 ng of RNA using the NZY First-Strand cDNA Synthesis Kit (MB12501, Nzytech, Lisboa, Portugal), following the manufacturer's instructions. Finally, real-time PCR was performed with the Power SYBR Green PCR Master Mix (ThermoFisher Scientific, Waltham, MA, USA). For each reaction, together with the SYBR Green mix, 10 ng of cDNA and 0.06 pmol of each primer (forward and reverse) were combined. Two thermocyclers were used, one for 96-well plates, the QuantStudio™ 3 Real-Time PCR System, 96-well, 0.1 mL Block (ThermoFisher Scientific, Waltham, MA, USA), and another for 384-well plates, the Real-Time PCR System (Bio-Rad, Hercules, CA, USA). The program was as follows: 10 min of denaturation at 95 °C, followed by 45 cycles of 10 s of denaturation at 95 °C, 30 s of annealing at 60 °C, and 30 s of extension at 60 °C. The expression of ACE2 (forward sequence: 5′-CATTGGTCTTCTGTCACCCGA-3′ and reverse sequence: 5′-CCCCAACTATCTCTCGCTTCAT-3′), ADAM17 (forward sequence: 5′-GGGCAGAGGGGAAGAGAGTA-3′ and reverse sequence: 5′-TGTGGAGACTTG-AGAATGCGA-3′), and NRP1 (forward sequence: 5′-GCCACAGTGGAACAGGTGAT-3′ and reverse sequence: 5′-ATGACCGTGGGCTTTTCTGT-3′) genes was evaluated. Importin 8 (IPO8) was used as a reference gene (forward sequence: 5′-CCAAGGGGTGGTTCATTCT-3′ and reverse sequence: 5′-TCTTGCCACAGCTCTTCATC-3′). Control reactions without reverse transcriptase were performed to exclude non-specific amplification of genomic DNA.

4.3. Western Blotting

Total proteins were extracted from 100 to 200 mg of AT with lysis buffer (50 mM Tris pH 7.6, 10 mM NaCl, 5 mM EDTA, 1 mM β-glycerophosphate, and 0.25% *v*/*v* Triton X-100) supplemented with protease (1:500) and phosphatase (1:250) inhibitors (Sigma-Aldrich, St. Louis, MO, USA). Lysis of AT was performed in a MagNA Lyser Instrument (Roche, Mannheim, Germany), with magnetic BulkBeads (Precellys, Montigny-le-Bretonneux, France). Then, samples were sonicated for 15 min in the BioruptorTM UCD-200 (Diagenode, Liège, Belgium) and centrifuged at 10,000× *g* for 10 min at 4 °C to discard the fat layers. The supernatant was stored at −20 °C for later protein quantification by the Bradford Reagent assay (Bio-Rad, Hercules, CA, USA). A total of 30 µg of protein was loaded per well and resolved in 10% SDS-PAGE gels (25 mA per gel). Proteins were then transferred into nitrocellulose membranes (Bio-Rad, Hercules, CA, USA) for 2 h at 30 V. All membranes were stained with Ponceau S and one image of each was recorded in a Chemi-DocTM XRS instrument (Bio-Rad, Hercules, CA, USA). Then, membranes were blocked with 5% *w*/*v* non-fat dry milk (Molico®) dissolved in 0.1 M Tris-buffered saline with 0.1% Tween 20 (TBST), except the ones that were incubated with anti-TMPRSS2 antibody, that were blocked with 5% *w*/*v* bovine serum albumin (BSA) in TBST. After this, membranes were probed with primary antibodies (dissolved in 5% BSA, 0.1% NaN3 in TBST) specific for ACE2 (1:1000, abcam 108252, Cambridge, United Kingdom), TM-PRSS2 (1:1000, abcam 109131, Cambridge, United Kingdom), ADAM17 (1:000, Santa Cruz Biotechnology sc-390859, Dallas, TX, USA), and NRP1 (1:1000, abcam 81321, Cambridge, United Kingdom), for three nights for ACE2 and one night for the remaining antibodies, at 4 °C. Subsequently, membranes were washed with TBST and incubated with the respective anti-rabbit (1:10,000, A16096, Invitrogen, Waltham, MA, EUA) or anti-mouse (1:5000, 715-035-150, Jackson Immuno Research, West Grove, PA, USA) horseradish-conjugated secondary antibodies diluted in 5% non-fat dry milk or 5% BSA (for TMPRSS2) in TBST for 1 h at room temperature with stirring. Finally, after washing the membranes with TBST, proteins were incubated with the Clarity ECL Western blotting substrates (Bio-Rad, Hercules, CA, USA). ECL signals and Ponceau S-stained membranes were captured in the Chemi-DocTM XRS instrument (Bio-Rad, Hercules, CA, USA) and quantified by densiometric analysis with the Image Lab 6.1 software (Bio-Rad, Hercules, CA, USA). Protein values were normalised by determination of the ratio of each ECL signal with the respective Ponceau S lane quantification. A Western blot analysis of internal loading controls, such as tubulin, actin, or Lamin A/C expression, was also performed. However, their protein expression levels linearly correlated with BMI, compromising their suitability as stable loading controls (Figure S2). This observation is consistent with previous reports [74,75]. Therefore, total protein staining with Ponceau S was used as protein loading normaliser, which is a widely accepted alternative when stable internal controls are not available [76].

Some membranes were re-used for incubation with other antibodies. For this, a 20 min stripping protocol with warm 10% SDS was performed, followed by washes with TBST, blocking, and incubation with specific primary and secondary antibodies.

4.4. Statistical Analysis

All statistical analyses were conducted in GraphPad Prism 9 (version 9.5.1). The correlation experiments were analysed by the Pearson correlation coefficient. The remaining analyses were performed using one-way ANOVA, followed by post-hoc analyses using Tukey's multiple comparisons test. All results are presented as mean ± standard error of the mean (SEM) and p-values < 0.05 were considered statistically significant. Outliers were removed after identification by the ROUT method test using the GraphPad Software, setting Q to 1.0%, in the correlation experiments.

5. Conclusions

In conclusion, this study provides novel insights into the expression patterns of SARS-CoV-2 cell entry mediators in AT under conditions of obesity, ex-obesity, or advanced age. ACE2, TMPRSS2, ADAM17, and NRP1 are more expressed in visceral AT, with obesity significantly influencing their protein expression. This may enhance the susceptibility to SARS-CoV-2 infection and contribute to the severe inflammatory responses observed in COVID-19. Although ageing also contributes to increased expression of the receptors and co-receptors in AT, ACE2 and NRP1 expression is much more upregulated in obesity. Importantly, persistent TMPRSS2 and ADAM17 expression was shown, even after significant weight loss, reflecting the long-lasting impact of obesity on adipose tissue biology. Nevertheless, normal weight individuals with previous obesity may have lower COVID-19 risk since present reduced ACE2 and NRP1 protein levels, receptors that bind directly to the SARS-CoV-2 virus. These results underscore the critical need to continue to investigate the molecular mechanisms driving the expression of SARS-CoV-2 receptors and their implications for metabolic health and infectious disease susceptibility. Such efforts could ultimately lead to targeted therapeutic strategies aimed at mitigating the impact of adipose tissue and obesity on COVID-19 outcomes and broader metabolic complications.

Supplementary Materials: The following supporting information can be downloaded at https://www.mdpi.com/article/10.3390/ijms26157313/s1.

Author Contributions: M.S. and A.M.G. conceived and designed the study and planned experiments; H.A., A.R.R. and A.M.G. supervised the study; S.M.O., F.S., C.P., M.J., I.C.-S., I.F.-P. and A.L.-M. provided biological samples and acquired anthropometric and biochemical data; M.S. and M.F. performed experiments and analysed data; A.R.R. and A.M.G. validated experimental methods and results; M.S. performed the figures and wrote the original draft of the manuscript; S.M.O., F.S., C.P., I.F.-P., D.N., H.A., A.R.R. and A.M.G. made manuscript revisions. All authors have read and agreed to the published version of the manuscript.

Funding: Maria Salazar (DOI 10.54499/2020.10112.BD) and Adriana Rodrigues (DOI 10.54499/DL-57/2016/CP1355/CT0009) are supported by the European Union and the Portuguese Foundation for Science and Technology (FCT). Cardiovascular R&D Center is supported by FCT–Portuguese Foundation for Science and Technology, under the scope of the UnIC (UIDB/00051/2020 and UIDP/00051/2020).

Institutional Review Board Statement: The study was conducted after approval by the Ethics Committees of Centro Hospitalar Universitário São João (CHUSJ) (references CES 236/2021 and 109/2022) and Instituto Português de Oncologia do Porto (IPO-Porto) (reference CES IPO 38/2018).

Informed Consent Statement: Informed consent (World Medical Association Declaration of Helsinki–Ethical Principles for Medical Research Involving Human Subjects) was obtained from the enrolled patients before the surgical procedure.

Data Availability Statement: The data are not publicly available due to privacy and ethical restrictions but are available from the corresponding author by reasonable request.

Conflicts of Interest: The authors declare that they have no conflicts of interest.

Abbreviations

The following abbreviations are used in this manuscript:

ACE2	Angiotensin-converting enzyme 2
AD	Abdominal deep adipose tissue
ADAM17	Protease A disintegrin and metalloproteinase 17
AS	Abdominal superficial adipose tissue
AT	Adipose tissue
BMI	Body mass index
CHUSJ	Centro Hospitalar Universitário São João
COVID-19	Coronavirus disease 2019
E	Epiploon adipose tissue
HDL	High-density lipoprotein
IPO-Porto	Instituto Português de Oncologia do Porto
IPO8	Importin 8
LDL	Low-density lipoprotein
NRP1	Neuropilin-1
S	Spike
SARS-CoV-2	Severe acute respiratory syndrome coronavirus 2
T	Thigh
TMPRSS2	Transmembrane protease serine 2

References

1. Chen, Y.; Klein, S.L.; Garibaldi, B.T.; Li, H.; Wu, C.; Osevala, N.M.; Li, T.; Margolick, J.B.; Pawelec, G.; Leng, S.X. Aging in COVID-19: Vulnerability, immunity and intervention. *Ageing Res. Rev.* **2021**, *65*, 101205. [CrossRef]
2. Salazar, M.; Barochiner, J.; Espeche, W.; Ennis, I. COVID-19 and its relationship with hypertension and cardiovascular disease. *Hipertens. Riesgo Vasc.* **2020**, *37*, 176–180. [CrossRef]
3. Cao, H.; Baranova, A.; Wei, X.; Wang, C.; Zhang, F. Bidirectional causal associations between type 2 diabetes and COVID-19. *J. Med. Virol.* **2023**, *95*, e28100. [CrossRef]
4. Quarleri, J.; Delpino, M.V. The interplay of aging, adipose tissue, and COVID-19: A potent alliance with implications for health. *Geroscience* **2024**, *46*, 2915–2932. [CrossRef]
5. Rodrigues, A.R.; Salazar, M.J. and Gouveia, A.M. Obesity and Adipose Tissue Remodeling. In *Understanding Obesity: From Its Causes to Impact on Life*; Bentham Science: Sharjah, United Arab Emirates, 2020; pp. 55–80.
6. Steinberg, E.; Wright, E.; Kushner, B. In Young Adults with COVID-19, Obesity Is Associated with Adverse Outcomes. *West. J. Emerg. Med.* **2020**, *21*, 752–755. [CrossRef]
7. Lighter, J.; Phillips, M.; Hochman, S.; Sterling, S.; Johnson, D.; Francois, F.; Stachel, A. Obesity in Patients Younger Than 60 Years Is a Risk Factor for COVID-19 Hospital Admission. *Clin. Infect. Dis.* **2020**, *71*, 896–897. [CrossRef]
8. Deng, M.; Qi, Y.; Deng, L.; Wang, H.; Xu, Y.; Li, Z.; Meng, Z.; Tang, J.; Dai, Z. Obesity as a Potential Predictor of Disease Severity in Young COVID-19 Patients: A Retrospective Study. *Obesity* **2020**, *28*, 1815–1825. [CrossRef]
9. Zhang, F.; Xiong, Y.; Wei, Y.; Hu, Y.; Wang, F.; Li, G.; Liu, K.; Du, R.; Wang, C.Y.; Zhu, W. Obesity predisposes to the risk of higher mortality in young COVID-19 patients. *J. Med. Virol.* **2020**, *92*, 2536–2542. [CrossRef]
10. Dietz, W.; Santos-Burgoa, C. Obesity and its Implications for COVID-19 Mortality. *Obesity* **2020**, *28*, 1005. [CrossRef]

11. Li, M.Y.; Li, L.; Zhang, Y.; Wang, X.S. Expression of the SARS-CoV-2 cell receptor gene ACE2 in a wide variety of human tissues. *Infect. Dis. Poverty* **2020**, *9*, 45. [CrossRef]
12. Jia, X.; Yin, C.; Lu, S.; Chen, Y.; Liu, Q.; Bai, J.; Lu, Y. Two Things about COVID-19 Might Need Attention. *Preprints* **2020**, 2020020315. [CrossRef]
13. Al-Benna, S. Association of high level gene expression of ACE2 in adipose tissue with mortality of COVID-19 infection in obese patients. *Obes. Med.* **2020**, *19*, 100283. [CrossRef]
14. Saccon, T.D.; Mousovich-Neto, F.; Ludwig, R.G.; Carregari, V.C.; Dos Anjos Souza, A.B.; Dos Passos, A.S.C.; Martini, M.C.; Barbosa, P.P.; de Souza, G.F.; Muraro, S.P.; et al. SARS-CoV-2 infects adipose tissue in a fat depot- and viral lineage-dependent manner. *Nat. Commun.* **2022**, *13*, 5722. [CrossRef]
15. Martínez-Colón, G.J.; Ratnasiri, K.; Chen, H.; Jiang, S.; Zanley, E.; Rustagi, A.; Verma, R.; Chen, H.; Andrews, J.R.; Mertz, K.D.; et al. SARS-CoV-2 infection drives an inflammatory response in human adipose tissue through infection of adipocytes and macrophages. *Sci. Transl. Med.* **2022**, *14*, eabm9151. [CrossRef]
16. Basolo, A.; Poma, A.M.; Bonuccelli, D.; Proietti, A.; Macerola, E.; Ugolini, C.; Torregrossa, L.; Giannini, R.; Vignali, P.; Basolo, F.; et al. Adipose tissue in COVID-19: Detection of SARS-CoV-2 in adipocytes and activation of the interferon-alpha response. *J. Endocrinol. Investig.* **2022**, *45*, 1021–1029. [CrossRef]
17. Hoffmann, M.; Kleine-Weber, H.; Schroeder, S.; Krüger, N.; Herrler, T.; Erichsen, S.; Schiergens, T.S.; Herrler, G.; Wu, N.H.; Nitsche, A.; et al. SARS-CoV-2 Cell Entry Depends on ACE2 and TMPRSS2 and Is Blocked by a Clinically Proven Protease Inhibitor. *Cell* **2020**, *181*, 271–280.e8. [CrossRef]
18. Jocher, G.; Grass, V.; Tschirner, S.K.; Riepler, L.; Breimann, S.; Kaya, T.; Oelsner, M.; Hamad, M.S.; Hofmann, L.I.; Blobel, C.P.; et al. ADAM10 and ADAM17 promote SARS-CoV-2 cell entry and spike protein-mediated lung cell fusion. *EMBO Rep.* **2022**, *23*, e54305. [CrossRef]
19. Xiao, L.; Sakagami, H.; Miwa, N. ACE2: The key Molecule for Understanding the Pathophysiology of Severe and Critical Conditions of COVID-19: Demon or Angel? *Viruses* **2020**, *12*, 491. [CrossRef]
20. Lambert, D.W.; Yarski, M.; Warner, F.J.; Thornhill, P.; Parkin, E.T.; Smith, A.I.; Hooper, N.M.; Turner, A.J. Tumor necrosis factor-alpha convertase (ADAM17) mediates regulated ectodomain shedding of the severe-acute respiratory syndrome-coronavirus (SARS-CoV) receptor, angiotensin-converting enzyme-2 (ACE2). *J. Biol. Chem.* **2005**, *280*, 30113–30119. [CrossRef]
21. Heurich, A.; Hofmann-Winkler, H.; Gierer, S.; Liepold, T.; Jahn, O.; Pöhlmann, S. TMPRSS2 and ADAM17 cleave ACE2 differentially and only proteolysis by TMPRSS2 augments entry driven by the severe acute respiratory syndrome coronavirus spike protein. *J. Virol.* **2014**, *88*, 1293–1307. [CrossRef]
22. Cantuti-Castelvetri, L.; Ojha, R.; Pedro, L.D.; Djannatian, M.; Franz, J.; Kuivanen, S.; van der Meer, F.; Kallio, K.; Kaya, T.; Anastasina, M.; et al. Neuropilin-1 facilitates SARS-CoV-2 cell entry and infectivity. *Science* **2020**, *370*, 856–860. [CrossRef]
23. Daly, J.L.; Simonetti, B.; Klein, K.; Chen, K.E.; Williamson, M.K.; Antón-Plágaro, C.; Shoemark, D.K.; Simón-Gracia, L.; Bauer, M.; Hollandi, R.; et al. Neuropilin-1 is a host factor for SARS-CoV-2 infection. *Science* **2020**, *370*, 861–865. [CrossRef]
24. Ou, M.Y.; Zhang, H.; Tan, P.C.; Zhou, S.B.; Li, Q.F. Adipose tissue aging: Mechanisms and therapeutic implications. *Cell Death Dis.* **2022**, *13*, 300. [CrossRef]
25. Favre, G.; Legueult, K.; Pradier, C.; Raffaelli, C.; Ichai, C.; Iannelli, A.; Redheuil, A.; Lucidarme, O.; Esnault, V. Visceral fat is associated to the severity of COVID-19. *Metabolism* **2021**, *115*, 154440. [CrossRef]
26. Zipeto, D.; Palmeira, J.D.F.; Argañaraz, G.A.; Argañaraz, E.R. ACE2/ADAM17/TMPRSS2 Interplay May Be the Main Risk Factor for COVID-19. *Front. Immunol.* **2020**, *11*, 576745. [CrossRef]
27. Senapati, S.; Banerjee, P.; Bhagavatula, S.; Kushwaha, P.P.; Kumar, S. Contributions of human ACE2 and TMPRSS2 in determining host-pathogen interaction of COVID-19. *J. Genet.* **2021**, *100*, 12. [CrossRef]
28. Wilson, A.M.; Shao, Z.; Grenier, V.; Mawambo, G.; Daudelin, J.F.; Dejda, A.; Pilon, F.; Popovic, N.; Boulet, S.; Parinot, C.; et al. Neuropilin-1 expression in adipose tissue macrophages protects against obesity and metabolic syndrome. *Sci. Immunol.* **2018**, *3*, eaan4626. [CrossRef]
29. Cancello, R.; Tordjman, J.; Poitou, C.; Guilhem, G.; Bouillot, J.L.; Hugol, D.; Coussieu, C.; Basdevant, A.; Bar Hen, A.; Bedossa, P.; et al. Increased infiltration of macrophages in omental adipose tissue is associated with marked hepatic lesions in morbid human obesity. *Diabetes* **2006**, *55*, 1554–1561. [CrossRef]
30. Mayi, B.S.; Leibowitz, J.A.; Woods, A.T.; Ammon, K.A.; Liu, A.E.; Raja, A. The role of Neuropilin-1 in COVID-19. *PLoS Pathog.* **2021**, *17*, e1009153. [CrossRef]
31. Emilsson, V.; Gudmundsson, E.F.; Aspelund, T.; Jonsson, B.G.; Gudjonsson, A.; Launer, L.J.; Lamb, J.R.; Gudmundsdottir, V.; Jennings, L.L.; Gudnason, V. Serum levels of ACE2 are higher in patients with obesity and diabetes. *Obes. Sci. Pract.* **2021**, *7*, 239–243. [CrossRef]
32. Pietri, L.; Giorgi, R.; Bégu, A.; Lojou, M.; Koubi, M.; Cauchois, R.; Grangeot, R.; Dubois, N.; Kaplanski, G.; Valéro, R.; et al. Excess body weight is an independent risk factor for severe forms of COVID-19. *Metabolism* **2021**, *117*, 154703. [CrossRef]

33. Saul, H.; Gursul, D.; Piernas, C. People carrying excess weight have an increased risk of severe covid-19. *BMJ* **2022**, *376*, o141. [CrossRef]
34. Patel, V.B.; Mori, J.; McLean, B.A.; Basu, R.; Das, S.K.; Ramprasath, T.; Parajuli, N.; Penninger, J.M.; Grant, M.B.; Lopaschuk, G.D.; et al. ACE2 Deficiency Worsens Epicardial Adipose Tissue Inflammation and Cardiac Dysfunction in Response to Diet-Induced Obesity. *Diabetes* **2016**, *65*, 85–95. [CrossRef]
35. Couselo-Seijas, M.; Almengló, C.; Agra-Bermejo, R.M.; Luis Fernandez, Á.; Alvarez, E.; González-Juanatey, J.R.; Eiras, S. Higher ACE2 expression levels in epicardial cells than subcutaneous stromal cells from patients with cardiovascular disease: Diabetes and obesity as possible enhancer. *Eur. J. Clin. Investig.* **2021**, *51*, e13463. [CrossRef]
36. Frühbeck, G.; Catalán, V.; Valentí, V.; Moncada, R.; Gómez-Ambrosi, J.; Becerril, S.; Silva, C.; Portincasa, P.; Escalada, J.; Rodríguez, A. FNDC4 and FNDC5 reduce SARS-CoV-2 entry points and spike glycoprotein S1-induced pyroptosis, apoptosis, and necroptosis in human adipocytes. *Cell Mol. Immunol.* **2021**, *18*, 2457–2459. [CrossRef]
37. Steenblock, C.; Bechmann, N.; Beuschlein, F.; Wolfrum, C.; Bornstein, S.R. Do adipocytes serve as a reservoir for severe acute respiratory symptom coronavirus-2? *J. Endocrinol.* **2023**, *258*, e230027. [CrossRef]
38. Al Heialy, S.; Hachim, M.Y.; Senok, A.; Gaudet, M.; Abou Tayoun, A.; Hamoudi, R.; Alsheikh-Ali, A.; Hamid, Q. Regulation of Angiotensin- Converting Enzyme 2 in Obesity: Implications for COVID-19. *Front. Physiol.* **2020**, *11*, 555039. [CrossRef]
39. Taneera, J.; El-Huneidi, W.; Hamad, M.; Mohammed, A.K.; Elaraby, E.; Hachim, M.Y. Expression Profile of SARS-CoV-2 Host Receptors in Human Pancreatic Islets Revealed Upregulation of ACE2 in Diabetic Donors. *Biology* **2020**, *9*, 215. [CrossRef]
40. Da Eira, D.; Jani, S.; Ceddia, R.B. Obesogenic and Ketogenic Diets Distinctly Regulate the SARS-CoV-2 Entry Proteins ACE2 and TMPRSS2 and the Renin-Angiotensin System in Rat Lung and Heart Tissues. *Nutrients* **2021**, *13*, 3357. [CrossRef]
41. Li, L.; Spranger, L.; Soll, D.; Beer, F.; Brachs, M.; Spranger, J.; Mai, K. Metabolic impact of weight loss induced reduction of adipose ACE-2—Potential implication in COVID-19 infections? *Metabolism* **2020**, *113*, 154401. [CrossRef]
42. Soll, D.; Beer, F.; Spranger, L.; Li, L.; Spranger, J.; Mai, K. Effects of Weight Loss on Adipose and Muscular Neuropilin 1 mRNA Expression in Obesity: Potential Implication in SARS-CoV-2 Infections? *Obes. Facts* **2022**, *15*, 90–98. [CrossRef]
43. El-Sayed Moustafa, J.S.; Jackson, A.U.; Brotman, S.M.; Guan, L.; Villicaña, S.; Roberts, A.L.; Zito, A.; Bonnycastle, L.; Erdos, M.R.; Narisu, N.; et al. ACE2 expression in adipose tissue is associated with cardio-metabolic risk factors and cell type composition-implications for COVID-19. *Int. J. Obes.* **2022**, *46*, 1478–1486. [CrossRef]
44. Dou, H.; Feher, A.; Davila, A.C.; Romero, M.J.; Patel, V.S.; Kamath, V.M.; Gooz, M.B.; Rudic, R.D.; Lucas, R.; Fulton, D.J.; et al. Role of Adipose Tissue Endothelial ADAM17 in Age-Related Coronary Microvascular Dysfunction. *Arterioscler. Thromb. Vasc. Biol.* **2017**, *37*, 1180–1193. [CrossRef]
45. Voros, G.; Maquoi, E.; Collen, D.; Lijnen, H.R. Differential expression of plasminogen activator inhibitor-1, tumor necrosis factor-alpha, TNF-alpha converting enzyme and ADAMTS family members in murine fat territories. *Biochim. Biophys. Acta* **2003**, *1625*, 36–42. [CrossRef]
46. Khatib, M.; Zvibel, I.; Zelber-Sagi, S.; Varol, C.; Lahat, G.; Abu-Abeid, S.; Klausner, J.M.; Halpern, Z.; Fishman, S. Discriminatory metabolic and inflammatory parameters in serum and omental adipose tissue of obese patients with different insulin sensitivity. *J. Clin. Transl. Endocrinol.* **2014**, *1*, 115–119. [CrossRef]
47. Matsui, Y.; Tomaru, U.; Miyoshi, A.; Ito, T.; Fukaya, S.; Miyoshi, H.; Atsumi, T.; Ishizu, A. Overexpression of TNF-α converting enzyme promotes adipose tissue inflammation and fibrosis induced by high fat diet. *Exp. Mol. Pathol.* **2014**, *97*, 354–358. [CrossRef]
48. Yong, S.B.; Song, Y.; Kim, Y.H. Visceral adipose tissue macrophage-targeted TACE silencing to treat obesity-induced type 2 diabetes. *Biomaterials* **2017**, *148*, 81–89. [CrossRef]
49. McGrath, C.B.; Shreves, A.H.; Shanahan, M.R.; Guard, H.E.; Nhliziyo, M.V.; Pernar, C.H.; Penney, K.L.; Lotan, T.L.; Fiorentino, M.; Mucci, L.A.; et al. Etiology of prostate cancer with the TMPRSS2:ERG fusion: A systematic review of risk factors. *Int. J. Cancer* **2024**, *156*, 1898–1908. [CrossRef]
50. Schumacher, N.; Rose-John, S. ADAM17 orchestrates Interleukin-6, TNFα and EGF-R signaling in inflammation and cancer. *Biochim. Biophys. Acta Mol. Cell Res.* **2022**, *1869*, 119141. [CrossRef]
51. Cancello, R.; Zulian, A.; Gentilini, D.; Mencarelli, M.; Della Barba, A.; Maffei, M.; Vitti, P.; Invitti, C.; Liuzzi, A.; Di Blasio, A.M. Permanence of molecular features of obesity in subcutaneous adipose tissue of ex-obese subjects. *Int. J. Obes.* **2013**, *37*, 867–873. [CrossRef]
52. Kristem, L.; Recamonde-Mendoza, M.; Cigerza, G.C.; Khoraki, J.; Campos, G.M.; Mazzini, G.S. Roux-en-Y Gastric Bypass Downregulates Angiotensin-Converting Enzyme 2 (ACE2) Gene Expression in Subcutaneous White Adipose Tissue: A Putative Protective Mechanism Against Severe COVID-19. *Obes. Surg.* **2021**, *31*, 2831–2834. [CrossRef]
53. Cauwenberghs, N.; Prunicki, M.; Sabovčik, F.; Perelman, D.; Contrepois, K.; Li, X.; Snyder, M.P.; Nadeau, K.C.; Kuznetsova, T.; Haddad, F.; et al. Temporal changes in soluble angiotensin-converting enzyme 2 associated with metabolic health, body composition, and proteome dynamics during a weight loss diet intervention: A randomized trial with implications for the COVID-19 pandemic. *Am. J. Clin. Nutr.* **2021**, *114*, 1655–1665. [CrossRef]

54. Lim, S.; Zhang, M.; Chang, T.L. ACE2-Independent Alternative Receptors for SARS-CoV-2. *Viruses* **2022**, *14*, 2535. [CrossRef]
55. Alipoor, S.D.; Mirsaeidi, M. SARS-CoV-2 cell entry beyond the ACE2 receptor. *Mol. Biol. Rep.* **2022**, *49*, 10715–10727. [CrossRef]
56. Argyrou, M.; Pitsillou, E.; Hung, A.; El-Osta, A.; Karagiannis, T.C. Insights into the pathogenic mechanisms associated with the SARS-CoV-2 spike protein. *J. Struct. Biol.* **2025**, *217*, 108229. [CrossRef]
57. Melano, I.; Cheng, W.C.; Kuo, L.L.; Liu, Y.M.; Chou, Y.C.; Hung, M.C.; Lai, M.M.C.; Sher, Y.P.; Su, W.C. A disintegrin and metalloproteinase domain 9 facilitates SARS-CoV-2 entry into cells with low ACE2 expression. *Microbiol. Spectr.* **2023**, *11*, e0385422. [CrossRef]
58. Mahdi, M.; Kiarie, I.W.; Mótyán, J.A.; Hoffka, G.; Al-Muffti, A.S.; Tóth, A.; Tőzsér, J. Receptor Binding for the Entry Mechanisms of SARS-CoV-2: Insights from the Original Strain and Emerging Variants. *Viruses* **2025**, *17*, 691. [CrossRef]
59. Chen, N.; Decker, K.E.; Schulz, S.R.; Kempf, A.; Nehlmeier, I.; Moldenhauer, A.S.; Dopfer-Jablonka, A.; Behrens, G.M.N.; Stankov, M.V.; Manthey, L.; et al. Comparative Analysis of Host Cell Entry Efficiency and Neutralization Sensitivity of Emerging SARS-CoV-2 Lineages KP.2, KP.2.3, KP.3, and LB.1. *Vaccines* **2024**, *12*, 1236. [CrossRef]
60. Li, Z.; Lu, Q.; Zhang, R.; Xilifu, N.; Li, Y.; Huang, X.; Zhang, J.; Zang, S.; Jiang, G.; Liu, J. Metabolic memory in gestational diabetes enhances SARS-CoV-2 susceptibility in postpartum women: A prospective cohort study integrated with longitudinal metabolomics. *Endocr. Connect.* **2025**, *14*, e240681. [CrossRef]
61. Hinte, L.C.; Castellano-Castillo, D.; Ghosh, A.; Melrose, K.; Gasser, E.; Noé, F.; Massier, L.; Dong, H.; Sun, W.; Hoffmann, A.; et al. Adipose tissue retains an epigenetic memory of obesity after weight loss. *Nature* **2024**, *636*, 457–465. [CrossRef]
62. Wang, Y.; Wang, Y.; Luo, W.; Huang, L.; Xiao, J.; Li, F.; Qin, S.; Song, X.; Wu, Y.; Zeng, Q.; et al. A comprehensive investigation of the mRNA and protein level of ACE2, the putative receptor of SARS-CoV-2, in human tissues and blood cells. *Int. J. Med. Sci.* **2020**, *17*, 1522–1531. [CrossRef]
63. Zhang, L.; Chen, Y.; Li, C.; Liu, J.; Ren, H.; Li, L.; Zheng, X.; Wang, H.; Han, Z. RNA binding protein PUM2 promotes the stemness of breast cancer cells via competitively binding to neuropilin-1 (NRP-1) mRNA with miR-376a. *Biomed. Pharmacother.* **2019**, *114*, 108772. [CrossRef]
64. Wang, X.; Hu, H.; Liu, H. RNA binding protein Lin28B confers gastric cancer cells stemness via directly binding to NRP-1. *Biomed. Pharmacother.* **2018**, *104*, 383–389. [CrossRef]
65. Zhang, G.; Chen, L.; Khan, A.A.; Li, B.; Gu, B.; Lin, F.; Su, X.; Yan, J. miRNA-124-3p/neuropilin-1(NRP-1) axis plays an important role in mediating glioblastoma growth and angiogenesis. *Int. J. Cancer* **2018**, *143*, 635–644. [CrossRef]
66. Zhang, L.; Chen, Y.; Wang, H.; Zheng, X.; Li, C.; Han, Z. miR-376a inhibits breast cancer cell progression by targeting neuropilin-1 NR. *Onco Targets Ther.* **2018**, *11*, 5293–5302. [CrossRef]
67. Liu, Y.; Beyer, A.; Aebersold, R. On the Dependency of Cellular Protein Levels on mRNA Abundance. *Cell* **2016**, *165*, 535–550. [CrossRef]
68. Medina-Enríquez, M.M.; Lopez-León, S.; Carlos-Escalante, J.A.; Aponte-Torres, Z.; Cuapio, A.; Wegman-Ostrosky, T. ACE2: The molecular doorway to SARS-CoV-2. *Cell Biosci.* **2020**, *10*, 148. [CrossRef]
69. Jiang, X.; Eales, J.M.; Scannali, D.; Nazgiewicz, A.; Prestes, P.; Maier, M.; Denniff, M.; Xu, X.; Saluja, S.; Cano-Gamez, E.; et al. Hypertension and renin-angiotensin system blockers are not associated with expression of angiotensin-converting enzyme 2 (ACE2) in the kidney. *Eur. Heart J.* **2020**, *41*, 4580–4588. [CrossRef]
70. Zhang, X.J.; Qin, J.J.; Cheng, X.; Shen, L.; Zhao, Y.C.; Yuan, Y.; Lei, F.; Chen, M.M.; Yang, H.; Bai, L.; et al. In-Hospital Use of Statins Is Associated with a Reduced Risk of Mortality among Individuals with COVID-19. *Cell Metab.* **2020**, *32*, 176–187.e4. [CrossRef]
71. Nichols, J.H.; Smith, A.M.; Jonsson, C.B. The Intersection of SARS-CoV-2 and Diabetes. *Microorganisms* **2025**, *13*, 1390. [CrossRef]
72. Shukla, A.K.; Awasthi, K.; Usman, K.; Banerjee, M. Role of renin-angiotensin system/angiotensin converting enzyme-2 mechanism and enhanced COVID-19 susceptibility in type 2 diabetes mellitus. *World J. Diabetes* **2024**, *15*, 606–622. [CrossRef]
73. Al-Kuraishy, H.M.; Al-Gareeb, A.I.; El Kholy, A.A.; El-Khateeb, E.; Alexiou, A.; Papadakis, M.; Elekhnawy, E.; Alsubaie, N.; Hamad, R.S.; Batiha, G.E. The potential therapeutic effect of metformin in type 2 diabetic patients with severe COVID-19. *Eur. Rev. Med. Pharmacol. Sci.* **2023**, *27*, 11445–11456. [CrossRef]
74. Hansson, B.; Morén, B.; Fryklund, C.; Vliex, L.; Wasserstrom, S.; Albinsson, S.; Berger, K.; Stenkula, K.G. Adipose cell size changes are associated with a drastic actin remodeling. *Sci. Rep.* **2019**, *9*, 12941. [CrossRef]
75. Yang, W.; Guo, X.; Thein, S.; Xu, F.; Sugii, S.; Baas, P.W.; Radda, G.K.; Han, W. Regulation of adipogenesis by cytoskeleton remodelling is facilitated by acetyltransferase MEC-17-dependent acetylation of α-tubulin. *Biochem. J.* **2013**, *449*, 605–612. [CrossRef]
76. Sander, H.; Wallace, S.; Plouse, R.; Tiwari, S.; Gomes, A.V. Ponceau S waste: Ponceau S staining for total protein normalization. *Anal. Biochem.* **2019**, *575*, 44–53. [CrossRef]

International Journal of
Molecular Sciences

MDPI

Article

Regulation of Human Sortilin Alternative Splicing by Glucagon-like Peptide-1 (GLP1) in Adipocytes

Ashley Lui [1,†], Rekha S. Patel [2,†], Meredith Krause-Hauch [1], Robert P. Sparks [2,3] and Niketa A. Patel [1,2,*]

1 Department of Molecular Medicine, University of South Florida, Tampa, FL 33612, USA; ashleylui@usf.edu (A.L.); meredith.krause-hauch@va.gov (M.K.-H.)
2 Research Service, James A. Haley Veterans Hospital, Tampa, FL 33612, USA; rekha.patel1@va.gov (R.S.P.); robert.sparks@umassmed.edu (R.P.S.)
3 Department of Medicine, Division of Gastroenterology, UMass Chan Medical School, Worcester, MA 01655, USA
* Correspondence: niketa.patel@va.gov; Tel.: +1-(813)-972-2000
† These authors contributed equally to this work.

Abstract: Type 2 diabetes mellitus is a chronic metabolic disease with no cure. Adipose tissue is a major site of systemic insulin resistance. Sortilin is a central component of the glucose transporter -Glut4 storage vesicles (GSV) which translocate to the plasma membrane to uptake glucose from circulation. Here, using human adipocytes we demonstrate the presence of the alternatively spliced, truncated sortilin variant (Sort_T) whose expression is significantly increased in diabetic adipose tissue. Artificial-intelligence-based modeling, molecular dynamics, intrinsically disordered region analysis, and co-immunoprecipitation demonstrated association of Sort_T with Glut4 and decreased glucose uptake in adipocytes. The results show that glucagon-like peptide-1 (GLP1) hormone decreases Sort_T. We deciphered the molecular mechanism underlying GLP1 regulation of alternative splicing of human sortilin. Using splicing minigenes and RNA-immunoprecipitation assays, the results show that GLP1 regulates Sort_T alternative splicing via the splice factor, TRA2B. We demonstrate that targeted antisense oligonucleotide morpholinos reduces Sort_T levels and improves glucose uptake in diabetic adipocytes. Thus, we demonstrate that GLP1 regulates alternative splicing of sortilin in human diabetic adipocytes.

Keywords: type 2 diabetes; human adipocytes; human adipose tissue; glucose uptake; splicing minigene; antisense oligonucleotides; splice factors; TRA2B

1. Introduction

Type 2 diabetes mellitus (T2DM) continues to be a rising epidemic with more than 34.2 million Americans diagnosed with T2DM in 2020 and 88 million living with prediabetes [1]. The prevalence of T2DM has steadily increased since the mid-1900s and is projected to reach 60.7 million Americans by 2060 [2]. T2DM, a chronic disease with no cure, is characterized by hyperglycemia and insulin resistance and promotes metabolic syndrome and comorbidities such as cardiovascular disease, neuropathy, and nephropathy.

Adipose tissue is a major site of systemic insulin resistance contributing to T2DM progression and serves as a predictor for cardiovascular disease [3]. Nearly 90% of patients with T2DM are overweight or obese [1]. In adipocytes, the major glucose transporter, glucose transporter- 4 (Glut4), is tethered in the cytoplasm within Glut4 Storage Vesicles (GSVs) [4]. Postprandial signals translocate GSVs to the membrane, integrating Glut4 for glucose uptake from the circulation and GSVs are later recycled back through the trans-Golgi network [5]. Perturbed signaling in insulin-resistant adipocytes results in less translocation of GSVs to the plasma membrane and thus reduces glucose uptake, contributing to the diabetic pathology [6–8].

Components of GSVs have been identified [9–12] and of these, trafficking protein sortilin is an essential factor for GSV formation and function [13,14]. Sortilin is a multi-ligand sorting receptor ubiquitously expressed in the body. Its cellular role involves endocytosis of targets and trafficking of multiple vesicles through the trans-Golgi network as well as targeting to the lysosome for degradation. In adipocytes, sortilin's luminal domain binds Glut4 [15], and its motif rich cytoplasmic tail guides sortilin through the trans-Golgi network via specific post translational modifications [16,17] and interactions with various proteins [18–20].

Alternative splicing increases protein diversity by generating distinct proteins from the same gene via inclusion and exclusion of exons or alternate 5′ or 3′ splice site usage. Alternative splicing produces multiple products, often with different functions, from the same gene. Mis-regulation of alternative splicing may contribute to or cause a disease [21]. An alternatively spliced human sortilin variant was identified in neurons with inclusion of an exon, named exon 17b, inserted between exons 17 and 18 in sortilin pre-mRNA [22,23]. Importantly, the mouse and human sortilin pre-mRNA sequences are different, and it has been demonstrated [22–24] that a longer sortilin splice variant is present in mouse cells, which is not seen in human cells. In neurons, sortilin functions as a receptor for progranulin, while in adipocytes, sortilin is a component of the GSVs and functions as a sorting protein. However, presence of alternatively spliced variants of sortilin in human adipocytes, and their roles in glucose metabolism are not yet reported. Since sortilin is a component of GSVs in adipocytes and glucose uptake is reduced in diabetes, we evaluated human adipose tissue and human adipocytes from normal and type 2 diabetic individuals for the presence of sortilin splice variants. Our results demonstrate presence of the truncated splice variant (Sort_T) in human adipose tissue and adipocytes whose expression is increased in T2DM. Based on this discovery, we evaluated the role of the splice variant Sort_T in glucose uptake in T2DM adipocytes. The results demonstrate that Sort_T over-expression decreases glucose uptake in adipocytes. Glucagon-like peptide-1 (GLP1), an incretin hormone, and its analogs regulate glucose uptake and are widely used as therapeutics to manage T2DM. However, the effect of GLP1 on the expression of sortilin splice variants is not yet known. Hence, we hypothesized that GLP1 may regulate the alternative splicing of sortilin in human adipocytes. We have systematically evaluated the underlying alternative splicing mechanisms of sortilin pre-mRNA regulated by GLP1 in diabetic adipocytes.

2. Results

2.1. *Alternatively Spliced Sortilin Variants in Human Adipocytes*

Human visceral adipose stem cells (hASCs; from ZenBio™, Durham, NC, USA) from pooled nondiabetic or type 2 diabetic donors were used for experiments. We have previously shown that these hASCs maintain the patient's phenotype [25,26]. To evaluate the expression of sortilin, type 2 diabetic (T2DM) and nondiabetic (NDM) hASCs were differentiated to mature adipocytes, PCR was performed using primers targeting different domains of sortilin, and products were separated on a 1% agarose gel (Figure 1a). A single product was detected using primers that amplify sortilin's exons 4 to 6 in both NDM and T2DM adipocytes. Two products were detected in T2DM adipocytes using primers that amplify sortilin's exons 16–20. To determine the identity of these two PCR products, bands from the gel were purified and sequenced. Sequencing results (Figure 1b) indicated an inclusion of 99bp in the upper band, which corresponded to the exon 17b and confirmed the presence of the alternatively spliced variant as a result of exon 17b inclusion between exons 17 and 18 in human adipocytes. Further analysis of this sequence revealed an in-frame stop codon TAG (indicated by bold, underlined). Sequences were compared to exonic and intronic sequences from NCBI (Reference Sequence NM_002959.7, Gene ID: 6272 in *Homo sapiens*). The constitutively expressed product is referred to as Sort_FL, while the alternative variant is referred to as Sort_T. Our results demonstrate alternative splicing of sortilin in human adipocytes and increased levels of the alternatively spliced variant Sort_T in T2DM adipocytes.

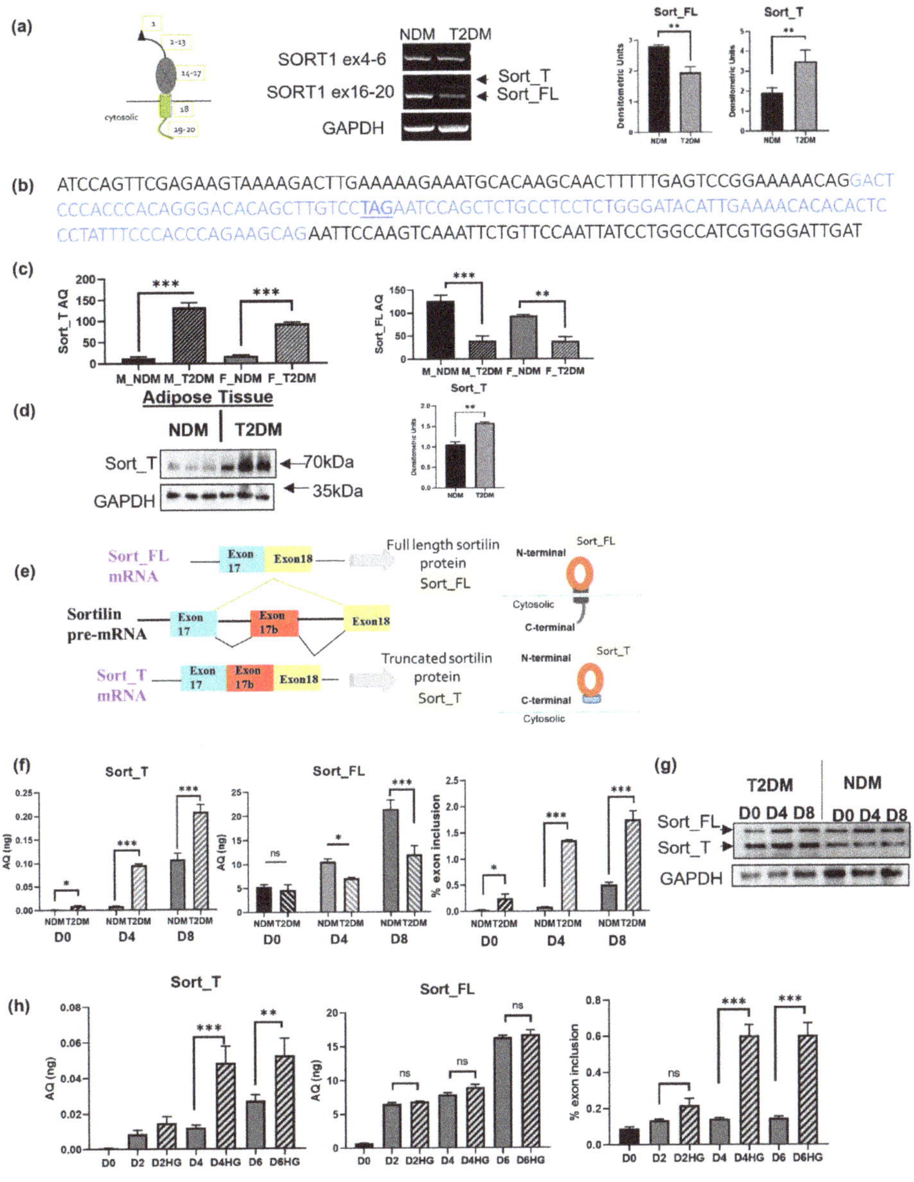

Figure 1. Sortilin alternatively spliced variants in diabetic human visceral adipocytes and adipose tissue. (**a**) HASCs from diabetic (DM) and nondiabetic (T2DM) donors were differentiated to mature adipocytes, RNA was isolated, and PCR performed using primers amplifying the region between exon 4 to 6 or the region between exon 16 to 20 (exon numbers of sortilin indicated in schematic). Products were separated on a 1% agarose gel and visualized using ethidium bromide staining. Graphs show densitometric analysis of individual PCR products normalized to GAPDH (n = 4). Statistical analysis was performed using *t*-test and ** $p < 0.01$. (**b**) The PCR products obtained using primers to sortilin cytoplasmic tail on exon 16 (sense) to exon 20 (antisense) were purified and sequenced. Sequencing of top band revealed inclusion of a 99bp sequence (as indicated by blue font), named exon 17b, between exons 17 and 18 of sortilin mRNA with an in-frame stop codon (bold, underlined).

(**c**) Adipose tissue from male and female, nondiabetic and diabetic donors was homogenized, and RNA was isolated. Real-time QPCR was performed using primers specifically within exon 17b detecting Sort_T and primer on the exon 17 to exon 18 junction to detect Sort_FL. A standard curve for each sortilin variant was included with the qPCR assay along with samples analyzed in triplicate. Absolute quantification (AQ, ng) for Sort_FL and Sort_T was calculated by normalizing the values to GAPDH absolute levels. Statistical analysis was performed using *t*-test between T2DM and NDM within each gender, ** $p < 0.01$ and *** $p < 0.001$. (**d**) Adipose tissue from T2DM and NDM donors were homogenized and whole-cell lysates were harvested. Western blot was performed and immunoblotted using Sort_T specific antibody as indicated. Graphs show densitometric analysis of Sort_T normalized to GAPDH (n = 3). Statistical analysis was performed using two-tailed Student's *t*-test and ** $p < 0.01$. (**e**) Schematic of sortilin pre-mRNA alternative splicing generating Sort_FL mRNA and Sort_T mRNA which produces Sort_FL protein (with C-terminal cytosolic tail) and Sort_T (truncated C-terminal). (**f**) T2DM and NDM hASCs were differentiated and harvested every four days (Day 0 preadipocytes: D0, Day 4 differentiated adipocytes: D4, Day 8 mature adipocytes: D8) to determine sortilin variant levels during adipogenesis. Real-time SYBR Green qPCR was performed using primers specifically for Sort_T and Sort_FL. A standard curve for each sortilin variant was included with the qPCR assay along with samples analyzed in triplicate. Graphs represent absolute quantification (AQ, ng) for Sort_FL and Sort_T calculated by normalizing to GAPDH AQ. Percent exon 17b inclusion was calculated using the following equation: %exon inclusion = $\frac{\text{Sort\_T}}{\text{Sort\_T+Sort\_FL}} * 100\%$. Statistical analysis was performed via multiple *t*-tests between diabetic and nondiabetic samples for each sortilin variant at different time points (n = 3). * $p < 0.05$, *** $p < 0.001$, ns = not significant. (**g**) Western blot was performed on NDM and T2DM D0, D4 and D8 whole-cell lysates using antibodies to N-terminal sortilin (detects both Sort_FL and Sort_T simultaneously) and GAPDH. Experiments repeated thrice. (**h**) NDM hASCs were differentiated in the presence of 25 mM glucose and harvested every 48 h along with a control sample. Real-time SYBR Green qPCR was performed. Graphs represent absolute quantification (AQ, ng) for Sort_FL and Sort_T calculated by normalizing to GAPDH and percent exon 17b inclusion was calculated using the following equation: %exon inclusion = $\frac{\text{Sort\_T}}{\text{Sort\_T+Sort\_FL}} * 100\%$. Statistical analysis was performed via multiple *t*-tests between control and high-glucose (HG) samples for each sortilin variant at different time points (n = 3). ** $p < 0.01$, *** $p < 0.001$, ns = not significant.

2.2. *Sort_T Is Increased in Diabetic Visceral Adipose Tissue*

Human visceral adipose tissue from T2DM and NDM donors (de-identified, male and female, n = 8 each, NHSR determination) were obtained to determine the presence of sortilin variants. Real-time qPCR was performed using primers specifically on exon 17b detecting Sort_T and a primer on the exon 17 to exon 18 junction to detect Sort_FL. The results demonstrate a marked increase in Sort_T levels in T2DM adipose tissue with concurrent reduced Sort_FL levels compared to NDM adipose tissue. This pattern occurred in both male and female adipose tissue (Figure 1c). Interestingly, total sortilin levels are lower in females compared to males. To confirm protein levels of truncated sortilin variant in adipose tissue, a Western blot was performed (Figure 1d). Lysates from NDM and T2DM visceral adipose tissue were analyzed via Western blot using Sort_T specific antibody. The results show increased Sort_T protein expression in T2DM adipose tissue. Figure 1e shows a schematic of sortilin pre-mRNA alternative splicing generating Sort_FL mRNA and Sort_T mRNA, which produces Sort_FL protein (full length with C-terminal cytosolic tail) and Sort_T (truncated C-terminal, missing cytosolic tail), respectively.

2.3. *Differential Sortilin Splice Variant Expression in Nondiabetic and Diabetic Adipocytes during Adipogenesis*

Next, we sought to elucidate the expression of sortilin splice variants during adipogenesis. The NDM and T2DM hASCs were differentiated and harvested as pre-adipocytes (D0), terminally differentiated adipocytes (D4), and mature adipocytes (D8). RNA was processed using real-time qPCR using primers specific for Sort_T and Sort_FL. The percent of Sort_T in total sortilin was calculated as percent exon inclusion for D0, D4 and D8 (Figure 1f).

Western blot analysis was performed on whole-cell lysates and the results show high Sort_T expression levels during adipogenesis in T2DM compared to NDM (Figure 1g). These results demonstrate that Sort_T is significantly increased during adipogenesis in T2DM adipocytes.

2.4. Hyperglycemia Increases Expression of Sort_T during Adipogenesis

To elucidate the effect of hyperglycemia (hallmark of diabetes) on sortilin alternative splicing, NDM hASCs were grown and differentiated in the presence of 25 mM high glucose (HG) and collected every 48 h. Real-time qPCR was performed using primers specific for Sort_T and Sort_FL and results showed a marked increase in Sort_T levels and percent exon 17b inclusion in response to hyperglycemia (Figure 1h).

2.5. Molecular Dynamics and Intrinsic Disorder Analysis of Sortilin Splice Variants

Available structures of sortilin in the literature and the Protein Data Bank only consist of the VPS10p luminal domain. With the improvement of molecular modeling, especially de novo prediction, we sought to model the complete conformation of sortilin splice variants (Figure 2a) using AlphaFold2 (https://alphafold.ebi.ac.uk/), an artificial intelligence (AI) software by DeepMind [27]. Following 3D structure prediction, the lowest energy folding compositions for Sort_T and Sort_FL were equilibrated for 100 ns using molecular dynamics (MD) performed with NAMD 2.12 [28] using the CHARMM36m force field [29]. Root mean square deviation (RMSD) for sortilin splice variants (Figure 2b) indicates stability of protein conformations over time [30,31].

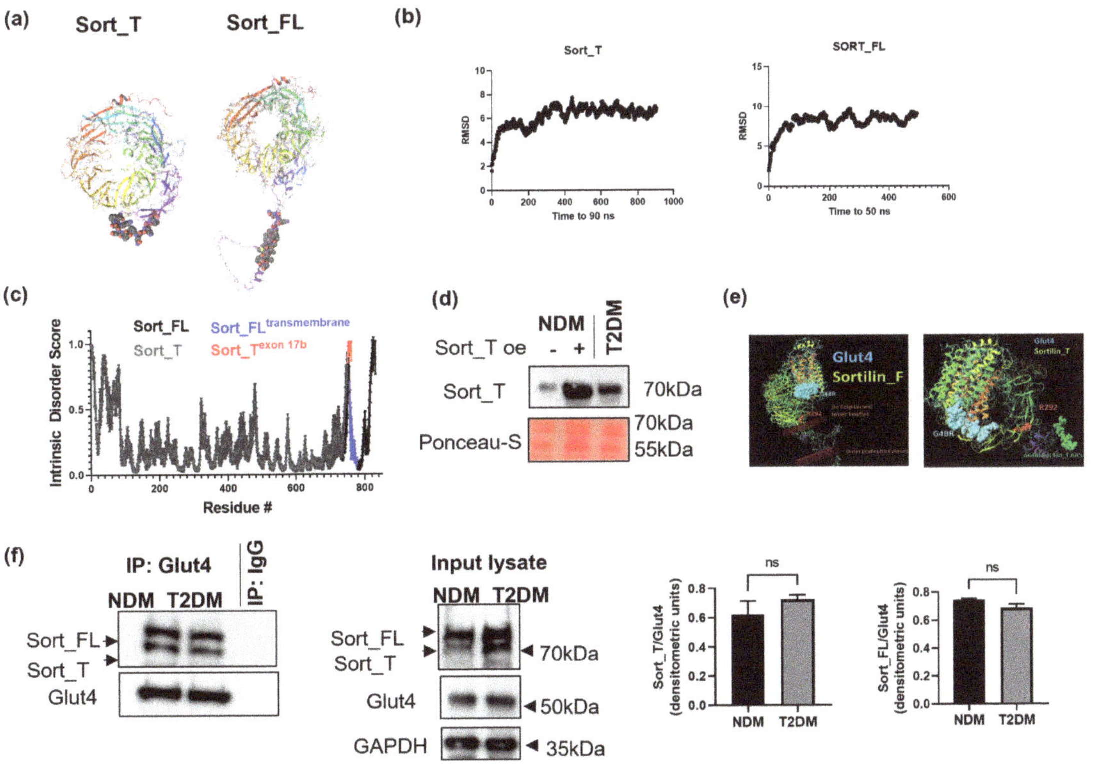

Figure 2. Molecular modeling and association of sortilin variants with Glut4. (**a**) De novo structure of sortilin splice variants was modeled using AlphaFold software. (**b**) Molecular dynamics (MD) was

performed on the lowest energy structure for Sort_T and Sort_FL using Schrodinger software. RMSD was calculated, indicating stability over 100 ns MD minimization. (**c**) Intrinsic disorder was predicted using PONDR-Fit software (http://original.disprot.org/pondr-fit.php) using sequences for each sortilin variant. Scores over 0.5 are predicted to be intrinsically disordered. (**d**) NDM adipocytes were transfected with Sort_T for 48 h. Conditioned media was collected from NDM, T2DM, or NDM over-expressing Sort_T, methanol extracted and analyzed via Western blot using the Sort_T specific antibody. Ponceau-S staining was used to visualize equal loading of lysates in the lanes. (**e**) Using the protein folding server I-TASSER (https://zhanggroup.org/I-TASSER/) and Schrodinger, models of Sort_FL, Sort_T and Glut4 were generated and equilibrated. Visualization and analysis were carried out using VMD 1.9.3. Characterization of protein–protein interactions between sortilin variants and Glut4 was carried out using ClusPro. Sort_FL was embedded into the membrane at TM domain (Shown as brown bars). Glut4 binding region (G4BR) is shown in blue. Arg (R292, in red) is on end of exon 17 and serves as an orientation point. The additional amino acids on Sort_T are shown in green. (**f**) Coimmunoprecipitation assay was performed by using Glut4 antibody for immunoprecipitation. Western blot analysis was then performed on the immunoprecipitated (IP) samples using antibodies against sortilin (against the N-terminal domain recognizing both variants) or Glut4 as indicated in the figure. Graphs show densitometric analysis of individual sortilin bands normalized to Glut4 (n = 3). Simultaneously, the input lysate was analyzed via Western blot using antibodies against sortilin or Glut4 (n = 3). Statistical analysis was performed via two-tailed *t*-test, ns = not significant.

Since insertion of exon 17b occurs between the disordered 10CC motif [32] and the ordered transmembrane domain, we evaluated whether the sortilin exon 17b insertion introduced ordered or amplified the already intrinsically disordered 10CC region. Intrinsically disordered regions (IDRs) have increased solvent exposure due to reduced or absent secondary and tertiary structure. This open conformation may account for more post-translational modifications and potential binding partners, but may affect the overall protein's function if the domain is necessary for protein–ligand affinity [33]. Using the meta predictor PONDR-Fit (http://original.disprot.org/pondr-fit.php), protein sequences for Sort_FL and Sort_T were analyzed. PONDR-Fit calculates an average disorder score and standard error per residue and thus predicts disorder indicated by scores over 0.5 [33,34]. The results (Figure 2c) show that Sort_FL contains the ordered structure of the transmembrane region (blue) and the intrinsic disordered cytoplasmic domain (ID score over 0.5). The results also demonstrate that the 10 amino acids of Sort_T (marked in red) contribute to the 10CC intrinsic disordered region.

2.6. Sort_T Is Secreted from Adipocytes

Since Sort_T lacks the transmembrane domain, we evaluated whether sortilin is present extracellularly in adipocytes. NDM adipocytes were transfected with Sort_T for 48 h. Conditioned media were collected from NDM, T2DM, or NDM over-expressing Sort_T, methanol extracted and analyzed via Western blot using the Sort_T specific antibody. The results (Figure 2d) show increased Sort_T in media from T2DM and in NDM over-expressing Sort_T compared to NDM adipocytes.

2.7. Computational Modeling Predicts Sort_T Maintains Ability to Bind Glut4

The lowest energy conformation structures for Sort_FL, Sort_T, and Glut4 were generated as described above. Previous studies indicate sortilin's VPS10p domain binds the first loop of Glut4's luminal domain [15]; thus, computational modeling was used to dock each sortilin splice variant to the first luminal loop of Glut4 using the ClusPro [35] server (https://cluspro.org/). The results (Figure 2e) predict similar Glut4 interaction for both Sort_T and Sort_FL at a site distant from the well-studied inner propeller binding pocket indicated by the Arg-292. The similar interaction conformations suggest the retained binding ability of Sort_T to Glut4.

2.8. Glut4 Binds to Sort_FL and Sort_T

To verify binding of Sort_T to Glut4, a co-immunoprecipitation assay was performed in NDM and T2DM adipocytes. Glut4 antibody was used to immunoprecipitate associated proteins and a Western blot was performed. Antibody targeting sortilin's N-terminal domain was used to visualize both sortilin variants. The results (Figure 2f) show Glut4 binds both Sort_FL and Sort_T splice variants in NDM and T2DM adipocytes.

2.9. Effects of Sortilin Splice Variants on Insulin Stimulated Glucose Uptake

Next, we evaluated the physiological outcome of increased Sort_T levels in adipocytes. Sort_T was overexpressed in NDM adipocytes, then treated with 100 nM insulin followed by glucose uptake assay. Over-expression of Sort_T was verified via real-time qPCR (Figure 3a). Impaired insulin stimulated glucose uptake was seen in NDM adipocytes overexpressing Sort_T compared to control NDM (Figure 3b).

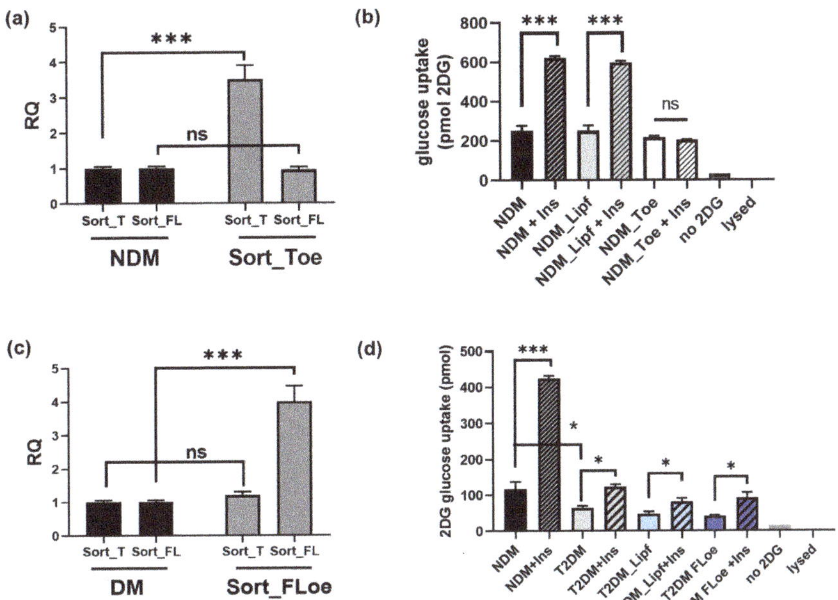

Figure 3. Sort_T overexpression decreases insulin stimulated glucose uptake. (**a**) Sort_T was transfected into NDM adipocytes, and RNA was harvested to confirm overexpression of Sort_T by qPCR analysis. *** $p < 0.001$, ns = not significant. (**b**) Simultaneously Sort_T was overexpressed in NDM adipocytes and glucose uptake assay was performed and uptake of 2 deoxyglucose-6-phosphate: (2DG) was measured; lipofectamine mock transfection, no DG and lysed cells were used as controls (n = 5). Statistical analysis was performed by multiple t-tests between basal and insulin treated samples; *** $p < 0.001$, ns = not significant. (**c**) Sort_FL was over-expressed in T2DM adipocytes, and RNA was harvested to confirm Sort_FL overexpression via qPCR analysis. Statistical analysis was performed by multiple t-tests between DM and Sort_FLoe samples; *** $p < 0.001$, ns = not significant. (**d**) Simultaneously, glucose uptake assay was performed on T2DM adipocytes overexpressing Sort_FL as described above (n = 3). Controls included NDM adipocytes, lipofectamine mock transfection, no 2DG, and lysed cells. Statistical analysis was performed via multiple t-tests between basal nondiabetic and diabetic adipocytes, and basal and insulin treated samples; * $p < 0.05$, *** $p < 0.001$, ns = not significant.

To determine if increasing Sort_FL levels in T2DM adipocytes could enhance insulin stimulated glucose uptake, Sort_FL was overexpressed in T2DM adipocytes and glucose uptake assay performed. Over-expression of Sort_T was verified via real-time qPCR (Figure 3c). The results (Figure 3d) demonstrate no change in insulin-stimulated glucose uptake with Sort_FL overexpression in T2DM adipocytes compared to control T2DM.

2.10. GLP1 Decreases Exon 17b Inclusion in Sortilin Pre-mRNA

We evaluated therapeutics developed for T2DM management and sought to determine if they may affect alternative splicing of sortilin. To test this, T2DM adipocytes were treated with metformin, rosiglitazone, insulin, or glucagon-like peptide-1 (GLP1). RNA was isolated and qPCR was performed using primers specific to Sort_T or Sort_FL. The results (Figure 4a,b) show all treatments reduced Sort_T expression; however, GLP1 was the most effective at significantly reducing Sort_T levels and increasing Sort_FL in T2DM adipocytes.

Based on these results, we determined the effect of GLP1 on sortilin splice variants by performing a dose and time curve. T2DM adipocytes were treated with increasing doses of GLP1 for 30 min and qPCR was performed. The absolute quantification and percent exon inclusion results (Figure 4c) demonstrate that 8 nM of GLP1 is an optimal dose for reducing exon 17b inclusion into sortilin pre-mRNA in T2DM adipocytes. This was followed by a time curve using 8 nM GLP1 concentration in T2DM adipocytes. Absolute quantification and percent exon inclusion was determined, and the results (Figure 4d) demonstrate a time-dependent decrease in Sort_T expression levels. Next, T2DM adipocytes were treated with GLP1 (8 nM) for 30 min. Cells were simultaneously harvested for Western blot analysis of protein expression of sortilin variants and gel PCR for sortilin splice variant levels. Additionally, RNA from T2DM adipocytes treated with 8 nM or 30 nM GLP1was also evaluated simultaneously for sortilin splice variant expression using real-time qPCR. The results (Figure 4e) show decrease in Sort_T in GLP1 treated T2DM adipocytes.

2.11. GLP1R Agonist Liraglutide Decreases Exon 17b Inclusion in Sortilin Pre-mRNA

GLP1 has a short half-life of two minutes and GLP1R agonists have been developed with modifications to increase stability of the peptide. Liraglutide has a 97% sequence homology to GLP1 and has an inclusion of a fatty acid chain and a Lys34Arg substitution, which increases the half-life to 13 h in vivo [36–38]. Hence, we evaluated liraglutide's effects on sortilin's pre-mRNA alternative splicing. The T2DM adipocytes were treated with 8 nM or 30 nM liraglutide for 10 min. The results (Figure 4f) show a dramatic reduction of Sort_T with a concurrent increase in Sort_FL.

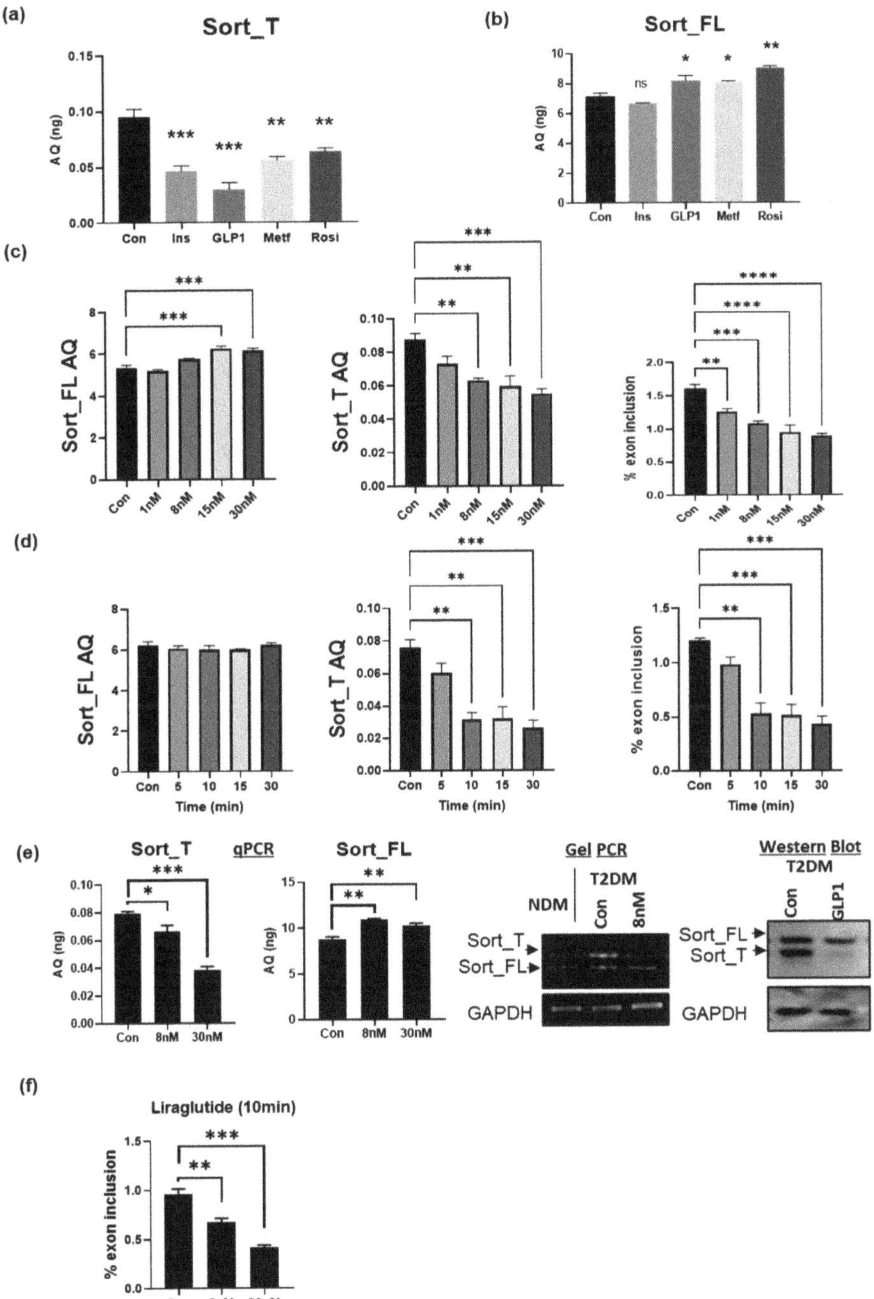

Figure 4. GLP1 reduces Sort_T levels in diabetic adipocytes. T2DM adipocytes and treated with 10 μM metformin (24 h), rosiglitazone (24 h), 100 nM insulin (30 min) or 30 nM GLP1 (30 min). RNA was isolated and qPCR was performed using primers specific to (**a**) Sort_T or (**b**) Sort_FL. To determine absolute quantities (AQ, ng) of each splice variant, a standard curve was included and normalized to GAPDH expression (n = 3). Statistical analysis was performed using one-way ANOVA, comparing each

treatment to control. * $p < 0.05$, ** $p < 0.01$, *** $p < 0.001$, ns = not significant. (**c**) T2DM adipocytes treated with increasing doses of GLP1 for 30 min. RNA was isolated and qPCR was performed using primers specific to Sort_T and Sort_FL. A standard curve for each sortilin splice variant was included and absolute quantification (AQ) was determined and normalized to GAPDH levels (n = 3). Percent exon inclusion was determined from absolute quantification of each splice variant using the following equation: $\%\text{exon inclusion} = \frac{\text{Sort\_T}}{\text{Sort\_T+Sort\_FL}} * 100\%$. Statistical analysis was performed via one-way ANOVA using untreated T2DM adipocytes as control, ** $p < 0.01$, and *** $p < 0.001$, **** $p < 0.0001$. (**d**) T2DM adipocytes were treated with 8 nM GLP1 and RNA was harvested at different time points (n = 3). QPCR was performed using primers specific to Sort_T and Sort_FL. A standard curve for each sortilin splice variant was included and absolute quantification quantified and normalized to GAPDH levels followed by quantification of percent exon inclusion. Statistical analysis was performed via one-way ANOVA using untreated diabetic adipocytes as control, ** $p < 0.01$, and *** $p < 0.001$. (**e**) T2DM adipocytes were treated with 8 nM GLP1 for 30 min. Whole-cell lysates were analyzed via Western blot using antibodies against sortilin (against the N-terminal domain recognizing both variants) or GAPDH (n = 3) or total RNA was used in PCR with primers spanning exon 16 to exon 20 (detecting both variants). Products were separated on 1% agarose gel and visualized using ethidium bromide (n = 3). Separately, real-time qPCR was performed using primers specific to Sort_T and Sort_FL. A standard curve for each sortilin splice variant was included and absolute quantification was determined and normalized to GAPDH levels (n = 3). Statistical analysis was performed via one-way ANOVA using untreated T2DM adipocytes as control, * $p < 0.5$, ** $p < 0.01$, and *** $p < 0.001$. (**f**) T2DM adipocytes were treated with 8 nM or 30 nM Liraglutide, a GLP1 receptor agonist, for 10 min. RNA was isolated and qPCR was performed using primers specific to Sort_T and Sort_FL (n = 3). Percent exon inclusion was determined from absolute quantification of each splice variant using the following equation: $\%\text{exon inclusion} = \frac{\text{Sort\_T}}{\text{Sort\_T+Sort\_FL}} * 100\%$. Statistical analysis was performed via one-way ANOVA using untreated T2DM adipocytes as control. ** $p < 0.01$, *** $p < 0.001$.

2.12. Predicted Splice Factors Regulating Sortilin Alternative Splicing in Diabetic Adipocytes

Splice factors are trans-acting proteins that are components of the spliceosome. The binding of splice factors to their conserved cis sequences along the pre-mRNA regulates the inclusion or exclusion of the exon by the spliceosome during pre-mRNA splicing. The dynamic interaction of splice factors regulates the pattern of alternative splicing to produce different mRNAs from a single gene. To predict the splice factors involved in sortilin's alternative splicing, sortilin exon 17b (99bp) and 200, base pairs upstream and downstream (totaling 499 nucleotides), were computationally analyzed using ESEfinder [39] and SFmap [40]. The analysis revealed strong consensus sequences for splice factors hnRNPA1, SRSF5 (aka SRp40), TRA2B (aka Tra2β1), TDP43, and SRSF1 (aka SF2/ASF) (schematic in Figure 5a).

Since GLP1 treatment decreased the expression of Sort_T, we sought to determine if the expression levels of any of the identified splice factors were affected by GLP1 treatment in T2DM adipocytes in concurrence with the changes in Sort_T expression. T2DM adipocytes were treated with 8 nM GLP1 for 30 min. Real-time qPCR was performed, and the results show increased levels of Sort_T in T2DM adipocytes and a decrease in Sort_T levels in response to GLP1 treatment (Figure 5b). Samples were further analyzed via qPCR for expression levels of hnRNPA1, SRSF5, TRA2B, TDP43, and SRSF1 to identify the splice factor whose expression levels correlated with Sort_T expression levels in GLP1-treated T2DM adipocytes. The results (Figure 5c) demonstrate that TRA2B is increased in T2DM adipocytes and treatment with GLP1 decreases its levels. While the levels of all these splice factors differed between NDM and T2DM adipocytes, the results indicate TRA2B (hereafter referred to as Tra2β1 protein) as the splice factor regulated by GLP1 treatment whose expression pattern changed in concurrence with Sort_T expression levels.

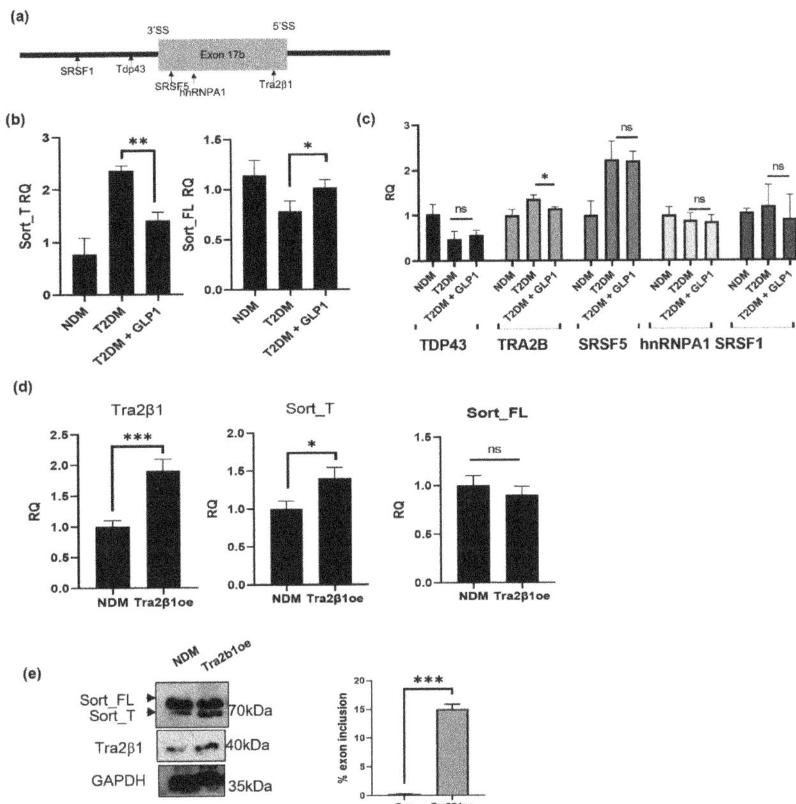

Figure 5. Tra2β1 mediates GLP1-regulated sortilin alternative splicing in diabetic adipocytes. (**a**) Schematic of bioinformatics-predicted splice factor binding on human sortilin exon 17b sequence and upstream intronic sequences. T2DM adipocytes were treated with 8 nM GLP1 for 10 min. RNA was harvested and qPCR was performed (**b**) using primers specific to Sort_T and Sort_FL or primers specific to (**c**) splice factors TDP43, TRA2B, SRSF5, hnRNAA1, and SRSF1. Results were normalized to GAPDH, relative quantification (RQ) was calculated by setting NDM as reference (n = 3). Statistical analysis was performed via two-tailed t-test between T2DM adipocytes and T2DM GLP1-treated sample. * $p < 0.05$, ** $p < 0.01$ and ns = not significant. (**d**) NDM adipocytes were transfected with Tra2β1 plasmid (Tra2β1oe) or lipofectamine mock transfection for 48 h. RNA was harvested, and qPCR performed using primers specific to Tra2β1, Sort_FL and Sort_T. Results were normalized to GAPDH, relative quantification (RQ) was calculated by setting NDM as reference (n = 3). Statistical analysis was performed via two-tailed Student's t-test. *** $p < 0.001$, * $p < 0.05$, and ns = not significant. (**e**) Cell lysates were harvested from NDM adipocytes overexpressing Tra2β1 (Tra2β1oe) and mock transfected control cells (NDM). Western blot was performed and immunoblotted using antibodies indicated (n = 3). Graph represents percent exon inclusion from densitometric analysis of blot using the equation: $\%\text{exon inclusion} = \frac{\text{Sort\_T}}{\text{Sort\_T}+\text{Sort\_FL}} * 100\%$. Statistical analysis was performed via two-tailed Student's t-test, *** $p < 0.001$.

2.13. *Tra2β1 Regulates Sortilin Alternative Splicing in Diabetic Adipocytes*

Next, we evaluated if Tra2β1 directly increases expression of Sort_T. For Tra2β1 overexpression, NDM adipocytes were transfected with Tra2β1 plasmid for 48 h. The QPCR (Figure 5d) and Western blot analysis (Figure 5e) results demonstrate that Tra2β1 overexpression significantly promotes percent sortilin exon 17b inclusion and an increase in Sort_T protein expression.

2.14. GLP1 Decreases Inclusion of Sortilin Exon 17b in a Splicing Minigene

The heterologous splicing minigene is a tool used to study alternative splicing without the influence of other endogenous factors. Hence, to evaluate if Tra2β1 could regulate the inclusion of sortilin's exon 17b directly, a sortilin exon 17b minigene was cloned. The pSPL3 splicing vector has two constitutive splicing exons: splice donor (SD) and splice acceptor (SA) exons, with a multiple cloning sites between them. Sortilin exon 17b and 200 bp upstream and downstream (499 bp) were cloned into the multicloning site of the pSPL3 vector and the resulting sortilin splicing minigene (MG) was verified via sequencing. Primer positions (Figure 6a) are depicted by arrows on SD and SA exons. Upon splicing, two products are obtained: a constitutive splicing of SD–SA and a product of 99 base pairs, larger due to the inclusion of sortilin exon 17b between SD and SA exons. The constitutive splicing of SD to SA provides an in-sample control for comparison with exon 17b inclusion.

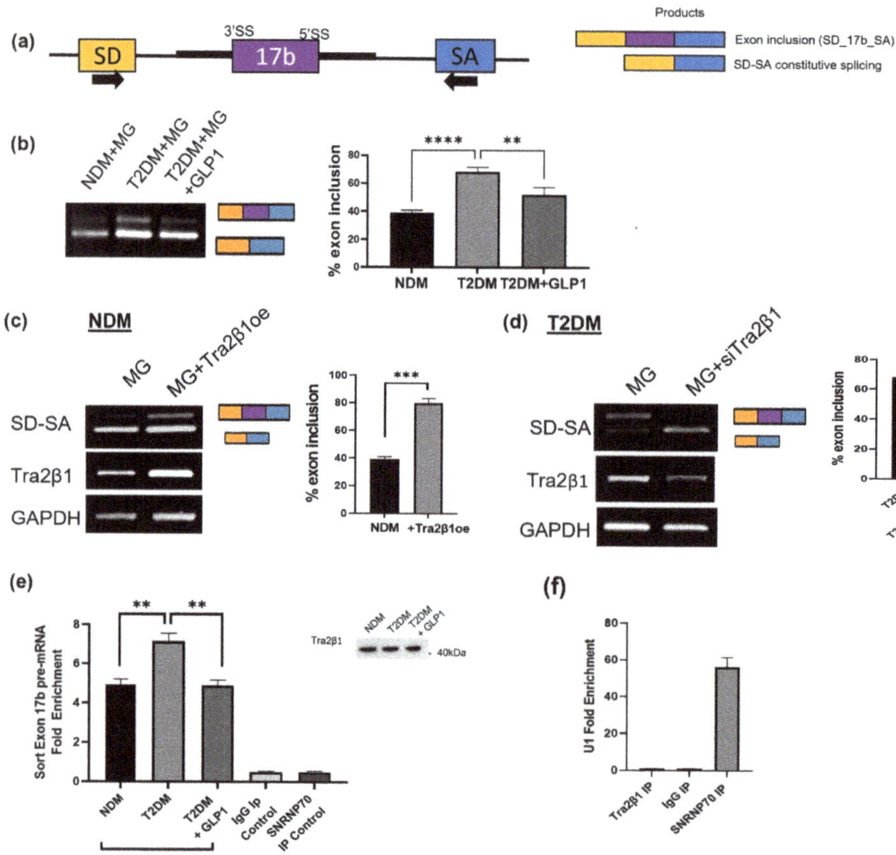

Figure 6. Human sortilin splicing minigene exon 17b inclusion is regulated by GLP1 and Tra2β1 (**a**) Schematic representing the sortilin splicing minigene (MG) with sortilin exon 17b (99 bp) and 200 bp of its flanking 3′ and 5′ introns cloned between exons SD (splice donor) and SA (splice acceptor) on the pSPL3 heterologous splicing vector. The arrows indicate primer positions to generate product with exon 17b inclusion (SD spliced to exon 17b spliced to exon SA) or exon 17b exclusion (constitutive splicing of SD to SA). (**b**) MG was transfected into NDM and T2DM adipocytes followed by GLP1 treatment (8 nM, 10 min) in T2DM (n = 3). RNA was isolated, and PCR was performed using primers on exons SD and SA to yield two products (n = 3). Densitometric analysis was performed, and

percent exon inclusion was calculated using the following equation: %exon inclusion = $\frac{SD\_17b\_SA}{SD\_17b\_SA+SD\_SA} * 100\%$. Statistical analysis was performed using one-way ANOVA. ** $p < 0.01$ and **** $p < 0.0001$. (**c**) NDM adipocytes were transfected with Tra2β1 plasmid (Tra2β1oe) for 48 h followed by MG transfected for 24 h, RNA was isolated, and PCR was performed using primers on exons SD and SA to yield two products (n = 3). Densitometric analysis was performed, percent exon inclusion was calculated, and statistical analysis was performed using two-tailed Student's *t*-test, *** $p < 0.001$. (**d**) T2DM adipocytes were transfected with Tra2β1 siRNA (siTra2β1) for 48 h followed by MG transfected for 24 h, RNA was isolated, and PCR was performed using primers on exons SD and SA to yield two products (n = 3). Densitometric analysis was performed, percent exon inclusion was calculated, and statistical analysis was performed using two-tailed Student's *t*-test, ** $p < 0.01$. (**e**) T2DM adipocytes were treated with 8 nM GLP1 for 10 min. RIP assay was performed by immunoprecipitating with Tra2β1 antibody, IgG antibody as a negative control, and SNRNP70 antibody as a positive control. RNA was eluted from antibodies and qPCR was performed using primers specific for sortilin exon 17b and Western blot was performed on input (shown in inset) using Tra2β1 antibody. ** $p < 0.01$. (**f**) U1, a spliceosome snRNP with known binding to SNRNP70 protein was analyzed by qPCR as a positive control. Analysis to calculate fold enrichment was performed using Ct values and Excel template provided by SigmaAldrich (St. Louis, MO, USA). Statistical Analysis was performed using one-way ANOVA, comparing NDM vs. T2DM, and T2DM vs. T2DM + GLP1.

NDM and T2DM adipocytes were transfected with MG for 24 h and T2DM+MG were treated with 8 nM GLP1 for 10 min. RNA was isolated and PCR was performed using SD and SA primers. The results show an increase in exon 17b inclusion in T2DM adipocytes compared to nondiabetic controls and a decrease in exon 17b inclusion with GLP1 treatment (Figure 6b).

2.15. Tra2β1 Increases Inclusion of Sortilin Exon 17b in a Splicing Minigene

To determine the role of Tra2β1 on sortilin exon 17b inclusion, NDM adipocytes were transfected with Tra2β1 plasmid for 48 h, followed by MG transfection for 24 h. PCR was performed using SD and SA primers. The results show an increase in exon 17b inclusion in the MG splicing with Tra2β1 overexpression compared to the lipofectamine control (Figure 6c).

In separate experiments, Tra2β1 was depleted using siRNA in T2DM adipocytes for 48 h followed by MG transfection for 24 h. RNA was isolated and PCR was performed. The results show a decrease in sortilin exon 17b inclusion in the MG splicing with reduced Tra2β1 levels (Figure 6d).

2.16. RIP Assay Shows Decreased Tra2β1 Binding to Sortilin Pre-mRNA with GLP1 Treatment

To evaluate Tra2β1 binding to exon 17b on sortilin pre-mRNA, an RNA immunoprecipitation assay (RIP) was performed. NDM and T2DM adipocytes were treated with 8 nM GLP1 for 10 min, then harvested. RIP assay was performed using the Tra2β1 antibody to immunoprecipitate and pull down associated sortilin pre-mRNA. An SNRP70 antibody was used as a positive control and an IgG antibody served as a negative control. A portion of input was removed to confirm equal Tra2β1 protein immunoprecipitation (inset) and RNA was eluted. QPCR was performed using primers on sortilin exon 17b and fold enrichment was calculated (Figure 6e). Fold enrichment for U1 binding (binds SNRNP70 protein) was calculated (Figure 6f) as a positive control. The results show that T2DM adipocytes have increased Tra2β1 binding to sortilin pre-mRNA, which is reduced with GLP1 treatment.

2.17. Morpholino Antisense Oligonucleotides Decrease Exon 17b Inclusion and Improve Glucose Uptake in Diabetic Adipocytes

Morpholinos are antisense oligonucleotides with a modified morpholine backbone, resulting in the ability to influence splicing events without degradation of the pre-mRNA transcript. Morpholino ASOs were designed spanning the Tra2β1 consensus sequence within exon 17b, such that the ASO inhibits Tra2β1 binding to sortilin pre-mRNA. The optimal 25-mer ASO (ASO_{1152}) was blasted using an NCBI Nucleotide Blast to confirm uniqueness and minimize off target effects within the human genome. T2DM adipocytes were treated with 1 μM or 10 μM ASO_{1152}, or 10 μM of a nonspecific control ASO (ASO_{NS}), or Endo-Porter transfection control (EP) for 48 h. The results (Figure 7a) show reduced Sort_T expression with no change in Sort_FL expression levels with ASO. Western blot analysis (Figure 7b) showed decrease in Sort_T protein levels. Apoptosis assay (Figure 7c) demonstrated no toxicity with ASO_{1152}. Next, T2DM adipocytes were treated with ASO_{1152}, and glucose uptake assay was performed. The results (Figure 7d) show improved insulin-stimulated glucose uptake with the morpholino ASO_{1152} treatment.

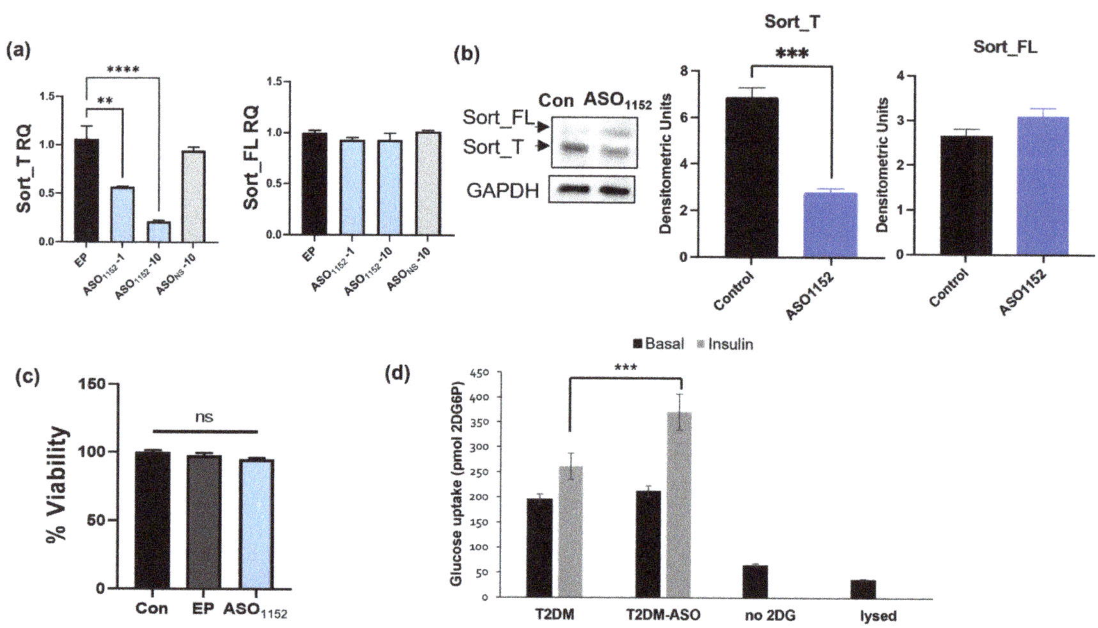

Figure 7. Morpholino antisense oligonucleotides mask Tra2β1 binding site on sortilin pre-mRNA. A Morpholino antisense oligonucleotide was developed to target the Tra2β1 consensus sequence and named ASO_{1152}. (**a**) T2DM adipocytes were treated with ASO_{1152} at 1 μM and 10 μM, mock Endo-Porter transfection (EP), and nonspecific antisense oligo control (ASO_{ns}) at 10 μM for 48 h. RNA was isolated and qPCR performed using primers specific to Sort_FL and Sort_T and normalized to GAPDH levels and Endo-Porter mock transfection control (n = 3). ** $p < 0.01$, **** $p < 0.0001$. (**b**) Simultaneously, Western blot analysis was performed using antibodies to N-terminal sortilin (detects both Sort_FL and Sort_T simultaneously) and GAPDH (n = 3). Statistical analysis was performed using one-way ANOVA. *** $p < 0.001$, and ns = not significant. (**c**) Acridine orange (AO) and propidium iodide (PI) viability assay was performed on untreated T2DM adipocytes, Endo-Porter mock transfection control, and morpholino ASO_{1152} treated T2DM adipocytes. Stained cells were analyzed in a cellometer (n = 5). Statistical analysis was performed using one-way ANOVA. ns = not significant. (**d**) T2DM adipocytes were treated with ASO_{1152}. Insulin-stimulated glucose uptake assay was performed and uptake of 2 deoxyglucose-6-phosphate: (2DG6P) was measured; no DG and lysed cells were used as controls (n = 5). Statistical analysis was performed via multiple *t*-tests between basal and insulin-treated samples; *** $p < 0.001$, ns = not significant.

3. Discussion

Sortilin is a multiligand receptor and trafficking protein that is constitutively expressed in multiple organs. Previously, Prudencio et al. described a truncated sortilin variant in the brain and showed that in human neuronal cells, the truncated sortilin is a nonfunctional receptor that does not internalize progranulin [22,41]. However, in adipocytes, sortilin functions as a trafficking protein and binds to Glut4 in GSVs. Here, we report that in addition to the constitutively expressed full-length sortilin (Sort_FL), human adipose tissue and adipocytes express the alternatively spliced, truncated sortilin (Sort_T) splice variant. We demonstrate that Sort_T expression is increased in human adipocytes from T2DM individuals. Our data also show that Sort_T is found in the conditioned media of diabetic adipocytes and non-diabetic adipocytes over-expressing Sort_T, which is in concurrence with previous reports [22,41–43]. These results suggest that Sort_T is either secreted directly from the trans Golgi network or is included in the GSV budding and held in the cytoplasm. Sort_FL is comprised of a large and highly conserved VPS10p domain (10-propeller structure) at its amino terminus, followed by a single-pass transmembrane domain and a short motif rich C-terminal tail. The 10CC site (comprised of 10 cysteines and 5 disulfide bonds) is directly below the VPS10p domain. The 10CC domain serves as an allosteric site, affecting ligand binding and protein stability [44]. In the alternatively spliced Sort_T variant, exon 17b inserts immediately after the 10CC site, introducing 10 novel amino acids prior to the stop codon, resulting in a truncated protein that lacks the cytosolic tail (schematic shown in Figure 1e). The C-terminal cytosolic tail of Sort_FL binds to retromer complex proteins and regulates movement of GSVs within the trans-Golgi network [18–20].

Since Sort_T lacks the cytosolic tail, we sought to determine the role of Sort_T in GSVs in adipocytes, which is not yet known. Artificial-intelligence-based structure prediction and molecular modeling indicated that Sort_T retained its ability to bind Glut4, and our coimmunoprecipitation assay results verified this interaction. The overexpression of Sort_T in nondiabetic adipocytes reduced glucose uptake while the over-expression of Sort_FL could not rescue T2DM glucose uptake. Based on these observations, we posit that Sort_T may affect GSV stability and trafficking in adipocytes. In diabetic adipocytes, the lack of sortilin's C-terminus results in the loss of guiding interactions, including post-translational modifications needed for endosomal sorting. Consequentially, Sort_T and its associated proteins may be transported to the lysosome for degradation, contributing to decreased glucose uptake. We are undertaking experiments such as pulse chase assays and measurements of UPP and lysosomal proteases to evaluate if Sort_T goes through lysosomal degradation with or without Glut4 in T2DM. This is beyond the scope of this manuscript, which is focused on the GLP1-mediated alternative splicing of sortilin. GSV has several components such as IRAP and LRP1, and the interaction of Sort_T with other GSV proteins is unknown and is being investigated by our laboratory. It is also possible that the absence of the cytoplasmic tail of Sort_T may not be solely responsible for aberrant Glut 4 trafficking, as other GSV proteins may compensate for it. It is likely that multiple defects contribute to impaired Glut4 trafficking and the degradation of GSVs in diabetes in adipocytes.

GLP1 is an incretin hormone normally found in blood at low concentrations (50 pmol/L), whose concentration is rapidly increased (peak levels of 200 pmol/L) following a meal [45]. Basal and postprandial levels of GLP1 are reduced in T2DM despite preservation of GLP1 receptor sensitivities and pathways. GLP1 was identified as a major contributor to the success of bariatric surgery in alleviating insulin resistance [46,47], and the GLP1 response is suggested as a predictor of success for remediation of T2DM via bariatric surgery [48–52]. GLP1 receptor agonists (GLP1Rs), GLP1 analogs, and DPP4 inhibitors behave similarly by preserving or mimicking GLP1 action. This family of T2DM medications shows cardio-protectivity, aids in weight loss via delayed gastric emptying and satiety, and has rare instances of hypoglycemia. 3T3L1 adipocytes treated with GLP1 demonstrate increased adipogenesis, resulting in advantages such as increased levels of Glut4, insulin receptor substrate-1 (IRS1), and insulin receptor β (IRβ) [53–55]. Contrasting results are seen in human adipocytes differentiated from human adipose stem cells (hASCs). Human adipocytes treated with GLP1 show a reduction of adipogenesis via a decreased expression of PPAR-γ, CEBP-α, and GSK-3 [56–58].

Alternative splicing is a post-transcriptional process resulting in the expansion of the transcriptome and proteome of a cell. This dynamic method is orchestrated by a collection of trans factors such as RNA binding proteins and spliceosome components (about 200 proteins associate/dissociate with the spliceosome) recognizing conserved cis sequences on pre-mRNA. Alternative splicing events are unique to tissues, developmental stages, sex, and disease states. The results of alternative splicing are pervasive, affecting every function of the cell through a multitude of means. Serine-arginine rich splice factors (SR proteins) are splicing enhancers known to act as a tether between pre-mRNA transcripts and the spliceosome complex, thus increasing exon inclusion Studies have implicated SR proteins in 5′ splice site selection, binding U1 snRNP, and aiding sequence recognition for splicing [59–61]. Tra2β1 is an SR-like protein, also known as SFRS10 or RA301, and is composed of two arginine-serine-rich (RS) domains with an RNA recognition motif (RRM) in between. The TRA2B gene was initially discovered in Drosophila and was named transformer 2β homolog and later renamed SFRS10 with a new SR splice factor nomenclature introduced by Manley and Krainer in 2010 [62]. Tra2β1/SFRS10 is not to be confused with SR splicing factor SRSF10 (also known as TASR1/SRp38/SRrp40). Increased Tra2β1 is associated with circulatory thickening and lesions [63] and is increased with inflammation [64] and nerve injury [65], while Tra2β1 is decreased in liver and muscle where it regulates the splicing of LPIN1 in obese humans and mice [66]. In adipocytes, Tra2β1 affects adipogenesis and enhances lipolysis [67]. Several genes are alternatively spliced in insulin resistance and T2DM [68–70]; notably, the insulin receptor gene itself is alternatively spliced and regulated by its own hormone ligand, insulin [71,72]. The data in Figure 5b,c show that the splice factor, TDP43, is reduced in T2DM, concurrent with increased Sort_T levels. Our unpublished data show that over-expressing TDP43 reduces Sort_T levels in adipocytes, which is in concurrence with previous reports on neurons [22,41]. However, GLP1 treatment did not affect TDP43 levels but only affected Tra2β1 levels as shown in Figure 5c. Hence, we pursued the evaluation of Tra2β1 as the splice factor regulated by GLP1 treatment in T2DM adipocytes, as we sought to evaluate the mechanism of the GLP1-mediated alternative splicing of sortilin in adipocytes.

Here, we demonstrate that the incretin hormone, GLP1, regulates the alternative splicing of sortilin in human adipocytes. Hormones such as insulin and GLP1 and insulin-sensitizers such as metformin and rosiglitazone affect kinases and phosphatases that phosphorylate/dephosphorylate trans-factors in the spliceosome, thereby changing their interaction (promote or inhibit) with the cis-element (consensus sequences) on the pre-mRNA. These often affect common signaling pathways such as PKC, PKA, PI3K/AKT that cross-talk to mediate the biological outcome. GLP1 mediates its effects via a G-protein-coupled receptor and signals through the PI3K/AKT, PKA/cAMP, and PKC pathways [73,74]. Previous studies have indicated GLP1 inducing rapid changes (<15 min) in the ERK/MEK pathway in adipocytes compared to slower changes in the PI3K/AKT signaling (2 h) [75]. GLP1 receptor (GLP1R) expression is higher in visceral adipose tissue compared to subcutaneous adipose tissue [76]. GLP1-GLP1R signaling in adipocytes is also shown to be mediated via the PKA and MAPK pathways in adipocytes [75,77]. Suppression of GLP1R expression reduces proliferation and differentiation while inducing apoptosis of pre-adipocytes [75]. GLP1 has also been shown to mediate its effects through the IL6 pathway in human adipocytes [78]. We are currently evaluating whether GLP1 signaling may also affect kinases that phosphorylate splicing trans-factors and whether it may potentially affect alternative splicing events through the signaling cascades in adipocytes.

A widely used method to alter alternative splicing is the use of antisense oligonucleotides (ASOs) to mask an enhancer or silencer binding site, resulting in exon skipping. Morpholinos ASOs are comprised of a morpholine backbone to protect the ASO from

RNAse H recognition and cleavage [79,80]. An outcome of modified backbones is the ability to modulate exon inclusion/exclusion without the degradation of the whole pre-mRNA transcript in the nucleus [81,82]. Our results demonstrate that morpholino ASO, targeted to mask the Tra2β1 binding site on exon 17b, reduces Sort_T expression and improves glucose uptake. This may be developed into a new therapeutic option for the management of T2DM.

The in vivo environment in adipose tissue in individuals with type 2 diabetes is, however, varied due to the polygenic nature of the disease. Hence, the limitation of this study is that a direct effect of GLP1, modulating sortilin splicing and culminating in improved glucose uptake in patients with type 2 diabetes, is yet to be demonstrated in vivo. Additionally, Sort_T may influence the formation or trafficking of GSV, which may be affected when sortilin splice variant levels are changed. The signaling pathway of GLP1 as well as the levels and mechanisms of association with its receptor (GLP1R) or signaling via other receptors, to regulate alternative splicing of sortilin, remains to be established in human adipocytes. In summary, we have identified sortilin alternative splicing as an important regulatory process in human adipocytes, and our data indicate that high levels of truncated sortilin substantially contribute to the manifestation of comorbidities associated with type 2 diabetes. Using robust tools and assays, we have demonstrated that GLP1 regulates the alternative splicing of sortilin. The study has established that truncated human sortilin has physiological consequences in glucose uptake in adipose tissue and may be a viable target in the management of type 2 diabetes.

4. Materials and Methods

4.1. Cell Culture

Human-visceral-adipose-derived stem cells (HASCs) and media were purchased from Zenbio™ (Durham, NC, USA) and passaged as pre-confluent cultures in pre-adipocytes media at 37 °C and 5% CO_2. Normal (HbA1c < 5.7%), and diabetic hASCs (HbA1c > 7%) were from pooled donors and were BMI matched. On day 2, media were changed to differentiation media, and cells were harvested on day 6 (4 days of differentiation) as differentiated adipocytes for experiments. For mature adipocytes, cells were kept in differentiation media for 6 days and changed to adipocyte media for 48 h before harvesting on day 10.

4.2. Human Adipose Tissue

Visceral adipose tissue from male/female nondiabetic/diabetic donors (de-identified, discarded samples from elective surgeries, BMI-matched, diabetic HbA1c > 6.5, nondiabetic HbA1c < 6.5, IRB #Pro00020295: non-human subject research (NHSR)) was homogenized in a bead homogenizer. Whole-cell lysates were harvested or RNA was isolated using Trizol (ThermoFisher Scientific, Waltham, MA, USA) per manufacturer's instructions.

4.3. Plasmid Transfections

Human sortilin plasmids for Sort_FL and Sort_T were designed by our lab and purchased from VectorBuilder (Chicago, IL, USA). HASCs were grown to confluency and differentiated for 2 days in differentiation media, 1 µg of plasmid per 1 mL of media, or lipofectamine mock transfection control was performed following manufacturer's instructions (Lipofectamine 3000, ThermoFisher #L3000015) for 48 h in differentiation media, total 4 days of differentiation.

4.4. Polymerase Chain Reaction

RNA was isolated using Trizol™ (ThermoFisher # 15596026) and cDNA was synthesized using 1 µg RNA (260/230 > 1.8 and 260/290 > 1.8) using iScript™ (Bio-Rad #1708891). Target amplification was performed with 1 µL cDNA was and JumpStart RED-Taq ReadyMix Reaction Mix (Sigma-Aldrich #P0982, St. Louis, MO, USA). Products were separated on a 1% agarose gel, visualized using ethidium bromide, and imaged on the

ProteinSimple FluorChem™. Densitometric analysis was performed using AlphaView Software (Version 3.3.0).

Human primers used included hSortilin exon 16 S to hSortilin exon 20 AS to see both sortilin splice variants simultaneously. For qPCR analysis and absolute quantification, human primers used included hSortilin exon 17b S and exon 17b AS to detect Sort_T or hSortilin, exon 17 and exon 17/18 AS to detect Sort_FL. See Table 1 for all human primer sequences. SYBR Green qPCR was performed on the ViiA 7 (ABI) and primer concentrations were optimized for a single melt curve and consistent amplification. Plate set-up included a standard series, no template control, and no reverse transcriptase control. A standard curve was generated and used to calculate absolute quantities of target expression using primers as per experiment. Samples were normalized to GAPDH.

Table 1. Human Primers with sequence used for PCR analysis and sequencing.

Gene	Species	Forward and Reverse Primer Sequences (5′-3′)
β-Actin	Human	CTCTTCCAGCCTTCCTTCCT AGCACTGTGTTGGCGTACAG
Sortilin exon 16–20	Human	AAGGGCCACGACCTGGAGTT CAATCTCGAGCTATTCCAAGAGGTCCTCATCTG
Sort_T	Human	TTGTCCTAGAATCCAGCTCTGC TCTGGGTGGGAAATAGGGAGT
Sort_FL	Human	TGGGGTAAATCCAGTTCGAG GACTTGGAATTCTGTTTTTCCGGAC
GAPDH	Human	GATCATCAGCAATGCCTCCT TGTGGTCATGAGTCCTTCCA
TRA2B	Human	TGTCTACTCGCAGGCGTCATGT AGACACATCGGCAATGGGACCA
SD-SA	N/A	GTGAACTGCACTGTGACAAGC CACCTGAGGAGTGAATTGGTC

Relative quotient (RQ) was determined using the comparative method (ΔΔCT). Primers sequences are shown in Table 1 below. Primers were either purchased from Origene (Rockville, MD, USA) or designed using the primer-designing tool Primer BLAST (NCBI) including verification of primer specificity, followed by synthesis by Eurofins Genomics (Louisville, KY, USA). Next, the optimal primer concentration for forward and reverse primers was determined from a range of 50–900 nM. The final concentration was selected to ensure efficiency and specificity for its target, based on the dissociation curve that showed a single sharp peak, indicating that the primers amplify one specific target.

4.5. Western Blot Analysis

Cell lysates were harvested using lysis buffer (Cell Signaling 9803S, Danvers, MA, USA) + 10% protease/phosphatase inhibitor (Pierce A32957, A32953 purchased from ThermoFisher Scientific, Waltham, MA, USA). Conditioned media were collected from cells, mixed with 4 times the volume of methanol, and vortexed well. A volume of chloroform was added and vortexed followed by centrifugation of fat at 9000× g for 5 min. The interface layer was carefully separated, mixed with 3 times the volume of methanol, followed by centrifugation of fat at 9000× g for 10 min at 4 °C. The supernatant was decanted and the pellet was reconstituted in lysis buffer. Lysates were kept on ice or stored in the freezer for an hour, then sonicated briefly. Cell lysates (80 µg) were separated on a 7.5% SDS-PAGE gel, wet-transferred to nitrocellulose membranes, blocked with 5% nonfat dried milk in Tris-Buffered Saline with 0.05% Tween 20 (TBST). Membranes were probed with Sortilin N- terminal antibody to visualize both splice variants (R&D Systems MAB3154; RRID: AB_2192601, Minneapolis, MN, USA), Sortilin C- terminal antibody to visualize full length sortilin (Abcam #ab16640; RRID: AB_2192606, Cambridge, UK), Glut4 (Cell

Signaling #2213; RRID: AB_823508, Danvers, MA, USA), and GAPDH (Santa Cruz #sc25778; RRID: AB_10167668, Dallas TX, USA). The Sort_T antibody was custom made by Pacific Immunology (Ramona, CA, USA) against a synthetic peptide, CSPEKQDSHPQGHSLS, in New Zealand white rabbits. The serum was passed through a peptide affinity column to purify the monoclonal antibody. The specificity for Sort_T antibody was determined using ELISA with pre-immune serum as a control. It was verified via a competition assay with the peptide epitope followed by Western blot analysis.

Secondary HRP antibodies for rabbit (Biorad #5196-2504, Hercules, CA, USA) and mouse (Biorad #0300-0108P, Hercules, CA, USA) were used with chemiluminesence reagent (Pierce #32109 purchased from ThermoFisher Scientific, Waltham, MA, USA) for detection. Images were digitally captured using ProteinSimple FluorChem^{TM} and densitometric analysis was performed using AlphaView Software (version 3.3.0).

4.6. Coimmunoprecipitation Assay

Human nondiabetic and diabetic pre-adipocytes were grown to confluency and differentiated for 4 days in differentiation media. Cells were harvested and 200 µg in 200 µL volume of lysate was used for coimmunoprecipitation assay. Lysates were rocked with Protein A/G Agarose (Santa Cruz sc2003, Dallas, TX, USA) for 30 min at 4 °C to clear nonspecific binding then centrifuged at 2000 rpm for 1 min to remove protein beads. Immunoprecipitation was performed using 2 µg Glut4-agarose antibody (Santa Cruz sc53566 AC, Dallas TX, USA) or IgG (Santa Cruz sc-51993, Dallas, TX, USA) as control and rocked overnight at 4 °C. Lysate was then centrifuged, and pellet was washed with PBS and resuspended in Laemmli buffer. Western blot analysis was performed as described above.

4.7. Glucose Uptake Assay

Human nondiabetic and diabetic pre-adipocytes were grown to confluency in a 96-well plate and differentiated for 2 days in differentiation media; 1µg plasmid per 1 mL of media for Sort_T, Sort_FL or lipofectamine control was overexpressed for 48 h in differentiation media. Glucose uptake was performed (ProMega #TM467, Madison, WI, USA) by measuring luciferase luminescence. Briefly, cells were serum-starved in Krebs-Ringer Phosphate HEPES buffer (KRPH) for an hour before insulin treatment (100 nM, 10 min). Cells were washed and treated with 1 mM of 2-deoxyglucose 6-phosphate (2DG6P) for 10 min. Cells were washed again in KRPH buffer and lysed. A standard curve was generated, and luminescence was measured on a Biotek Synergy Mx microplate reader (Agilent Technologies, Santa Clara, CA, USA) using a 1 s integration time and 2DG6P quantification of samples were calculated.

4.8. Computational Modeling

Sequences for full-length human Glut4 and Sort_FL were obtained from UniProt[232] and Sort_T alternatively spliced variant sequence was derived via translation of Sort_T nucleotide sequence and alignment with Sort_FL. Sequences were submitted to Alphafold 2.2.0 for protein folding and structure prediction. Folded sortilin variants and Glut4 proteins were minimized in Schrodinger protein preparation wizard (2019-1: Maestro, Schrödinger, LLC, New York, NY, USA). Molecular dynamics simulations were performed using NAMD 2.12 and CHARMM36m force field at a 100 ns time scale. The system was prepared using the CHARMM-GUI solution builder, with a salt concentration of 150 mM NaCl. Visualization and analysis were carried out using Visual Molecular Dynamics (VMD 1.9.3). The system was equilibrated for 100 ns, restraining the Cα atoms of the protein (1.0 kcal/mol/A2) to allow for solvation. Root mean square deviations (RMSD) using backbone Cα atoms were calculated using VMD RMSD trajectory tool (VMD 1.9.3)for Sort_FL, and Sort_T. Protein–protein docking was performed using the last frame from the equilibrated structures of Sort_T, Sort_FL, and Glut4 on the ClusPro server, targeting the first luminal loop of Glut4 interaction to Sort_FL and Sort_T.

4.9. Disorder Analysis

Sortilin protein sequences were analyzed using PONDR-Fit (www.disprot.org accessed on May 2019), a multi-residue meta-predictor of intrinsic disorder in protein sequences. PONDR-Fit uses an 8-fold cross-validation method to calculate a disorder score and standard error per residue. The results were exported from the PONDR-Fit server; prediction scores above 0.5 are considered disordered.

4.10. Cloning Human Sortilin Exon 17b-Splicing Minigene

The human-sortilin-splicing minigene (MG) was cloned by isolating genomic DNA from human adipose stem cells (hASCs) and amplifying exon 17b $\pm$ 200 bp of the upstream and downstream introns using primers with designed SacI and NheI restriction sites. The product and the pSPL3 splicing vector were digested with NheI and SacI restriction enzyme, followed by ligating the insert into the pSPL3 splicing minigene. The orientation of insert was verified via PCR and sequencing of the minigene.

4.11. RNA-Immunoprecipitation Assay (RIP)

Imprint RNA Immunoprecipitation kit (Sigma-Aldrich #RIP, St Louis, MO, USA) was purchased and assay was carried out per manufacturer's instructions. Briefly, human nondiabetic and diabetic adipocytes were differentiated into adipocytes and diabetic adipocytes were treated with GLP1 (8 nM, 10 min); 2×10^6 cells were resuspended in 525 µL RIP Lysis buffer. Cells were harvested then rocked in protein A/G beads for 1 h to clear nonspecific binding. Immunoprecipitation was performed with 2 µg Tra2β1/SFRS10 antibody, SNRNP70 antibody (positive control, Millipore #03-103), or IgG antibody (as a negative control, included in kit). A portion was removed to confirm equal immunoprecipitation via Western blot. RNA was eluted, isolated, and treated with DNase to remove genomic DNA. SYBR Green qPCR was performed using primers specific to sortilin exon 17b and U1 (positive control for SNRNP70). The yield (percent input) and specificity (fold enrichment) were calculated using a provided RIP Excel template from Sigma-Aldrich (St. Louis, MO, USA).

4.12. Morpholino Treatment

Human diabetic pre-adipocytes were grown to confluency in a 12-well plate and differentiated for 2 days in differentiation media. Morpholinos were added to fresh differentiation media, to a final concentration of 1 µM or 10 µM, and dish was swirled, then Endo- Porter (Gene Tools, Philomath, OR, USA) 6 µL/mL of media was added and swirled. Cells were incubated for 48 h. Sequences used for the 3′ splice site are 5′-TGTGTGTATGATTGCACAGGACTCC-3′ and for the 5′ splice site, 5′-AGAAGCAGGTAGG TTACACGAAAAC-3′.

4.13. Statistical Analysis

Densitometric analysis of Western blots was carried out using AlphaView™ software (Version 3.3.0) from ProteinSimple™. Experiments were independently repeated three to five times for reproducibility. PRISM software (version 10.0.2) (GraphPad) was used for statistical analysis. Two-way analysis of variance (ANOVA) or Student's t test was used in the analysis, indicated by the figure legends (* $p < 0.05$, significant; ** $p < 0.01$, highly significant; *** $p < 0.001$, extremely significant). Each data set passed the Shapiro–Wilk normality test (alpha = 0.05).

Author Contributions: A.L. performed experiments, analyzed the data, and co-wrote the manuscript. R.S.P. performed the co-immunoprecipitation assays, Western blots, and PCR experiments. M.K.-H. performed the statistical analysis of the data. R.P.S. performed the AlphaFold modeling and molecular dynamics. N.A.P. developed the concept, designed the experiments, planned, obtained funding, and co-wrote the paper. All authors have read and agreed to the published version of the manuscript.

Funding: This work was funded by Department of Veterans Affairs VAMR I01BX003836 (N.A.P.), Department of Veterans Affairs VAMR I01BX005591 (N.A.P.), and Department of Veterans Affairs RCS IK6BX005387 (N.A.P.). This work does not reflect the views or opinions of the James A. Haley VA Hospital nor the US Government.

Institutional Review Board Statement: Not applicable.

Informed Consent Statement: Not applicable.

Data Availability Statement: Full size blots for all figures are uploaded with the paper. No new data were created or analyzed in this study. Data sharing is not applicable to this article.

Acknowledgments: We thank Wayne Guida (Department of Chemistry, USF) for the use of the Schrodinger, LLC software packages.

Conflicts of Interest: The authors declare no conflict of interest.

References

1. Center for Disease Control and Prevention. *National Diabetes Statistics Report;* Center for Disease Control and Prevention: Atlanta, GA, USA, 2020.
2. Lin, J.; Thompson, T.J.; Cheng, Y.J.; Zhuo, X.; Zhang, P.; Gregg, E.; Rolka, D.B. Projection of the future diabetes burden in the United States through 2060. *Popul. Health Metr.* **2018**, *16*, 9. [CrossRef] [PubMed]
3. Berezin, A.E.; Berezin, A.A.; Lichtenauer, M. Emerging Role of Adipocyte Dysfunction in Inducing Heart Failure among Obese Patients with Prediabetes and Known Diabetes Mellitus. *Front. Cardiovasc. Med.* **2020**, *7*, 583175. [CrossRef]
4. Rea, S.; James, D.E. Moving GLUT4: The biogenesis and trafficking of GLUT4 storage vesicles. *Diabetes* **1997**, *46*, 1667–1677. [CrossRef] [PubMed]
5. Hashiramoto, M.; James, D.E. Characterization of insulin-responsive GLUT4 storage vesicles isolated from 3T3-L1 adipocytes. *Mol. Cell Biol.* **2000**, *20*, 416–427. [CrossRef]
6. Thurmond, D.C.; Kanzaki, M.; Khan, A.H.; Pessin, J.E. Munc18c function is required for insulin-stimulated plasma membrane fusion of GLUT4 and insulin-responsive amino peptidase storage vesicles. *Mol. Cell Biol.* **2000**, *20*, 379–388. [CrossRef] [PubMed]
7. Atkinson, B.J.; Griesel, B.A.; King, C.D.; Josey, M.A.; Olson, A.L. Moderate GLUT4 overexpression improves insulin sensitivity and fasting triglyceridemia in high-fat diet-fed transgenic mice. *Diabetes* **2013**, *62*, 2249–2258. [CrossRef]
8. Bai, L.; Wang, Y.; Fan, J.; Chen, Y.; Ji, W.; Qu, A.; Xu, P.; James, D.E.; Xu, T. Dissecting multiple steps of GLUT4 trafficking and identifying the sites of insulin action. *Cell Metab.* **2007**, *5*, 47–57. [CrossRef]
9. Jedrychowski, M.P.; Gartner, C.A.; Gygi, S.P.; Zhou, L.; Herz, J.; Kandror, K.V.; Pilch, P.F. Proteomic Analysis of GLUT4 Storage Vesicles Reveals LRP1 to Be an Important Vesicle Component and Target of Insulin Signaling. *J. Biol. Chem.* **2010**, *285*, 104–114. [CrossRef]
10. Pilch, P.F. The mass action hypothesis: Formation of Glut4 storage vesicles, a tissue-specific, regulated exocytic compartment. *Acta Physiol.* **2008**, *192*, 89–101. [CrossRef] [PubMed]
11. Morris, N.J.; Ross, S.A.; Lane, W.S.; Moestrup, S.K.; Petersen, C.M.; Keller, S.R.; Lienhard, G.E. Sortilin Is the Major 110-kDa Protein in GLUT4 Vesicles From Adipocytes. *J. Biol. Chem.* **1998**, *273*, 3582–3587. [CrossRef]
12. Bogan, J.S.; Kandror, K.V. Biogenesis and regulation of insulin-responsive vesicles containing GLUT4. *Curr. Opin. Cell Biol.* **2010**, *22*, 506–512. [CrossRef] [PubMed]
13. Shi, J.; Kandror, K.V. Sortilin is essential and sufficient for the formation of Glut4 storage vesicles in 3T3-L1 adipocytes. *Dev. Cell* **2005**, *9*, 99–108. [CrossRef] [PubMed]
14. Huang, S.; Czech, M.P. The GLUT4 Glucose Transporter. *Cell Metab.* **2007**, *5*, 237–252. [CrossRef] [PubMed]
15. Kim, J.Y.; Kandror, K.V. The first luminal loop confers insulin responsiveness to glucose transporter 4. *Mol. Biol. Cell* **2012**, *23*, 910–917. [CrossRef]
16. McCormick, P.J.; Dumaresq-Doiron, K.; Pluviose, A.-S.; Pichette, V.; Tosato, G.; Lefrancois, S. Palmitoylation Controls Recycling in Lysosomal Sorting and Trafficking. *Traffic* **2008**, *9*, 1984–1997. [CrossRef] [PubMed]
17. Li, J.; Matye, D.J.; Li, T. Insulin Resistance Induces Posttranslational Hepatic Sortilin 1 Degradation in Mice. *J. Biol. Chem.* **2015**, *290*, 11526–11536. [CrossRef] [PubMed]
18. Pan, X.; Zaarur, N.; Singh, M.; Morin, P.; Kandror, K.V. Sortilin and retromer mediate retrograde transport of Glut4 in 3T3-L1 adipocytes. *Mol. Biol. Cell* **2017**, *28*, 1667–1675. [CrossRef] [PubMed]
19. Seaman, M.N.J. The retromer complex—Endosomal protein recycling and beyond. *J. Cell Sci.* **2012**, *125*, 4693–4702. [CrossRef]
20. Nielsen, M.S.; Madsen, P.; Christensen, E.I.; Nykjaer, A.; Gliemann, J.; Kasper, D.; Pohlmann, R.; Petersen, C.M. The sortilin cytoplasmic tail conveys Golgi-endosome transport and binds the VHS domain of the GGA2 sorting protein. *EMBO J.* **2001**, *20*, 2180–2190. [CrossRef]
21. Tazi, J.; Bakkour, N.; Stamm, S. Alternative splicing and disease. *Biochim. Biophys. Acta* **2009**, *1792*, 14. [CrossRef]

22. Prudencio, M.; Jansen-West, K.R.; Lee, W.C.; Gendron, T.F.; Zhang, Y.-J.; Xu, Y.-F.; Gass, J.; Stuani, C.; Stetler, C.; Rademakers, R.; et al. Misregulation of human sortilin splicing leads to the generation of a nonfunctional progranulin receptor. *Proc. Natl. Acad. Sci. USA* **2012**, *109*, 21510–21515. [CrossRef]
23. Mohagheghi, F.; Prudencio, M.; Stuani, C.; Cook, C.; Jansen-West, K.; Dickson, D.W.; Petrucelli, L.; Buratti, E. TDP-43 functions within a network of hnRNP proteins to inhibit the production of a truncated human SORT1 receptor. *Hum. Mol. Genet.* **2016**, *25*, 534–545. [CrossRef] [PubMed]
24. Lui, A.; Sparks, R.; Patel, R.; Patel, N.A. Identification of Sortilin Alternatively Spliced Variants in Mouse 3T3L1 Adipocytes. *Int. J. Mol. Sci.* **2021**, *22*, 983. [CrossRef] [PubMed]
25. Shi, Y.; Parag, S.; Patel, R.; Lui, A.; Murr, M.; Cai, J.; Patel, N.A. Stabilization of lncRNA GAS5 by a Small Molecule and Its Implications in Diabetic Adipocytes. *Cell Chem. Biol.* **2019**, *26*, 319–330. [CrossRef] [PubMed]
26. Patel, R.S.; Carter, G.; El Bassit, G.; Patel, A.A.; Cooper, D.R.; Murr, M.; Patel, N.A. Adipose-derived stem cells from lean and obese humans show depot specific differences in their stem cell markers, exosome contents and senescence: Role of protein kinase C delta (PKCdelta) in adipose stem cell niche. *Stem Cell Investig.* **2016**, *3*, 2. [CrossRef] [PubMed]
27. Jumper, J.; Evans, R.; Pritzel, A.; Green, T.; Figurnov, M.; Ronneberger, O.; Tunyasuvunakool, K.; Bates, R.; Žídek, A.; Potapenko, A.; et al. Highly accurate protein structure prediction with AlphaFold. *Nature* **2021**, *596*, 583–589. [CrossRef] [PubMed]
28. Phillips, J.C.; Braun, R.; Wang, W.; Gumbart, J.; Tajkhorshid, E.; Villa, E.; Chipot, C.; Skeel, R.D.; Kale, L.; Schulten, K. Scalable molecular dynamics with NAMD. *J. Comput. Chem.* **2005**, *26*, 1781–1802. [CrossRef]
29. Huang, J.; Rauscher, S.; Nawrocki, G.; Ran, T.; Feig, M.; de Groot, B.L.; Grubmuller, H.; MacKerell, A.D., Jr. CHARMM36m: An improved force field for folded and intrinsically disordered proteins. *Nat. Methods* **2017**, *14*, 71–73. [CrossRef]
30. Maiorov, V.N.; Crippen, G.M. Significance of Root-Mean-Square Deviation in Comparing Three-dimensional Structures of Globular Proteins. *J. Mol. Biol.* **1994**, *235*, 625–634. [CrossRef]
31. Monzon, A.M.; Zea, D.J.; Fornasari, M.S.; Saldaño, T.E.; Fernandez-Alberti, S.; Tosatto, S.C.E.; Parisi, G. Conformational diversity analysis reveals three functional mechanisms in proteins. *PLoS Comput. Biol.* **2017**, *13*, e1005398. [CrossRef] [PubMed]
32. Quistgaard, E.M.; Groftehauge, M.K.; Madsen, P.; Pallesen, L.T.; Christensen, B.; Sorensen, E.S.; Nissen, P.; Petersen, C.M.; Thirup, S.S. Revisiting the structure of the Vps10 domain of human sortilin and its interaction with neurotensin. *Protein Sci.* **2014**, *23*, 1291–1300. [CrossRef] [PubMed]
33. Mollica, L.; Bessa, L.M.; Hanoulle, X.; Jensen, M.R.; Blackledge, M.; Schneider, R. Binding Mechanisms of Intrinsically Disordered Proteins: Theory, Simulation, and Experiment. *Front. Mol. Biosci.* **2016**, *3*, 52. [CrossRef] [PubMed]
34. Xue, B.; Dunbrack, R.L.; Williams, R.W.; Dunker, A.K.; Uversky, V.N. PONDR-FIT: A meta-predictor of intrinsically disordered amino acids. *Biochim. Biophys. Acta* **2010**, *1804*, 996–1010. [CrossRef] [PubMed]
35. Vajda, S.; Yueh, C.; Beglov, D.; Bohnuud, T.; Mottarella, S.E.; Xia, B.; Hall, D.R.; Kozakov, D. New Additions to the ClusPro Server Motivated by CAPRI. *Proteins* **2017**, *85*, 435. [CrossRef]
36. Crane, J.; McGowan, B. The GLP-1 agonist, liraglutide, as a pharmacotherapy for obesity. *Ther. Adv. Chronic Dis.* **2016**, *7*, 92–107. [CrossRef]
37. Fransson, L.; dos Santos, C.; Wolbert, P.; Sjöholm, Å.; Rafacho, A.; Ortsäter, H. Liraglutide counteracts obesity and glucose intolerance in a mouse model of glucocorticoid-induced metabolic syndrome. *Diabetol. Metab. Syndr.* **2014**, *6*, 3. [CrossRef]
38. Sjoholm, A. Liraglutide Therapy for Type 2 Diabetes: Overcoming Unmet Needs. *Pharmaceuticals* **2010**, *3*, 764. [CrossRef]
39. Cartegni, L.; Wang, J.; Zhu, Z.; Zhang, M.Q.; Krainer, A.R. ESEfinder: A web resource to identify exonic splicing enhancers. *Nucleic Acids Res.* **2003**, *31*, 3568–3571. [CrossRef]
40. Paz, I.; Akerman, M.; Dror, I.; Kosti, I.; Mandel-Gutfreund, Y. SFmap: A web server for motif analysis and prediction of splicing factor binding sites. *Nucleic Acids Res.* **2010**, *38*, W281–W285. [CrossRef]
41. Chen, Z.-Y.; Ieraci, A.; Teng, H.; Dall, H.; Meng, C.-X.; Herrera, D.G.; Nykjaer, A.; Hempstead, B.L.; Lee, F.S. Sortilin Controls Intracellular Sorting of Brain-Derived Neurotrophic Factor to the Regulated Secretory Pathway. *J. Neurosci.* **2005**, *25*, 6156–6166. [CrossRef]
42. Molgaard, S.; Demontis, D.; Nicholson, A.M.; Finch, N.A.; Petersen, R.C.; Petersen, C.M.; Rademakers, R.; Nykjaer, A.; Glerup, S. Soluble sortilin is present in excess and positively correlates with progranulin in CSF of aging individuals. *Exp. Gerontol.* **2016**, *84*, 96–100. [CrossRef] [PubMed]
43. Oh, T.J.; Ahn, C.H.; Kim, B.R.; Kim, K.M.; Moon, J.H.; Lim, S.; Park, K.S.; Lim, C.; Jang, H.; Choi, S.H. Circulating sortilin level as a potential biomarker for coronary atherosclerosis and diabetes mellitus. *Cardiovasc. Diabetol.* **2017**, *16*, 92. [CrossRef] [PubMed]
44. Trabjerg, E.; Abu-Asad, N.; Wan, Z.; Kartberg, F.; Christensen, S.; Rand, K.D. Investigating the conformational response of the Sortilin receptor upon binding endogenous peptide- and protein ligands by HDX-MS. *bioRxiv* **2018**. [CrossRef] [PubMed]
45. Vendrell, J.; El Bekay, R.; Peral, B.; García-Fuentes, E.; Megia, A.; Macias-Gonzalez, M.; Real, J.F.; Jimenez-Gomez, Y.; Escoté, X.; Pachón, G.; et al. Study of the Potential Association of Adipose Tissue GLP-1 Receptor with Obesity and Insulin Resistance. *Endocrinology* **2011**, *152*, 4072–4079. [CrossRef] [PubMed]
46. Webb, D.-L.; Abrahamsson, N.; Sundbom, M.; Hellström, P.M. Bariatric surgery—Time to replace with GLP-1? *Scand. J. Gastroenterol.* **2017**, *52*, 635–640. [CrossRef]
47. Madsbad, S.; Holst, J.J. GLP-1 as a Mediator in the Remission of Type 2 Diabetes after Gastric Bypass and Sleeve Gastrectomy Surgery. *Diabetes* **2014**, *63*, 3172–3174. [CrossRef]

48. Currie, A.; Bolckmans, R.; Askari, A.; Byrne, J.; Ahmed, A.R.; Batterham, R.L.; Mahawar, K.; Miras, A.D.; Pring, C.M.; Small, P.K.; et al. Bariatric-metabolic surgery for NHS patients with type 2 diabetes in the United Kingdom National Bariatric Surgery Registry. *Diabet. Med. J. Br. Diabet. Assoc.* **2023**, *40*, e15041. [CrossRef]
49. Hollanda, A.; Lecube, A.; Rubio, M.A.; Sanchez, E.; Vilarrasa, N.; Oliva, J.G.; Fernandez-Soto, M.L.; Salas-Salvado, J.; Ballesteros-Pomar, M.D.; Ciudin, A.; et al. New Metrics to Assess Type 2 Diabetes After Bariatric Surgery: The "Time-Within-Remission Range". *J. Clin. Med.* **2020**, *9*, 1070. [CrossRef]
50. Perry, Z.H.; Netz, U.; Tzelnick, S.; Berar, O.; Atias, S.; Lantsberg, L.; Avinoach, E.; Mizrahi, S. The Effect of Bariatric Surgery upon Diabetes Mellitus: A Proof of Concept by Using the Case of the Mid-Term Effect of Lap Adjustable Gastric Banding (LAGB) on Patients with Diabetes. *Metabolites* **2022**, *12*, 1236. [CrossRef]
51. Vilarrasa, N.; Goday, A.; Rubio, M.A.; Caixas, A.; Pellitero, S.; Ciudin, A.; Calanas, A.; Botella, J.I.; Breton, I.; Morales, M.J.; et al. Hyperinsulinemic Hypoglycemia after Bariatric Surgery: Diagnosis and Management Experience from a Spanish Multicenter Registry. *Obes. Facts* **2016**, *9*, 41–51. [CrossRef]
52. Xu, T.Q.; Kindel, T.L. The role of weight control in the management of type 2 diabetes mellitus: Bariatric surgery. *Diabetes Res. Clin. Pract.* **2023**, *199*, 110667. [CrossRef] [PubMed]
53. Lee, Y.-S.; Park, M.-S.; Choung, J.-S.; Kim, S.-S.; Oh, H.-H.; Choi, C.-S.; Ha, S.-Y.; Kang, Y.; Kim, Y.; Jun, H.-S. Glucagon-like peptide-1 inhibits adipose tissue macrophage infiltration and inflammation in an obese mouse model of diabetes. *Diabetologia* **2012**, *55*, 2456–2468. [CrossRef] [PubMed]
54. Hao, T.; Zhang, H.; Li, S.; Tian, H. Glucagon-like peptide 1 receptor agonist ameliorates the insulin resistance function of islet β cells via the activation of PDX-1/JAK signaling transduction in C57/BL6 mice with high-fat diet-induced diabetes. *Int. J. Mol. Med.* **2017**, *39*, 1029–1036. [CrossRef] [PubMed]
55. Gao, H.; Wang, X.; Zhang, Z.; Yang, Y.; Yang, J.; Li, X.; Ning, G. GLP-1 amplifies insulin signaling by up-regulation of IRbeta, IRS-1 and Glut4 in 3T3-L1 adipocytes. *Endocrine* **2007**, *32*, 90–95. [CrossRef]
56. Lee, H.M.; Joo, B.S.; Lee, C.H.; Kim, H.Y.; Ock, J.H.; Lee, Y.S. Effect of Glucagon-like Peptide-1 on the Differentiation of Adipose-derived Stem Cells into Osteoblasts and Adipocytes. *J. Menopausal Med.* **2015**, *21*, 93–103. [CrossRef] [PubMed]
57. El Bekay, R.; Coín-Aragüez, L.; Fernández-García, D.; Oliva-Olivera, W.; Bernal-López, R.; Clemente-Postigo, M.; Delgado-Lista, J.; Diaz-Ruiz, A.; Guzman-Ruiz, R.; Vázquez-Martínez, R.; et al. Effects of glucagon-like peptide-1 on the differentiation and metabolism of human adipocytes. *Br. J. Pharmacol.* **2016**, *173*, 1820–1834. [CrossRef]
58. Liu, H.; Zhan, Y.L.; Luo, G.J.; Zou, L.L.; Li, Y.; Lu, H.Y. Liraglutide and Insulin Have Contrary Effects on Adipogenesis of Human Adipose-Derived Stem Cells via Wnt Pathway. *Diabetes Metab. Syndr. Obes.* **2020**, *13*, 3075–3087. [CrossRef]
59. Hwang, D.Y.; Cohen, J.B. Base pairing at the 5′ splice site with U1 small nuclear RNA promotes splicing of the upstream intron but may be dispensable for slicing of the downstream intron. *Mol. Cell. Biol.* **1996**, *16*, 3012–3022. [CrossRef]
60. Zahler, A.M.; Roth, M.B. Distinct functions of SR proteins in recruitment of U1 small nuclear ribonucleoprotein to alternative 5′ splice sites. *Proc. Natl. Acad. Sci. USA* **1995**, *92*, 2642–2646. [CrossRef]
61. Freund, M.; Hicks, M.J.; Konermann, C.; Otte, M.; Hertel, K.J.; Schaal, H. Extended base pair complementarity between U1 snRNA and the 5′ splice site does not inhibit splicing in higher eukaryotes, but rather increases 5′ splice site recognition. *Nucleic Acids Res.* **2005**, *33*, 5112–5119. [CrossRef]
62. Manley, J.L.; Krainer, A.R. A rational nomenclature for serine/arginine-rich protein splicing factors (SR proteins). *Genes Dev.* **2010**, *24*, 1073–1074. [CrossRef] [PubMed]
63. Tsukamoto, Y.; Matsuo, N.; Ozawa, K.; Hori, O.; Higashi, T.; Nishizaki, J.; Tohnai, N.; Nagata, I.; Kawano, K.; Yutani, C.; et al. Expression of a novel RNA-splicing factor, RA301/Tra2beta, in vascular lesions and its role in smooth muscle cell proliferation. *Am. J. Pathol.* **2001**, *158*, 1685–1694. [CrossRef] [PubMed]
64. Takeo, K.; Kawai, T.; Nishida, K.; Masuda, K.; Teshima-Kondo, S.; Tanahashi, T.; Rokutan, K. Oxidative stress-induced alternative splicing of transformer 2β (SFRS10) and CD44 pre-mRNAs in gastric epithelial cells. *Am. J. Physiol.-Cell Physiol.* **2009**, *297*, C330–C338. [CrossRef] [PubMed]
65. Kiryu-Seo, S.; Matsuo, N.; Wanaka, A.; Ogawa, S.; Tohyama, M.; Kiyama, H. A sequence-specific splicing activator, tra2beta, is up-regulated in response to nerve injury. *Brain Res. Mol. Brain Res.* **1998**, *62*, 220–223. [CrossRef]
66. Pihlajamaki, J.; Lerin, C.; Itkonen, P.; Boes, T.; Floss, T.; Schroeder, J.; Dearie, F.; Crunkhorn, S.; Burak, F.; Jimenez-Chillaron, J.C.; et al. Expression of the splicing factor gene SFRS10 is reduced in human obesity and contributes to enhanced lipogenesis. *Cell Metab.* **2011**, *14*, 208–218. [CrossRef]
67. Li, H.; Cheng, Y.; Wu, W.; Liu, Y.; Wei, N.; Feng, X.; Xie, Z.; Feng, Y. SRSF10 regulates alternative splicing and is required for adipocyte differentiation. *Mol. Cell Biol.* **2014**, *34*, 2198–2207. [CrossRef]
68. Dlamini, Z.; Mokoena, F.; Hull, R. Abnormalities in alternative splicing in diabetes: Therapeutic targets. *J. Mol. Endocrinol.* **2017**, *59*, R93–R107. [CrossRef]
69. Petersen, M.C.; Shulman, G.I. Mechanisms of Insulin Action and Insulin Resistance. *Physiol. Rev.* **2018**, *98*, 2133–2223. [CrossRef]
70. Zhao, H.; Kong, H.; Wang, B.; Wu, S.; Chen, T.; Cui, Y. RNA-Binding Proteins and Alternative Splicing Genes Are Coregulated in Human Retinal Endothelial Cells Treated with High Glucose. *J. Diabetes Res.* **2022**, *2022*, 7680513. [CrossRef]
71. Renna, L.V.; Bosè, F.; Brigonzi, E.; Fossati, B.; Meola, G.; Cardani, R. Aberrant insulin receptor expression is associated with insulin resistance and skeletal muscle atrophy in myotonic dystrophies. *PLoS ONE* **2019**, *14*, e0214254. [CrossRef]

72. Norgren, S.; Zierath, J.; Wedell, A.; Wallberg-Henriksson, H.; Luthman, H. Regulation of human insulin receptor RNA splicing in vivo. *Proc. Natl. Acad. Sci. USA* **1994**, *91*, 1465–1469. [CrossRef] [PubMed]
73. Wan, W.; Qin, Q.; Xie, L.; Zhang, H.; Wu, F.; Stevens, R.C.; Liu, Y. GLP-1R Signaling and Functional Molecules in Incretin Therapy. *Molecules* **2023**, *28*, 751. [CrossRef] [PubMed]
74. Wei, L.; Mo, W.; Lan, S.; Yang, H.; Huang, Z.; Liang, X.; Li, L.; Xian, J.; Xie, X.; Qin, Y.; et al. GLP-1 RA Improves Diabetic Retinopathy by Protecting the Blood-Retinal Barrier through GLP-1R-ROCK-p-MLC Signaling Pathway. *J. Diabetes Res.* **2022**, *2022*, 1861940. [CrossRef]
75. Challa, T.D.; Beaton, N.; Arnold, M.; Rudofsky, G.; Langhans, W.; Wolfrum, C. Regulation of Adipocyte Formation by GLP-1/GLP-1R Signaling. *J. Biol. Chem.* **2012**, *287*, 6421–6430. [CrossRef] [PubMed]
76. Ejarque, M.; Guerrero-Perez, F.; de la Morena, N.; Casajoana, A.; Virgili, N.; Lopez-Urdiales, R.; Maymo-Masip, E.; Pujol Gebelli, J.; Garcia Ruiz de Gordejuela, A.; Perez-Maraver, M.; et al. Role of adipose tissue GLP-1R expression in metabolic improvement after bariatric surgery in patients with type 2 diabetes. *Sci. Rep.* **2019**, *9*, 6274. [CrossRef]
77. Chen, J.; Zhao, H.; Ma, X.; Zhang, Y.; Lu, S.; Wang, Y.; Zong, C.; Qin, D.; Wang, Y.; Yingfeng Yang, Y.; et al. GLP-1/GLP-1R Signaling in Regulation of Adipocyte Differentiation and Lipogenesis. *Cell. Physiol. Biochem. Int. J. Exp. Cell. Physiol. Biochem. Pharmacol.* **2017**, *42*, 1165–1176. [CrossRef]
78. Gutierrez, A.D.; Gao, Z.; Hamidi, V.; Zhu, L.; Saint Andre, K.B.; Riggs, K.; Ruscheinsky, M.; Wang, H.; Yu, Y.; Miller, C., 3rd; et al. Anti-diabetic effects of GLP1 analogs are mediated by thermogenic interleukin-6 signaling in adipocytes. *Cell Rep. Med.* **2022**, *3*, 100813. [CrossRef]
79. Ferguson, D.P.; Dangott, L.J.; Lightfoot, J.T. Lessons learned from vivo-morpholinos: How to avoid vivo-morpholino toxicity. *Biotechniques* **2014**, *56*, 251–256. [CrossRef]
80. Summerton, J. Morpholino antisense oligomers: The case for an RNase H-independent structural type. *Biochim. Biophys. Acta (BBA)-Gene Struct. Expr.* **1999**, *1489*, 141–158. [CrossRef]
81. Havens, M.A.; Hastings, M.L. Splice-switching antisense oligonucleotides as therapeutic drugs. *Nucleic Acids Res.* **2016**, *44*, 6549–6563. [CrossRef]
82. Havens, M.A.; Duelli, D.M.; Hastings, M.L. Targeting RNA splicing for disease therapy. *Wiley Interdiscip. Rev. RNA* **2013**, *4*, 247–266. [CrossRef] [PubMed]

MDPI AG
Grosspeteranlage 5
4052 Basel
Switzerland
Tel.: +41 61 683 77 34

International Journal of Molecular Sciences Editorial Office
E-mail: ijms@mdpi.com
www.mdpi.com/journal/ijms

www.ingramcontent.com/pod-product-compliance
Lightning Source LLC
Chambersburg PA
CBHW061628210926
51850CB00009B/2470
9783725873302